T0206089

Migration, Squatting and Radical Autonomy

This book offers a unique contribution, exploring how the intersections among migrants and radical squatters' movements have evolved over past decades. The complexity and importance of squatting practices are analyzed from a bottom-up perspective, to demonstrate how the spaces of squatting can be transformed by migrants. With contributions from scholars, scholar-activists, and activists, this book provides unique insights into how squatting has offered an alternative to dominant anti-immigrant policies, and the implications of squatting on the social acceptance of migrants. It illustrates the different mechanisms of protest followed in solidarity by migrant squatters and Social Center activists, when discrimination comes from above or below, and explores how different spatialities can be conceived and realized by radical practices.

Contributions adopt a variety of perspectives, from critical human geography, social movement studies, political sociology, urban anthropology, autonomous Marxism, feminism, open localism, anarchism and post-structuralism, to analyze and contextualize migrants and squatters' exclusion and social justice issues. This book is a timely and original contribution through its exploration of migrations, squatting and radical autonomy.

Pierpaolo Mudu is Professor in the Faculty of Urban Studies and Interdisciplinary Arts & Sciences at the University of Washington – Tacoma, USA.

Sutapa Chattopadhyay is Assistant Professor and Visiting Researcher at Maastricht University and United Nations University, the Netherlands.

Routledge Research in Place, Space and Politics Series
Series Edited by Professor Clive Barnett, Professor of Geography and Social Theory, University of Exeter, UK.

This series offers a forum for original and innovative research that explores the changing geographies of political life. The series engages with a series of key debates about innovative political forms and addresses key concepts of political analysis such as scale, territory and public space. It brings into focus emerging interdisciplinary conversations about the spaces through which power is exercised, legitimized and contested. Titles within the series range from empirical investigations to theoretical engagements and authors comprise of scholars working in overlapping fields including political geography, political theory, development studies, political sociology, international relations and urban politics.

Urban Refugees: Challenges in Protection, Service and Policy
Edited by Koichi Koizumi and Gerhard Hoffstaedter

Space, Power and the Commons: The Struggle for Alternative Futures
Edited by Samuel Kirwan, Leila Dawney, and Julian Brigstocke

Externalizing Migration Management: Europe, North America and the Spread of 'Remote Control' Practices
Edited by Ruben Zaiotti

Architecture and Space Reimagined: Learning from the Difference, Multiplicity, and Otherness of Development Practice
Richard Bower

Migration, Squatting and Radical Autonomy
Edited by Pierpaolo Mudu and Sutapa Chattopadhyay

Nation Branding and Popular Geopolitics in the Post-Soviet Realm
Robert A. Saunders

Political Street Art: Communication, Culture and Resistance in Latin America
Holly Ryan

Geographies of Worth: Rethinking Spaces of Critical Theory
Clive Barnet

Migration, Squatting and Radical Autonomy

Edited by Pierpaolo Mudu and Sutapa Chattopadhyay on behalf of SqEK (Squatting Europe Kollective)

LONDON AND NEW YORK

First published in paperback 2018

First published 2017
by Routledge
2 Park Square, Milton Park, Abingdon, Oxon OX14 4RN

and by Routledge
711 Third Avenue, New York, NY 10017

Routledge is an imprint of the Taylor & Francis Group, an informa business

British Library Cataloguing-in-Publication Data
A catalogue record for this book is available from the British Library

Library of Congress Cataloging-in-Publication Data
Names: Mudu, Pierpaolo, editor. | Chattopadhyay, Sutapa, editor.
Title: Migration, Squatting and Radical Autonomy / edited by Pierpaolo Mudu and Sutapa Chattopadhyay.
Description: Abingdon, Oxon ; New York, NY : Routledge, 2017. | Series: Routledge research in place, space and politics series
Identifiers: LCCN 2016004569 | ISBN 9781138942127 (hardback) | ISBN 9781315673301 (e-book)
Subjects: LCSH: Squatter settlements—Political aspects. | Squatters—Political activity. | Immigrants—Political activity. | Internal migrants—Political activity.
Classification: LCC HD7287.95 .M54 2017 | DDC 325—dc23
LC record available at https://lccn.loc.gov/2016004569

ISBN: 978-1-138-94212-7 (hbk)
ISBN: 978-1-138-49448-0 (pbk)
ISBN: 978-1-315-67330-1 (ebk)

Typeset in Times New Roman
by Apex CoVantage, LLC

Contents

Illustrations

Figures

Tables

Contributors

Thomas Aguilera is Phd in political science (2015) and postdoctoral researcher at Sciences Po Paris (Centre for European Studies) where he teaches political science, sociology and research methods. His research work concentrates on public policies towards urban illegalities, squats and informal slums in Europe, by crossing governance, social movements and urban sociology approaches in comparative studies. He also works on tourism policies and resistances to tourism patterns in Europe.

Bridget Anderson is Deputy Director and Senior Research Fellow at the Centre on Migration, Policy and Society (COMPAS). She is interested in labour migration and trafficking, migration, enforcement and the liberal state, and citizenship. Publications include *Us and Them? The Dangerous Politics of Immigration Control* (Oxford University Press, 2013), and *Doing the Dirty Work. The Global Politics of Domestic Labour* (Zed, 2000, winner of the Philip Abrams memorial prize). She has worked with a wide range of national and international organizations including the Trades Union Congress (TUC), the International Labour Organization (ILO), and the Organization for Security and Co-operation in Europe (OSCE).

Fulvia Antonelli is a researcher in anthropology at the University of Bologna, Italy. Her research interests include migration processes, urban ethnography, education, marginality and youth cultures. She published *Tranvieri: Etnografia di una palestra di pugilato* (Roma: Arcane, 2010); 'Dar El Baida mon amour' in M. Peraldi and M. Tozy (eds) *Casablanca, figures et scènes métropolitaines* (Paris: Karthala, 2011); 'Ethnographie de la 'malavita' mobile' and 'La rue ici, la rue ailleurs' in M. Peraldi (ed.) *Les Mineurs migrants non accompagnés.* Un defi pour les pays européens (Paris: Karthala, 2014). Her research is currently focused on issues around urban and social exclusion processes, migrations, and school drop-outs.

Kolar Aparna is currently pursuing her doctoral studies in Geography at the Nijmegen Center for Border Research, Radboud University, Netherlands. Her career has meandered across the arts and academia, from being a performing artist to a geographer. Her current research and activist interests fall at the intersections of artistic and scientific practices that help to address issues around power relations, spatial relations, in general, and transboundary or (cross)border relations; in particular her research project investigates the everyday relations of hos(ti)pitality towards and by 'asylum-seekers' and 'undocumented migrants' in the production of everyday space. She is also part of a collective called the 'Asylum University' in which academics and so-called 'undocumented migrants' work together towards critical production of knowledge on various topical issues.

Azozomox is a pseudonym of an activist who is based in Berlin, Germany and is influenced by anarchist and autonomist ideas.

Simone Beate Borgstede is a sociologist, historian and political activist. She teaches at Leuphana University Lüneburg. She received her PhD in Modern British History from University College London (2010) after doing MA in Sociology at the University of Essex and Social Economics in Hamburg. Her research interests include antisemitism, colonial racism and social movements. She has been working on transnational perspectives of refugee struggles, postcolonial theory and the history of racism. Her recent publication is ‚Geschichte ist immer offen: Denken und kämpfen mit Stuart Hall' in Dagmar Brunow (ed.) *Stuart Hall. Aktivismus, Pop und Politik* (2015). Her main publication is *All is Race. Benjamin Disraeli on race, nation and empire* (2011). She has also published on the social phenomenon of squatting in St. Pauli Hafenstrasse, Hamburg 1981–1987: *Der Kampf um die Herzen und Köpfe* in *Das Argument* (2010). Simone has lived in squatted houses of St. Pauli Hafenstraße for more than thirty years. From 1984–1985 she participated in a solidarity brigade for construction work in Nicaragua. She locates herself in the struggles of St. Pauli, such as the right to the city movement and the solidarity with the refugee group 'Lampedusa in Hamburg' campaign.

Florence Bouillon is socio-antropologist. She is Lecturer at the university Paris 8 and Researcher in the Center Norbert Elias (CNRS / EHESS). She looks into the marginalized forms of housing as being produced by institutional processes and to understand vulnerable city-dwellers' strategies of resistance to social disqualification. She has published three books and various articles on squats in France.

Andrew Burridge is an Associate Research Fellow at the University of Exeter in the UK, where he is part of a study into procedural consistency in asylum appeal hearings. His previous research has included a study of the humanitarian aid group 'No More Deaths' in the Mexico-US borderlands, and of the operation of the EU border agency Frontex. He is a former participant in the US and UK No Borders movement, has volunteered with No More Deaths, and has been involved in anti-detention and deportation campaigns. In 2012, Andrew was co-editor of the collection *Beyond Walls and Cages: Prisons, Borders, and Global Crisis* (UGA Press), with Jenna Loyd and Matt Mitchelson.

Calais Migrant Solidarity (CMS) is an international network of autonomous people involved in practical solidarity work with the migrant communities of Calais. CMS first got involved in migrant issues in 2009. This group has since been active with monitoring police activity, squatting for and with migrants, supporting migrant political protests and other direct action projects such as distributing food, clothing, water, blankets, and tents, doing outreach and publicizing the situation in Calais, and providing migrants with asylum and immigration information for the UK and other European countries.

Sara Casella Colombeau is a postdoctoral fellow at the LAMES (University of Aix-Marseille). In December 2013, she completed her PhD in political science from Sciences Po (Paris) on the transformation of the French border police in relation with the creation of the Schengen border zone. In 2014, she was a postdoctoral fellow at the International Centre for Comparative Criminology (ICCC) at University of Montreal (Canada).

Claudio Cattaneo is a Barcelona-based freelance researcher and activist. He is an active member with Research and Degrowth Barcelona/Università Carlo Cattaneo (Castellanza, Spain) and is currently involved in the squat Can Masdeu (Barcelona). He holds a Phd in Environmental Science on the ecological economics of urban squatters in Barcelona from the Institute of Environmental Science and Technology, Universitat Autónoma de Barcelona, where he is also a research associate. His research specializations are squatting, eco-communities and Degrowth. His relevant publications include: with G. d'Alisa, G. Kallis and C. Zografos: 'Degrowth Futures and Democracy' (*Futures* 44(6): 515–523, 2012); with G. d'Alisa: 'Household Work and Energy Consumption: A Degrowth Perspective. Catalonia's Case Study' (*Journal of Cleaner Production* 38, 71–79, 2013). 'The money-free autonomy of Spanish squatters' in A. Nelson and F. Timemrmann (eds) Life without money. The experience of urban squats in Collserola, Barcelona: what kind of degrowth? (*Journal of Cleaner Production*, 18(6), 581–589.) and with M. Martínez: *the Squatters' Movement in Europe: commons and autonomy as alternatives to capitalism* (London: Pluto Press, 2014).

Sutapa Chattopadhyay is a geographer and currently working at the University of Toronto. Previously she has worked as an Assistant Professor and researcher at the Universities of Minnesota-Duluth, Maastricht (Faculties of Health, Medicine and Life Sciences and School of Governance), United Nations. Her areas of interest are migrations, alternative development, indigeniety, anarcha-ecofeminism and radical autonomy. She is currently pursuing her own research on migrant incarceration and counter struggles in Rome, Italy and on indigenous food sovereignty in Andhra Pradesh, India. She is in the advisory board of ACME. She has published in ACME; *Gender, Place and Culture*; Population, Place and Space, Environment and Planning D, and Capitalism, Nature, Socialism on indigenous anti-colonial struggles, development-induced dislocation, colonial appropriation of nature, feminist research methods and border politics. She has several manuscripts in progress that focus on radical pedagogy, food sovereignty, gender-class struggles and militarization of borders. She is working on a book on wider indigenous struggles with Palgrave and on two journal special issues on eco-socialist pedagogy and discourses on the migrant position for two key radical journals.

Deanna Dadusc is a PhD candidate at the Universities of Kent (UK) and Utrecht (the Netherlands). Her research topic revolves around the criminalization of social movements. In particular, she combines her research activities with political activism in migrant and housing rights and squatting.

Cesare Di Feliciantonio is a PhD candidate in Geography at KU Leuven (Belgium) and Sapienza – University of Rome (Italy). His research interests include a wide range of issues from social and urban geography to critical political economy. Beyond academia, he is a militant in the queer leftist scene in Rome, and an active member of various collectives/groups, including the Roman node of Communia, a national network involving different forms of squatting and grassroots initiatives (communianet.org).

Romain Filhol is a PhD student in Geography at the Paris Est University, France. He is also affiliated to the Naples Federico II University, Italy. His interest in rurality and the social condition of food production has led him to study the case of the migrant agricultural workers and particularly their political struggles in Southern Italy. This research and fieldwork engagements has allowed him to combine his scientific work with political activism.

Federica Frazzetta is a PhD candidate in Sociology and Sociological Research at the University of Trento. In 2014 she completed her Masters in Political Science and International Relations from the University of Catania. In her undergraduate thesis, she analysed the case of No MUOS movement, a Sicilian antimilitarist movement that has been active since 2009. Also, she has been active in the students' movement, the 'Seconda Onda Studentesca' (Second Students' Wave), against school and university reform in 2010. She is a member of the Aleph Collective of Catania and No MUOS movement.

Stephania Grohman is an anthropologist and former squatter interested in the question how power and territoriality affect people's sense of being a 'self' inhabiting the 'space' of the body. She has recently obtained her doctorate based on long-term ethnographic research with squatters in the UK, exploring how the power of solidarity can offset the detrimental effect homelessness has on mental health. She is a qualified social worker, feminist and a long-time activist in the areas of housing and poverty. Currently, she works as a post-doctoral researcher at the Institute for Cognitive and Evolutionary Anthropology at the University of Oxford.

Duygu Gürsel is a PhD student at the Social Science Institute in Humboldt University of Berlin with funding from the Rosa Luxemburg Foundation. In 2013, she published an edited book volume entitled *Wer MACHT Demokratie?*

Kritische Beiträge zu Migration und Machtverhältnissen with Allmende e.V. and Zülfukar Cetin.

Serin D. Houston is an Assistant Professor of Geography and International Relations at Mount Holyoke College (USA). A variety of local and national grants have supported her research endeavours in two primary areas: 1) settlement, belonging, and social justice with migrant communities; and 2) encounters in and transformations of urban space. She has published in a range of peer-reviewed publications, including *Gender, Place, and Culture*, the *Geographic Bulletin*, *Progress in Human Geography*, *Qualitative Inquiry*, and *Social and Cultural Geography.*

International Women's Space (IWS) is a collective of women in an occupied building in Kreuzberg, Berlin (Germany) (see iwspace.wordpress.com/)

Vasiliki (Vaso) Makrygianni is a PhD researcher at the Department of Urban and Regional Planning at the Aristotle University of Thessaloniki (Greece). She also holds a diploma in Architecture. Her interests include critical urban theory, spatial analysis, feminism, mobilities, migration and urban social movements. She has taken part in architectural projects and has taught design. Currently she resides in Greece and is a member of the 'encounters and conflicts' group. Beyond that she is thinking of practising witchcraft in case she doesn't find another way to end capitalism.

Miguel Martínez is currently working as an Assistant Professor at City University of Hong Kong. He simultaneously holds a Senior Researcher position in the Department of Sociology II (Human Ecology and Population), University Computense of Madrid. His interests lie in urban sociology, social movements and participatory-activist methodologies. His latest publications are: *Okupaciones en movimiento. Derivas, estrategias y prácticas* (co-authored with M. Dominguez and E. Lorenzi; Madrid: Tierradenadie, 2010), 'The Citizen Participation of Urban Movements in Spatial Planning. A Comparison between Vigo and Porto' (*International Journal of Urban and Regional Research* 35(1): 147–171, 2011); co-editor with R. Adell: *Donde están las llaves? El movimiento okupa: prácticas y contextos sociales* (Madrid: La Catarata, 2004); '*Los movimientos sociales urbanos. Un analisis de la obra de Manuel Castells*' (*Revista Internacional de Sociología* 61(34): 81–106, 2003), *Okupaciones de viviendas y centros sociales* (Barcelona: Virus, 2002). He has been actively involved in various squatted social centres of Madrid. For more information see: www.miguelangelmartinez.net.

Pierpaolo Mudu is a geographer collaborating with the faculties of Urban Studies and Interdisciplinary Arts & Sciences at the University of Washington in Tacoma (USA). He worked in England at the University of Oxford and at the University of Reading and he has been visiting scholar at the University of Washington in Seattle and at the EHESS in Paris. He is in the International Advisory Board of ACME. He has been involved in the Italian Social Centers movement and over the years he has extensively published on their diffusion and patterns in (radical) academic journals (such as Antipode, ACME, GeoJournal, Urban Geography) and book chapters. Some of his latest publications are: *Self-managed Social Centers and the right to urban space* in Clough Marinaro I., Thomassen B. (eds): Global Rome, Changing Faces of the Eternal City (Bloomington & Indianapolis: Indiana University Press, 2014); *Ogni sfratto sarà una barricata: squatting for housing and social conflict in Rome* in SqEK, Cattaneo C. and Martínez M. (eds.) The squatters movement in Europe (London: Pluto Press, 2014); with A. Membretti: *Where global meets local: Italian social centres and the alterglobalization movement* in Flesher Fominaya C. and Cox L. (eds) Understanding European Movements (New York: Routledge, 2013). He is also the editor of a forthcoming special symposium for Antipode on the Italian squatting and "occupy" practices.

Nadia Nur is a Sociologist and holds her PhD in 'Territorial Policies and local project' from Roma Tre University (Italy). Her research interests are migration, urban movements and the right to the city.

Mimmo Perrotta is Assistant Professor in Sociology of Cultural Processes at the University of Bergamo, Italy, and co-editor of the journal Etnografia e ricerca qualitativa / Ethnography and Qualitative Research. His research interests include migration processes, the ethnography of work, the nexus between culture and power and oral history. He published a volume on Romanian workers in the Italian construction sector entitled *Vite in cantiere. Migrazione e lavoro dei rumeni in Italia* (Bologna, Il Mulino, 2011). His research is currently focused on migrant labour in agriculture and on a comparison between Southern and Northern Italy's agro-industrial supply chains. On these topics, he has published a number of articles in Italian and international journals.

Hans Pruijt teaches Sociology of Organizations and Organizing, and research skills at Erasmus University, Rotterdam (the Netherlands). He has a longstanding interest in processes of empowerment, both in urban movements and in the organization of work and working life policies. Recent publications include: with C. Roggeband, 'Autonomous and/or Institutionalized Social Movements? Conceptual Clarification and Illustrative Cases' (*International Journal of Comparative Sociology* 55(2), 144–165, 2014); with M.A. Yerkes, 'Empowerment as Contested Terrain. Employability of the Dutch Workforce' (*European Societies* 16(1) 48–67, 2014); 'Culture Wars, Revanchism, Moral Panics and the Creative City. A Reconstruction of a Decline of Tolerant Public Policy: The Case of Dutch Anti-squatting Legislation' (*Urban Studies* 50(6): 1114–1129, 2013).

Alejandro Sethman is a Phd candidate at Universidad de San Martín (Argentina) and Università degli studi di Roma La Sapienza (Italy). He is working on Italian housing issues, segregation in urban spaces, urban governance and migrations.

Tina Steiger holds a Bachelor degree in Political Science from the University of Florida, and a Masters in Urban Studies from the UNICA 4 Cities Urban Studies Program. Based in Denmark, she has worked as a graduate assistant at the HafenCity Universität Hamburg, as well as an external lecturer at Copenhagen University's Department of Modern Culture and Literature. She is involved in a number of projects engaging in cultural exchanges and radical politics in Copenhagen and Hamburg.

Henk van Houtum is the Head of the Nijmegen Centre for Border Research, Radboud University Nijmegen (the Netherlands) and Research Professor in Geopolitics of Borders, University of Bergamo (Italy). He has published extensively on the philosophy of b/ordering and othering, the EU border regime with regard to immigration, global injustices and critical cartography. His most recent book is *Borderland: Atlas, Essays and Design, History and Future of the Border landscape* (Blauwdruk: Wageningen, 2013). For more information, see: www.henkvanhoutum.nl.

Foreword

Bridget Anderson

How exciting: a volume that traces the web of connections, both empirical and theoretical, between different strategies of state resistance, using the same lens to explore struggles around living spaces and the struggles of migrants. Squatting and migration are rarely considered together. Migration is often used as a tool for division and for limiting the radical potential of grassroots struggles. Nowhere is this clearer than in the case of housing. Housing for those who are not wealthy is increasingly treated as a scarce resource to be rationed out to the deserving. The basic infrastructure of living – healthcare, food, a minimum income – is under sustained assault more generally as the European financial downturn is used to deliver swingeing cuts to minimal support and provisions. What is particular about housing, however, is that it is very much rationed at the local level, and the imagination of the neighbourhood as well as the nation is used as an exclusionary mechanism. Thus deservingness is not only about being a national citizen, but also belonging locally, and those excluded may be citizens from the wrong part of the country. Perhaps it should not be surprising then that so often housing campaigns have steered away from questions of immigration. 'What are you going to do about illegalized migrants?' has the potential to undermine people's responses and the ways forward they attempt to devise.

In a parallel move, migrants have often sought to distance themselves from 'criminals' and (non-immigration) forms of criminal activity. Indeed, the Good Migrant is often keen to present themselves as hard-working and honourable, contrasting themselves with the criminal and the terrorist. These may be citizens, but in practice, the implication is, migrants are morally superior. Slogans like: 'Refugees are not criminals' attempt to embed non-citizens in a community that assumes that those criminalized by the state are morally reprobate. The non-citizen is often under particular pressure to demonstrate a certain kind of good citizenship in order to be accepted. This is illustrated in the formal requirements of naturalization, which often require that an applicant for citizenship does not have a criminal record, is of 'good character', and often has a professional or otherwise respectable person to vouch for them.

These kinds of politics mean that squatters and migrants have been differentiated from each other, yet in practice, as we see from the contributions to this volume, squatters can be migrants, migrants are involved in struggles over housing, and there are spaces of political solidarity as well as productive conflicts and debates. This is the basis for deeply political connections, linking the local and the global, offering the opportunity for the production of new subjectivities and types of relations.

Viewing migration and squatting through the same lens foregrounds the very direct consequences for everyday life of state attempts to control mobility, often through the use of the law. These consequences are not only experienced by migrants. State, mobility control and law are inextricably intertwined. As John Torpey has argued, legitimate control over movement is the *sine qua non* of the modern state: a state must 'know' its population in order to be able to govern it (Torpey 2000). Yet this runs counter to what used to be the fundamental ideal of liberal freedom, 'the right of locomotion'. In his commentaries on the law of England, eighteenth-century jurist William Blackstone commented,

> the personal liberty of individuals … consists in the power of locomotion, of changing situation, or removing one's person to whatsoever place one's own inclination may direct; without imprisonment or restraint, unless by due course of law. (Blackstone, Commentaries 1979, 120–41)

The centrality of locomotion meant that 'the subject at the core of liberal theory has a corporeal dimension: the capacity of locomotion' (Kotef 2015). Kotef argues that mobility is inherently a political concept. Yet despite the recognition of this in classic liberal political theory (e.g., 'when the words Free, and Liberty, are applied to anything but Bodies they are abused; for that which is not subject to Motion is not subject to Impediment' (Hobbes 1651, 262)), the nineteenth-century erasure of the corporeal dimension of the liberal subject means that the centrality of mobility to freedom was lost, and Kotef notes that it is not even mentioned in Rawls' basic liberties (1971).

The 'freedom' to move can be granted only to those who know how to manage it properly. It might be thought then that this is the prerogative of citizens, and it is usually assumed that what Torpey (2000) calls 'state monopoly over the legitimate means of movement' results in surveillance and obstruction for non-citizens, but not for citizens. It is of course migrants who are subject to immigration controls and deportation, and whose entry into a state of which they are not a citizen may be refused. Indeed, it is supposed to be a defining feature of citizens that they are not subject to immigration controls, and in Europe this is also true of European Union citizens. Yet this does not mean that states protect their citizens from forced movement – extradition and extraordinary rendition evidence that under some circumstances states are prepared to allow citizens to be forcibly removed from their territory. The thousands of daily deportations not only suggest that states are not prepared to intervene to protect their citizens from other state's enforcement, but also that they will generally admit citizens being forced to return even against those citizens' wishes. That is, citizenship must be understood in terms of a global regime that enables state control over movement, rather than as a relation between an individual and a state. In this way 'citizenship' does not entail protection from enforced mobility.

Attention to squatting reveals that states may also attempt to directly shape the mobility of their citizens at the scale of the local, and that local and city governance structures can be important mechanisms for so doing. Citizens may be forced to 'move on' and vacate their living spaces, they may be banned from public spaces as nuisances and undesirables, they may be criminalized for sleeping in the street, and for begging in public squares, ending up incarcerated – the most direct form of control over mobility – because they cannot pay fines. It is imperative to bring these different oppressions and struggles into conversation with one another.

Squatters are strongly associated with vagrants: people who refused to accept the rule of feudal masters, and, later the rule of the market, or people who were excluded from society in multiple ways. There is a long history of fear of vagrancy, fear of 'masterless men' who threatened to disrupt the social order by not being in the right place. The first passports in England were issued, not to foreigners, but to people 'wandering abroad' outside the community where they were supposed to stay. If they were on their master's business and needed to travel, they needed to have a 'passport' that was stamped with the king's seal. By the sixteenth century, a good set of false papers cost over two pennies. The reason for this urge to control the mobility of the poor has not been connected to immigration controls whose origins are more likely to be located in the early twentieth century than in the vagrancy statutes of the fourteenth century.

Considering squatting and migration together offers an opportunity to explore ways in which we can fight against the positioning of migrants and citizens as competitors for the privileges of membership. This is extremely valuable, particularly since these privileges are increasingly poor. In Europe, rights and resources that were formerly distributed on the basis of universalism – albeit the universalism of the white, male, able-bodied subject – are now distributed on the basis of work. It is the 'worker citizen' or the taxpayer who deserves these rights. The rise of the worker citizen has seen the development of two types of undeservingness: idleness (the unemployed citizen) and not belonging (the migrant). It is politically crucial to overcome the division this makes, and shared living projects are critical fora for so doing.

The shared struggles of squatters and migrants – and the struggles between them – will be of increasing importance in years to come, and it is vital that we analyse and learn from them now. For there are political shifts afoot, manifest in Europe in the growing discontent with the politics of austerity, and the public anxiety about the treatment of people at Europe's borders. Who in 2014 would have anticipated the demonstrations of summer 2015, proclaiming 'Refugees are welcome here'? However, unless anti-austerity politics address migrants' rights, and support for refugees extends to those marginalized by the cuts to welfare, both movements will be fatally weakened.

It has become commonplace for politicians of all parties to recite standard references to Europe's respect for human rights, democracy and history of welcoming refugees as a prelude to introducing ever harsher immigration and asylum laws. As people gathered at the borders of Europe and were confronted with tear gas, batons and barbed wire, such claims began to ring increasingly hollow. Yet there was at the same time an unprecedented show of public support for new arrivals, with people promising to open their houses to refugees from Syria in particular. However, it is possible, but not easy, to live with strangers who it may turn out one does not particularly get on with, and who can end up sitting at home all day, not able to find work, under stress about their longer term prospects. It is even harder to live with them as equals.

Levels of poverty are increasing all over Europe and not only for asylum seekers. How will the promises of support extended to Syrian refugees be received by the hundreds of thousands of people who have had their benefits stopped or capped, who are sofa surfing, scraping by on the minimum wage? Or the people on housing lists or going to food banks who see that Syrians are accommodated but not them? If we are to avoid a competition between marginalized and impoverished groups, it is necessary to make the argument that better services for migrants must mean better services for everybody. This takes supporters of refugees off the terrain of humanitarian responses and demands that they argue for common interests rather than special cases. We can all agree that the current situation needs bold thinking and new paradigms. I would suggest that connecting the discontent with austerity and support for migrants is a critical first step, and this is precisely what we are seeing in the kinds of projects and struggles outlined in this volume.

References

Blackstone, W. (1979) *Commentaries on the Laws of England: A Facsimile of the First Edition of 1765–1769*. Chicago: University of Chicago Press.

Hobbes, T. (1651) *Leviathian: On the Liberty of Subjects*. Clarendon: Oxford University Press, Chapter 21: 262.

Kotef, H. (2015) *Movement and the Ordering of Freedom: On Liberties of Governances of Mobility*. Durham: Duke University Press.

Rawls, J. (1971) *A Theory of Justice*. Harvard: Harvard University Press.

Torpey, J. (2000) *The Invention of the Passport. Surveillance, Citizenship and the State*. Cambridge: Cambridge University Press.

Acknowledgements

Putting together this book would not have been possible without the collective collaboration, support and initiatives of the Squatting Europe Kollective. Additionally, many individuals gave us a crucial support to finalize different versions of the book.

Jessica Newman provided invaluable support for the coherence in English of the entire manuscript. Her patient reading of the chapters and the introduction has made this project a solidarity initiative, alongside the initiatives of the authors, reviewers and supporters. Besides relentlessly reading and editing the chapters, we have to admit that this book could not have been written this way without the constructive and perceptive peer review comments provided by Baptiste Colin, Laurence Cox, Giulio D'Errico, E.T.C. Dee, Kris Forkaseiwicz, Eliseo Fucolti, Andrej Holm, Flemming Mikkelsen, Federico Oliveri, Linus Owens, Gianni Piazza, Giovanni Picker, Luisa Rossini, Pietro Saitta, Galvão Santos, Margot Vedier, Tommaso Vitale and Judith Watson.

We are thankful to Andrea Aureli, Emily Gilbert, Heather Merill, Bart van der Steen and Alex Vasudevan's detailed critique prompting more appropriate and clearer ways of expressing and re-thinking the introduction of this volume. Margit Mayer gave us very useful advice to reconsider some critical parts of the first draft of the introduction.

Many others likewise have contributed through critiques, suggestions, and support from the proposal stage, on the introduction, the conclusions and on various chapters. Among them are Harald Bauder, Michelle Buckley, Carlo Cellamare, Pierluigi Cervelli, John Clark, Isabella Clough-Marinaro, Nick Dines, Salvatore Engel-Di Mauro, Anthony Falit-Baiamonte, Lucy Finchett-Maddock, Nik Heynen, Larry Knopp, Cristina Mattiucci, Peter Nyers, Eliot Tretter, Jim Tyner, Serge Weber and Djemila Zeneidi. We also thank the following people for their support: Bridget Anderson, George Caffentzis, Silvia Federici, Levi Gahman, Rich Heyman, Matt Meyers, Alan Moore, Laura Pulido, Christy Petropoulou, Sujata Ramachandran, Shyama Ramani, Nandita Sharma, Marina Sitrin and Cynthia Wright. We would like to express our particular thanks to Angela Davis and Noam Chomsky.

We are grateful for the welcoming atmosphere provided by Routledge and especially to Faye Leerink who worked with us closely from autumn 2014. Two anonymous reviewers gave encouraging comments and constructive suggestions which improved the final outcome. We also thank the Routledge team (Emma Chappell, Daniel Bourner, Michelle Antrobus and Fabienne Pedroletti) for finalizing many aspects of the book without which this volume would be incomplete.

Finally, we acknowledge all the resources and all the people known and unknown who have sustained us and given us space and time to write this book.

Introduction

Migration, squatting and radical autonomy

Pierpaolo Mudu and Sutapa Chattopadhyay

This book is a collective effort by a group of activists and researchers, some of whom are part of the Europe Kollective (SqEK).[1] We have combined our knowledge and experience to represent current social trends from the point of view of those who occupy and squat in places in order to oppose oppression, injustice, and lack of autonomy enforced by dominant relations to ensure benefits to a few privileged groups. This book testifies to the level of conscious struggles here and now in Europe. The analyses, references and websites[2] constitute a mine of information, otherwise scattered, for those who want to read the untold stories of migrants, Romas and refugees struggling through squatting, and take action. Organizing the knowledge of self-managed squats is a difficult task because the available information is oftentimes biased on the side of mainstream actors, as it is all too often produced by self-serving politicians, police reports, right-wing repressive campaigns, and mainstream academia and media. The information generated by squatter activists is often published in the form of zines, monographs and blogs, and published in native languages through local activist outlets. It is also a difficult task because keeping together a collective formed by individuals with very different origins and political practices is not simple (SqEK 2013). Summarizing research findings and praxis of scholars who are critical and engage in radical struggles is a challenging undertaking (Chatterton et al. 2010; Ruddick 2004; SqEK 2010). Added to this, we are aware of the role of language and the link between knowledge and power (Paasi 2015). Presenting this book in English or deducing every analysis in one language does not do justice to the rich diversity and the enormous wealth of knowledge presented in this manuscript that is related to hundreds of micro and macro social conflicts but still allows a wide circulation of stories that many readers could not be aware of.

We think it is relevant to put together our experiences, thoughts and challenges for two vital reasons: firstly, to show the unique and long-term resistance to close borders and to repressive policies; and secondly, to document the active participation of migrants in the squatter movement and in our societies at large.

Objectives and aims of this book

In this book, we put forth a collective effort towards understanding migrations *and* squatting in Western Europe and North America. We do not focus on either migration or squatting exclusively. Consequently, there are two points to be noted. Firstly, we are fully aware that migrations and squatting are important phenomena outside Europe and North America. In Europe, Social Centers are spaces, usually originated in the squatting of an abandoned place, where people experiment with non-institutional action and association through self-management. They can be ascribed to the long-term fractured tradition of communism and anarchism, obviously filtered by the new radical trends, for example feminism or autonomism (Gaillard 2013; Mudu 2012). The history of Social Centers and squatting has been mainly documented as occurring in Western Europe, but this is changing and can also be traced in Eastern European countries (Piotrowski 2011). Social Centers are spaces originally squatted, and several are legalized for organizing social activities. The first examples can be traced back to the 1960s, first in Amsterdam and then France (Pechu 2010), Germany (Vasudevan 2011), Denmark (Mikkelsen and Karpantshof 2001), Italy (Mudu 2004) and Switzerland (Pattaroni 2007). The UK experienced a different pattern: for years, squatting was mainly linked to housing unavailability, but in the last century Social Centers were established (Common Place 2008). We aim to further more detailed discussions based on the analyses of European and North American studies, following the peculiar nature of repressive policies and mechanisms that prevent and control migrations from the most impoverished regions of the world. Secondly, although the book is focused on migrants and squatters, we must be aware of the fact that we are dealing with two heterogeneous groups, convenient for some general discourses but too vague when addressing, in particular, the trends that combine migration and squatting; therefore we explore their meanings in detail in many chapters.

The book has two major objectives. First, we explore how the intersections between migrants and the radical squatter movements have evolved over the past decades and describe how the policies and discourses on the nexus of victimage (migrants as powerless victims) or security (migrants as dangerous security treats) and precarity (migrants occupying menial or illegal jobs) are resisted. Following up with the aforementioned aspects, we analyze how squatted spaces can be transformed by migrants in different European cities. We illustrate, for example, the different mechanisms of solidarity protests by migrant-squatters and Social Center activists, elaborating on: resurgence when discrimination comes from above or below; the productivities/arts of struggles against deep exclusion; and different kinds of collaborative strategies of struggle in context-specific and situated histories. This leads us to investigate how different spatialities are conceived and realized by radical practices, and to discuss the difficulties and critical issues that emerge when there is a real attempt to build and run self-managed, horizontal structures by heterogeneous subjects like migrants and radical activists. Eventually, we explore how the double repressive device of criminalizing migrants and squatting can be challenged.

Organization of this book

In the following paragraphs, we introduce the reader to the main themes and questions raised by the authors of the book. The book is organized in five sections. In Part I, we introduce the global context of bordering and frontiering that constructs a very articulated topography of the denial for many humans to be "subjects of lives"; this happens through a range of violent tactics that aim to dehumanize, criminalize and victimize migrants instead as subjects that threaten undeniably genuine rights-holders. Borders and courtrooms are two of the main criminalization infrastructures. In the US, Operation Streamline is a fast-track programme, costing between US$7–10 million per month that removes judicial discretion and allows prosecution of individuals apprehended crossing the Mexico-US border. Andrew Burridge, in his chapter, describes the Operation Streamline dragnet and resistance to it. Across the ocean, Frontex is a European agency with a 114-million-euro budget, in 2015, to reinforce border control and surveillance at the European borders. Sara Casella Colombeau introduces readers to the expanding and invasive growth of Frontex. All the huge b/ordering apparatuses are part of the "strategies of spatializations" that create "Undocumented Territories" – territories that are created by and for human beings who are illegalized by states. Henk van Houtum and Kolar Aparna investigated these "Undocumented Territories". When the border is concentrated in one place and crossing border movements are denied, we find people trapped, as is the case of Calais that "divides" France from the UK. The case of Calais represents one of the places of denial of the whole official rethoric on human rights circulating in Europe. But the solidarity squatting practices set up by the Calais Migrant Solidarity Group are one of the most interesting because illegalized migrants are able to resist extreme violent police tactics. The Calais Migrant Solidarity Group reflects on how different squatted spaces function in Calais. They introduce us to the issues of lack of shelter and housing faced by "non-citizen" migrants, similar to redlined minority "citizens".

Part II of the book focuses on struggles around housing and housing policies that prevent a large proportion of the population, migrants and squatters from occupying a place to live. Migrants are confined in low quality houses and to the lowest level of the housing market. They dwell in degraded private apartments and, sometimes, in subsidized social houses. What happens to those who have no access to any subsidized (social) housing? The cases of large cities such as Marseille in France, Rome in Italy and Berlin in Germany offer many examples of struggles where migrants have become active producers of their spaces of life. Florence Bouillon discusses the essential political questions that squats, used for housing purposes and mainly inhabited by migrants in France, pose to the societies in which they are located. She does so using her ten years of experience squatting in Marseille and Paris. Nadia Nur and Alejandro Sethman describe the participation of migrants within Rome's "Right to Inhabit" movement and analyze its implications on the expansion of rights for non-native Italians. Their description provides us with a new viewpoint on the development of housing patterns for migrants, migrant participation in squats and the political dimension of migrant housing activism that configures an emergent *urban citizenship*. Also small cities present interesting cases, such as the one of Catania, described by Federica Frazzetta, where a collective mostly composed of students has supported the squatting of a building by Roma migrants. The shortage of housing also affects students, in particular migrant students, who become fundamental political players in many major cities. Cesare Di Feliciantonio describes this effectively through student narratives. The concluding chapter of this section illustrates squatting experiences of Turkish migrant women in the early 1980s in Berlin. Azozomox and Duygu Gürsel, in their intervention, raise several vital questions. How did the struggle of migrants get marginalized in the narrative of urban struggles? How does the squatting of migrant women reveal the limits and the possibilities of the squatting movement? How do the untold stories of migrant squatting change our understanding of migration and the squatting movement?

Part III expands on diverse issues of exclusion, criminalization and precarity. The migrant-squatter combination is discussed in detail by Stephania Grohman. The migrant-squatter combination is of crucial importance, not only for the UK. In fact, while neither squatters nor migrants enjoy much support in public discourse, the convergence of the two groups, in the figure of the "migrant-squatter", combines two distinct modes of exclusion into an unparalleled image of threat to the territorial control of citizens. Severe modes of exclusion experienced by the Roma introduce us to the "Roma question" and the creation of slums. The French case is analyzed by Thomas Aguilera, who starts his analysis with two questions: how informal settlements and their inhabitants have been racialized since the 1960s by state policies and the media, and how this process has impeded the disruptive use of squatting by these groups and their supporters. He next explores how slum dwellers are able to resist and exploit resources to survive. The resistance Aguilera refers to is subversive in the sense that it challenges the rules of housing as well as social and urban policies. The same subversive resistance can be traced in Italy. For almost three years between 2003 and 2005 in Bologna, Romanians (Roma, for the most part) took part in an occupation that represented a complex political experience, participated in and analyzed by Fulvia Antonelli and Domenico Perrotta. The Bologna occupation meant not only a place of shelter and organization for migrants in transit or intending to settle in Italy, but also an opportunity for migrants and activists to build common pathways towards making claims for the right to a home, free mobility of persons, and a decent job. The claims of these rights are part of the struggle of the refugee group "Lampedusa in Hamburg" in Germany and the solidarity campaign organized between 2013 and 2015. Simone Borgstede shows how the solidarity campaign successfully forged and involved various spaces of conflict, including the St. Pauli neighborhood, and questions who belongs to where and what it means for "a community" to host people living without recognized equal rights and access to its resources. After the first three sections that have analyzed examples of the harshest conflicts and difficult struggles we have in Europe, we expand our reflections on the tough situations in which migrants, squatters and migrant-squatters find themselves.

Diverse strategies of squatting are encompassed in Part IV, exploring the asymmetry that exists between migrants and their supporters, and the issue of gender roles.[3] The first case of asymmetry that we explore is related to the "right to sanctuary". Serin Houston illuminates the processes whereby migrants seeking sanctuary are involved in squatted spaces provided by religious groups in the US. The right to seek refuge has opened new terrains of struggle that show the complexities of immigration policies, migrant experiences and strategies to support squatting. Based on his experience in Madrid, Spain, Miguel Martínez reflects on these complexities and the relationship between migrants and native political activists in their practices of squatting. This relationship took different forms in time, and different specific dynamics of autonomy, solidarity, engagement, and empowerment are identifiable. These dynamics can be outlined when analyzing squatting experiences in detail. This is what azozomox and the International Women's Group show in Berlin. Migrant stories can be self-transformative through a process of collective and mutual learning. The reasons for squatting, the discussions on gender issues, paternalistic approaches by leftist parties an inherent lack of comprehension of "the migrant" women by native Germans are covered in an extensive manner from the personal narratives of four women activists of the International Women's Space, interviewed by azozomox.

Part V highlights several autonomous struggles, mainly carried out and adopted by Social Centers and migrants, to carve a niche in their neighborhoods and among native communities, while at the same time adopting often contradictory tactics to oppose state repression. Tina Steiger uses the case of Trampolinehuset, an autonomous Center for refugees in Copenhagen, to shed light on the diverse and broad set of actors involved to actively challenge repressive asylum policies. Romain Filhol, by analyzing the movement of migrants and refugees in Caserta (Southern Italy), introduces one of most contradictory issues at stake in these struggles involving migrants and radical squatters. The issue is the need to provide legal papers for migrants and, at the same time, refuse to negotiate with institutions that are responsible for the state of affairs, institutions that are often corrupt and even run in open support of mafia activities. Greece offers a case where negotiations with institutions have not represented a characteristic of the autonomous groups. Vasiliki Makrygianni describes and analyzes how several spaces of solidarity, resistance and struggles have recently emerged, and Athens has become a privileged field of struggle. She highlights how much squatting as praxis of struggle, resistance and re-appropriation of the deprived means of production and reproduction is gaining ground. The permanent state of economic crisis faced by many countries leads us to rethink economic models. Claudio Cattaneo does this by drawing our attention to how to sustain an economically efficient, ecologically effective and socially just strategy for degrowth to contrast the capitalist imperative of infinite growth. He adequately distinguishes left and right degrowth proposals. From a short case study in New York, in the 1980s, Hans Pruijt offers readers a different take on the migrant-squatter analysis.

He unfolds potentially contradictory issues related to the "class", "belonging", "positionality", "hierarchy" and "difference" among squatters. This chapter presents the difficulties of being accepted in deprived neighborhoods and the potential to export experiences from one country, the Netherlands, to another, the US. The Netherlands have had the longest tradition of squatting in Europe. Deanna Dadusc investigates a series of issues that are fundamental to the struggles in the Netherlands and globally. She documents how migrants use squatting as a tool of protest and to gain visibility, as well as to open collective spaces to organize their struggles in a systematic way. She also addresses the contradiction of struggling to get legal papers for migrants and activism to create a world of "no papers at all".

In the next sections we discuss some general implications of experiences and practices that are addressed in detail in the chapters that compose the five sections of this book.

Why migration, squatting and radical autonomy?

The three components of the title of our book have no obvious sequence or single definition.

"Migrant" is perceptibly the most difficult word to use, and we cannot ignore questions of language, definitions, theories, labels or controversies. For many years it has often been combined with adjectives to distinguish good integrated migrants from criminal, malevolent, irregular, illegal, bad migrants. The destabilization of the meta-narratives around the binary of "integrated" opposed to "criminals" finds materiality in that area which seems to be its own cornerstone: land (in particular social construction of space) and law (in particular citizenship) (Benjamin 2008). Migration comes from the Latin word *migratio*, i.e., "to move from one place to another", probably related to the Greek verb *ameibein*, "to change". We consider migration an individual and collective experience and we prefer to use migrants in the plural, not in the singular form, because of the insurmountable plurality that exists and must be accounted for. Is "migrant", in the context of squatting, the right word to use? This question extends to all the other definitions attached to migrants, such as foreigners, newcomers, settlers, outsiders, expatriates, exiled and so on. Reconsidering all these definitions and adjectives, which basically can be applied to the entire population, for migrants, is not to deny their existence. Indeed it means taking into account the dynamics of the construction of space of migrants and squatters, the categories that are used and contain them, and the way they are disarticulated. What if unexpectedly migrants, the objects of rights, impossible citizens, take the risk to struggle? Who is the outsider or insider when people (mainly illegalized migrants) join to struggle against the securitization of citizenship? Yet, if migrants are global "external security threats" to the established order, then there are also "internal threats" that are represented by real ongoing practices such as racism, sexism, homophobia and fascism. Resisting these practices has been at the core of many squatting movements.

Also the label "squat" is questionable in various countries; for example, in Spain and Italy, the term "occupy" is used in a broad way that includes the term "squatting". Squatter is translated as *kraaker* in the Netherlands, *besætter* in Denmark and not used in Greece (van der Steen et al. 2014). Instead of naming these spaces "squats", adjectives like "recuperated" and "liberated" for buildings or spaces are preferred (Martínez's chapter). Squatting is usually recognized as an action of occupying a piece of land, a building or an apartment without legal property rights. This general definition is of no use if it is not put in the context of the different places where it happens and of the different struggles that are concomitant (SqEK 2013). In Europe, there is a long cycle of squatting for housing mainly related to migrations and to setting up intentional communities (Martínez 2012; SqEK, Cattaneo and Martinez 2014). Squatting has occurred mainly in big cities, but small urban cases also exist and are important in setting up forms of struggle (see Filhol's chapter).[4] Social Centers have also been related to innovative actions, such as critical mass bicycle rides in cities (Lorenzi 2012). At the core of these experiences, there is the self-management of many activities and direct action to reclaim spaces denied under the capitalist regimes. The direct actions performed in the occupied vacant properties and city spaces are a variety of creative and self-sustaining activities such as housing, guerrilla city greening projects, autonomy over food production, educational and artistic workshops, libraries, counter pedagogies, discussion forums, etc. (Mudu 2014; Moore and Smart 2015). In many cities, squatters have constituted some of the strongest opposition to urban renewal projects (Holm and Kuhn 2011). The Social Centers of Europe have vehemently protested the repressive state-capitalist trends of social exclusion that have multiplied in Europe over the past several decades (SqEK 2013; SqEK, Cattaneo and Martinez 2014). Nonetheless, these positive endeavours/initiatives to reclaim autonomy, self-liberation and self-determination by reusing/recycling resources that were previously left unused or squandered have been repeatedly contested and stopped by hideous surveillance and restrictive mechanisms by state police and judicial apparatuses, which is a matter of continual concern for squatters. Squatting is not a marginal social practice in Europe and has been linked to the shortage of housing after World War II and mass migration flows between and within European countries. But the historical and theoretical analyses or relationships between migration and squatting are omitted, ignored, or at best, overlooked. Sometimes the dialectics are placed in the discursive framework of illegality, precarity, unhealthy living conditions and empathy. One of the relationships that we want to highlight is with the concept, and its application, of autonomy.

The concept of "autonomy" is at the base of a demand for a better life that originated in ancient Greece (Castoriadis 1991). Autonomy is at the core of any project of democracy that advocates participation, responsibility and critical engagement with the political life, the life of the *polis*. Is this the current pattern of western democracies? Several social movements have devoted their energies to addressing this "rhetorical" question and re-defining and affirming autonomous subjects for new models of democracy and international relations.

Autonomy can be defined as either a process of labor self-valorization, negation of state power, or the rejection of colonial domination (Böhm et al. 2010). Regarding the first definition it is worth citing the Italian autonomist tradition that configured autonomy (*autonomia*) as independence of the workers from the general interest of the capitalist class without political mediation of parties or trade unions (Katsiaficas 1997). Autonomia was carried out by a conscious violation of laws and rejection of rules (Berardi 2007). The second definition of autonomy is related to a struggle for negation, the ability to say 'no' to existing forms of power and domination (Böhm et al. 2010; Holloway 2010). A third discourse on autonomy is related to post-development theories, calling for self-determination and self-organization of people against the imaginary of development (Böhm et al. 2010; Escobar 1992). When we link it to migrations, the concept of autonomy acquires a fourth meaning. In fact, the process of becoming autonomous is related to a series of interventions, around which a distance from the country of origin, relatives and the known environment are laid, allowing the distanced person to problematize who and where they can be and how they can be political subjects. The reasons that underlie migrations vary greatly, and we know that in many cases migrations are forced or induced by circumstantial and political changes or other people's activities, such as environmental disasters, poverty, sexual emancipation, war and persecution, and the like (Klein 2007). Most of the time different reasons to move create patterns that distance migrants from their place of origin or force them to "escape" their unsafe domicile. An analysis of these actions of autonomy is therefore fundamental in order to better understand migrations as a "social movement" and a necessity depending on individuals' will. This analysis is also worth studying because autonomy is historically specific, highly contextual and contested, and variably used within various political traditions, and "such flexibility in usage and interpretation makes it a dangerously fuzzy concept" (Pickerill and Chatterton 2006: 4). Projects of autonomy are also attributable to diverse migration trajectories that see migrants running their private business (ethnic economies, for example) often celebrated for their autonomous entrepreneurial spirit, but autonomy in this sense obviously offers very limited room for critical engagement against injustice, segregation and racism (Mudu 2007). This leads us to not underestimate the incorporation of autonomy within oppressive projects characterized by discourses of creativity and independence, or development by autarkic local practices and self-determination that imply "closed and patrolled frontiers".

The authors of this book have been looking for a definition of autonomy that is collective, invented by different subjects, originated by different perspectives, placed at the intersection of the occupation of spaces and social radical struggles. Realistically, migrants are autonomous when they squat alone without the initial support of native radical squatters, although some cooperation may occur later on (detailed in Martínez's chapter). At the same time, we delve into other expressions of autonomy which are radical and not built on the acceptance of injustices, but constitute a continuous intervention against the roots of power relations and a challenge to the functioning of corporate-driven capitalist societies. We can also conceive autonomy as a real non-hierarchical association among people (Katsiaficas 1997), or autonomy as the "right to self-government" (explained in Steiger's chapter). Borgstede (in Chapter 13) simply defines radical autonomy as: "you do not ask for what you need, you know what you need and try to gain with others, you have the right to have rights". Autonomy contains all these meanings together and it would be limiting to impose a narrow definition which would not do justice to the variety and variability of the different contexts and actors involved in practically defining it. At the same time these different claims around what autonomy might be open up spaces of tension and struggle around what autonomy might mean (Böhm et al. 2010). Eventually, autonomy means opening up frontiers of resistance and change towards radical practices, self-management and an equal society (Böhm et al. 2010).

In our framework, we highlight the urgency to re-think the space of the "newcomers", and this implies questioning the space between property regimes and citizenry. What is turned into a political contention is the fact that unequal function of the property regime and unequal access to national citizenship are in direct conflict with the right to decent and affordable housing, on the one hand, and the right to be granted a decent migrant or asylum status, on the other (Martínez and Grohman's chapters and Anderson's foreword). To outline the relationship between squatting, for housing or for Social Centers, and migrants, we have adopted a set of mixed theoretical and empirical frameworks. The real experiences that tie migrants to squatting and radical politics answer the methodological question on how to articulate these three enormous and cumbersome concepts chosen for the title. We explore how squatting offers an alternative to dominant and repressive anti-immigration regularization policies, and their implications on the social acceptance of migrants. This alternative is not a single successful pattern to be replicated and exported, but it is arranged around many attempts and failures to host, live and struggle with people that speak different languages, and have different social origins. The interaction between migrants and radical squatters is always full of surprises, frustrations, uncertainties, mistakes, passions, joy and fear. The relation between national and migrant squatters can even reproduce "colonial" relationships, backed by the "dream" behind the choice of the nationals to engage with migrants as part of an international proletariat that the activists must organize. Migrants are not usually anti-capitalist or autonomous. The intersection of migrants, radical struggles and squatting reveal an incredible set of multiscalar mechanisms that call into question the manufactured consensus of "who belongs where", as well as the prevailing configuration of housing and cultural rights. Questioning belonging mechanisms aims at building explicit politics of scale to contest and reconfigure the particular differentiations and hierarchies that shape citizenship and prevent the intersection of migrants and squatters (Smith 1996). This intersection is not invisible to authorities and, in fact, there is an invention of the "migrant-squatter" by police and the media, but it is related to the fear of the hybrid creature that results from their combination, a creature that is deemed an invader of space (see Grohman's chapter). But the migrant-squatter has demonstrated a long-term ability to unsetttle the re-proposition of the old Athenian way to re-produce *metics*.

In the fifth and fourth centuries BC, metics (freed slaves, artists and merchants) designed the status of foreigners, with a word originating from *metá* meaning change, and *oîkos* meaning house, but also eco is defined as economy or ecology. As often happens, one word that has several meanings makes us reflect on the condition of social change, related to where people settle and how resources are used, when privileges are enforced to create "inferior" status for metics, who in old Athens could not own property or marry a citizen. The inferior status and the "disturbance" brought by the migrant-squatter leads us to a different way of inventing cosmopolitanism, a fourth case other than the three identified by Harvey as "out of philosophical reflection", "out of an assessment of practical requirements and basic human needs" and "out of the ferment of social movements" (Harvey 2009). The first two cases point to the idea of imagining and claiming rights, and to the idea of nations as the main counterpart. Although Harvey is skeptical on cosmopolitanism as the exclusive preserve of the elite, he envisions both worker cosmopolitanism and ecological cosmopolitanism (Gidwani 2006). The third case looks for cosmopolitanism built out of social movements that work on long-term perspectives; but migrants are not a social movement in the popular academic discourse but a movement for existence (Fominaya and Cox 2013) – a movement for existence with a spatio-temporal frame of "here and now" (see chapter by Dadusc). The cosmopolitanism invented by migrant-squatters and squatters building their politics of scale and steering to autonomous radical subjects is a fourth kind. As soon as a cosmopolitanism which protects the right to freedom of mobility comes into being, there arises a new social duty which is clearly different from the institutional one to regulate mobility. The latter is conditioned by and mined with repression and oppression. Migrants and squatters represent themselves as subjects who have a "duty" to take action against the lottery of citizenships, begging for rights, the "reason of state" and the new war against movement of people, we are reminded of the war against the marginalized, sexed and racialized people (Federici 2004, Nyers 2003). This duty comes from answering in practice several questions that are disseminated and brought forward in this introductory part. Migrants and squatters work out a duty out of a hypothesis of rights; that is, rights that must be socially justified and proved. This duty has to be connected to the existing and desired social obligations of solidarity and hospitality. This duty is exerted in the grey area of the almost uncodifiable right of resistance, of disobedience to oppression and illegal acts carried out in states of need.

The right to be "here and now"

The short-term view of the migratory process that characterizes immigration policies and popular media constructions of migrants impress on people that migrants are, on the whole, security threats, undermining the cultural homogeneity of host societies (Castles 2004; Golash-Boza 2009). Current economic crises are blamed on the increase in numbers of illegalized migrants. These policies are formulated on models that construct divisionary borders between migrants from western metropoles and those from southern nations, in particular Arab countries, who had colonial relations with Europe (Bigo 2006; van Houtum 2010).

A "border" is not simply a line or a benign construct but a set of apparatuses that racializes and marginalizes, and a sheath that segregates large parts of the world (Walia 2013; see Part I of the book). Borders not only signal the direction in which money flows, billions of dollars or euros or other currencies, but also how the economy is shaped socially, for example with the need of reactionary and violent mobilization of people to patrol on land and maritime borders or a dragnet that eliminates outsiders (see Burridge's and Casella Colombeau's chapters). It is outside the scope of this book, but it is important to not forget migrants within risk construction patterns that offer a reduction of their spaces simply as spaces of flows organized by criminals. There is a widespread recognition of the uselessness of border controls to stop migrants or the justification of the existence or creation of borders – the maintenance and security for which have increased the toll of deaths and imprisoned vulnerable people. What should be done when the legal b/ordering strategies are illegal or justified on unjust racist principles? Why do such apparatuses continue to exist? How do migrants counter them? What if the law prohibits asylum seekers from working or pursuing education for an uncertain amount of time, marked by Kafkaesque bureaucratic procedures?

If migration is analytically defined or perceived as mere movement of people across spaces and networks, the definition stays superficial and ignores many factors, reasons and contexts under which people move. Furthermore, following the histories of migration, colonialism, global trade, arms trade and militaristic-neocolonialist interventions, the bourgeoisie's over-consumption and over-accumulation (Luxemburg 1963; Chattopadhyay 2014), the pertinent question that becomes apparent is, where lies the legitimacy of the EU or North American governments in rejecting the environmental and war refugees and asylum seekers? Departing from the conventional discourses of migration that depict migrants as "victims" "trapped between state and capital" (see azozomox and Gürsel's chapter), if migration is re-analyzed from more radical perspectives (e.g., feminist or autonomist), then it stands as a peculiar social movement. Migrants during their trajectories need to reconsider gender roles and when involved in squatting they have to tackle directly, without the filter of any institution, populist views of women and how patriarchy is taken for granted or how the manifestations of patriarchy are overlooked even in "advanced" democracies.

In many cases, women squatter migrants want to make real the possibilities of lives that disarticulate patriarchal systems of capitalist exploitation. This disarticulation happens through the acceptance of a common gender condition, the refusal of difference between migrants and non-migrants, and eventually the rejection of the visions of the world that are imposed by privileged classes. However, because of prominent queer activism, the situation is often complicated by the fact that in many squats the political activities avoid the common fallacy of equating gender only with women. Consequently, simplified feminist approaches that insist on the primacy of gender, treating structures such as gender and race as mutually exclusive, thereby marginalizing racism and other structures of oppression, are challenged. Succinctly, if migration is considered the mere movement of people, we miss the most vital point – that it significantly glosses over a set of bio-political devices that aim to subjugate and constrict migrant bodies and souls by controlling their social and economic aspects of life (Foucault 1979). Migrations encompass the social and subjective dimensions of movement, challenges and experiences in resisting violence at borders, precarious working conditions and racism in neighborhoods, and highlight the creativity of migrants as autonomous individual subjects in discovering and self-managing commons in solidarity with native radicals. Radical movements, in this sense, are largely trajectories of people that provoke direct changes that have implications for personal, social and environmental future revolutions. Nevertheless, migrations have always represented: 1) challenges while establishing social orders, 2) conditions of generation or re-generation of societies, and 3) subversive acts for those crossing national borders without legal documents (Papadopoulos et al. 2008; Bojadžijev and Karakayalı 2010; Mezzadra 2011).

Although at a normative level the restrictions on migrations act as a fundamental factor to regulate the job market, migrants are treated as a dangerous social exception. Western economies have been and are dependent on a migrant workforce. Yet harsh, convoluted and expensive regularization processes keep the status of particular migrant workers illegitimate so their labour can be over-used and de-valued, and this is framed within a labour market that keeps the relevant workforce out. "In Europe, rights and resources that were formerly distributed on the basis of universalism – albeit the universalism of the white, male, able-bodied subject – are now distributed on the basis of work. It is the 'worker citizen' or the taxpayer who deserves these rights. The rise of the worker citizen has seen the development of two types of undeservingness: idleness (the unemployed citizen) and not belonging (the migrant)" (Anderson's foreword in this volume). Over the past few decades, with the movements of people on-the-margin to the richest parts of the world, border controls and migration management have become pivotal in the capitalist discourse and for capitalist exploitation for profit generation (Balibar and Wallerstein 1991). Histories of migrations clearly show the failure of global capitalism and prolonged repression of those marginalized (see Makrygianni's chapter) – an "apparent" failure because the real purpose of restrictive migration policies is not to prevent people from migrating, but rather to produce a disciplined (labor) army of people that accepts marginalization in the name of better chances for the future, amnesties and the like (Cornelius *et al.* 1994). The EU's migration policy is an examplar in the circular way they facilitate labour migration and mobility while discouraging settlement (Feldman 2011). The increase in people crossing borders without legal documents also contests the existence of the nation-state, national boundaries, national identities and inherited privileges by citizenship status (Bauder 2003; Hayter 2000). Then citizenship and legality nexus, which are built upon a broad array of juridical categories, are challenged by migrations. Squatting adds other dimensions to this challenge because several key social aspects are put into question: 1) the neoliberal project of accumulation by dispossession; 2) the efficiency of privatization and "free" market as optimum service providers; 3) the idea of private property as the cornerstone of capitalist sovereignty and 4) the "bourgeois" rule of law characterized by the contradiction between universal rights and their actual implementation through national legislation (Aureli and Mudu 2015). Let us concentrate on the fourth point in the following paragraphs of this section.

The tragic epilogues of policies against migrants are written time and again in global histories. There are famous universal principles, not legally binding, such as the ones stated by the United Nations General Assembly in 1948.

> Everyone has the right to freedom of movement and residence within the borders of each state... Everyone has the right to leave any country, including his own, and to return to his country... (Universal Declaration of Human Rights 1948: Article 13)

But mainstream ideologies also recognize these principles, and one does not need to be a radical to believe in them. So, a paradoxical issue would be that radical movements can merely be the critical consciousness of legal orders that regulate the movement of people. That is the reason why we need to describe a different cosmopolitanism instrumental in creating previously nonexistent subjectivities. But if these subjectivities are not able to modify other subjects and the institutions, freedom of movement becomes a mere statement of principle. Migration and legality controls are juxtaposed with racism, sexism and classism with a total exclusion of migrants and their families as "aliens", putting a ban on their cross-border movements at times of crises and on their threadbare lives in host nations (Agamben 2003; Chattopadhyay 2013). Although an increasingly salient aspect of western nations lies in the recognition of the fundamental respect of human rights, in reality, none of the 1948 Universal Declaration of Human Rights conventions are respected, as activists and scholars claim that racial discrimination, harassment, ghettoization, segregation and violence towards migrant populations persist across European Union (EU) member states (Balibar and Wallerstein 1991; Fekete 1997). It is established that the post-9/11 attacks in New York and Washington DC, post-3/11 attacks in Madrid and post-7/7 attacks in London had no connection with migrants, refugees or asylum seekers. Still, the increasing global apartheid geopolitics are deceptively juxtaposed with a global war against illegalized migrants, the mystifying global 'war on terror'. These events have resulted in repressive b/ordering mechanisms through stricter policies. This is why this book starts with a section dedicated to the b/ordering framework that creates the "migrants". Without borders there are no migrants. The geopolitics of borders has inscribed the circulation of people within new juridical hierarchies (Rigo 2007) fueled by a "differentiating machine" of unequal spaces of citizenship rights (Isin 2002). Reflecting on migration and squatting in conjunction offers an opportunity to explore ways in which we can fight against the positioning of migrants and citizens as competitors for the increasingly poor privileges of membership (Anderson's foreword in this volume) and against the regimes of permanent temporariness that are increasingly enforced (Picker and Pasquetti 2015). These fights are carried out through squatting and

Social Center activities where citizenship is conceived 'beyond the state' or rejected as the constitutive ground of the political (Tyler and Marciniak 2013; Aureli and Mudu 2015).

Squatting for Social Centers and housing has also meant an action, or a collective political tool/mechanism, to collaborate and build solidarity networks through mutual aid and horizontality to actively and directly resist detention, deportation, and unfair immigration policies, and to challenge the legitimacy of immigration law enforcement and profits made from the current enforcement regime. In reality, different spatialities have been conceived and realized by radical practices. Abandoned space is recuperated and named not by authorities but by squatters, territorial stigmatization is challenged, gentrification opposed and borders disarticulated. Readers will find themselves interpreting a range of spaces that include not only Social Centers or abandoned squatted buildings, but also churches (*kirkeasyl*), "jungles", *kharabas*, slums and sanctuaries. Multiple political scales are built, sometime clearly identifiable: transnational (see Borgstede's chapter), international (Burridge's and Casella Colombeau's chapters), national (van Houtum and Aparna chapter) and local (Antonelli and Perrotta's chapter). But, quite often, embedded scalarities are the norm (see the chapter by Calais Migrant Solidarity Group). A range of political issues, not usually discussed, emerge when relations and structures have to be built by heterogeneous subjects such as migrants and activists within all these "deviant" spaces. But it is an endless spatial struggle (Makrygianni's chapter).

This is not a book on a peculiar niche of social issues, on odd or rare events. On the contrary, we maintain that repressive policies that target migrants and squatters are likely to be extended to the rest of the population (having already extended to non-humans); we sincerely wish to address the difficulty in setting up forms of resistance and different paths oriented by/to self-management. Having described the defeats, misery and disillusionment of various struggles, nevertheless, still leads us to reaffirm the fact that these forms of resistance are the only routes to escape and to eke out a living for those committed to freedom of movement, social justice, redistribution of resources and alternative lifestyles.

Squatting for housing and migrants

In Europe, Social Centers and squatted houses have different national and social patterns. To give an example, squatting solely dedicated to art performances exists in France and the Netherlands but are negligible in Spain and Italy. Southern European countries lack a welfare state, so the discussion on forms of income related to the activities occurring in the Social Centers is a long-term one (Membretti 2007). In any case, squatting produces a different economy (SqEK, Cattaneo and Martinez 2014). The interactions of radical activists and migrants in struggles carried out through squatting, or from occupied spaces, deserve careful analysis regarding the contexts, political trends, and prevalent typologies (Pruijt 2012). As pointed out in the previous section, squatting, in particular for housing, entails key social aspects that are related to capital accumulation, privatization and property regime. Current capitalism is oriented toward creating a huge lack of housing for the population. Public housing projects have been abandoned in favour of "free" market housing controlled by big private speculators. Lower- and even, in many cities, middle-class individuals and families do not have access to the renting market, not to mention buying a property. But the house is not a commodity like any other; it is not a simple object, and housing policies have become a tool to marginalize people who are already in a precarious position. Right to ownership is more important than the right to housing. Right to ownership is regulated by a property regime based on and fueling a sharp class division that offers no alternatives except for segregation in suburban derelict buildings if not homelessness to those who are in lower classes. Squatting is a way to proclaim one's existence, directly, physically and materially (Bouillon and Nur and Sethman chapters).

Although ignored or taken for granted in many migration studies, segregation that enforces social hierarchies and class divisions are problematic issues. Racialization and depoliticization are some of the devices used to deny the right to housing (see Frazzetta's chapter), but this can create the conditions for disruptive practices and subversive resistance (detailed in Aguilera's chapter). Many cases described in this book also offer the possibility of adding an element regarding the so-called "integration" of migrants because the possibility for poor people, in this case under-privileged migrants, to aspire to decent living conditions is negotiated with the working class, and this is clear when considering housing and education, for example (see chapters by Frazzetta, Bouillon and azozomox and IWS). Real solidarity, factual negotiation on the desires and needs of migrants, happens both in positive and negative terms at the bottom of the social hierarchies. Roma migrants have been and are undoubtedly at the bottom of European social hierarchies. Their condition is full of stories of migrations, squatting and evictions. Rarely have they been involved or perceived as part of radical struggles. When this happens, their struggles offer several points for reflection (see Aguilera, Frazzetta, and Antonelli and Perrotta's chapters). In the case of Roma, nomadism is neither political nor romantic; it is related to evictions and movements from camp to camp. Nomadism does not match the idea of creating temporary autonomous zones nor provide a particular joy of homelessness (Bookchin 1995). Nomadism becomes a social condition that entraps those pushed to move within harsh and difficult environments, and nomads "may turn its meaning upside down and adopt the label in a positive sense to empower themselves, like the queer and hacker activists" (Martínez's chapter). But the evident risk in the process of collective experimentation to build autonomous spaces is that the process is ultimately more transformative and empowering than the resulting structures (Brown 2007).

Migrant autonomy and radical squatting against racist trends

In this section, we analyze the broad issue of contrasting current racist and xenophobic trends and the questions and contradictions that we cannot escape if a successful intersection between migrants and radical squatters is expected. Patterns of migration to northern European countries have constituted the 'new' Europe after the destruction from World War II, while migration to southern European countries is a relatively new phenomenon as is the organization of repression against immigrants (Castles and Kosack 1973; Calavita 2005). For decades, to oppose the socialist bloc within the then European Community, the movement of people was presented with discourses on western open borders, freedom of movement and having a passport document denied in the Eastern bloc. With the collapse of the Soviet bloc, these discourses were soon superseded by a propaganda that sought to justify increasingly repressive measures against migrants and restrictions on the rights of asylum seekers for the sake of protecting the "purity" of western rich nations, thus transforming non-Europeans in an "underclass" to be exploited (Bhabha 1998; De Genova 2008; Chattopadhyay 2013). Repressive regularization processes, newspaper narratives, and state anti-immigrant, racist and sexist sentiments and discourses criminalized migrant subjectivities (see Dadusc's chapter). In many countries, these new patterns of racism (re)presented an important ingredient for the neo-fascist discourses (Pred 2000; Merrill 2006). The way to describe racism, its construction, and its critique is no longer valid, and new analysis to take countermeasures is needed. There is no racism without some form of violence and segregation, and counter-struggles to such violence (see azozomox and IWS's chapter). For instance, exclusion on the grounds of immigration restrictions can, and all too often does, cost lives (Düvell 2003). More than 20,000 people have died in the last twenty years trying to circumvent European entry restrictions (Harding 2012; Kingsley and Jones 2015). In confronting immigration (particularly its "illegal" construction) and racially discriminatory policies, many people have chosen to ignore or to violate laws in favour of assisting "out of status" migrants, refugees and asylum seekers in squatted religious spaces (see Borgstede, Houston, Steiger and Dadusc's chapters). Sanctuaries exist to secure immunity or survival for refugees (Derrida 2005).

In ancient Greek society and mythology, the right and duties related to hosting beggars and strangers were one of the tests to measure civilizations, and *The Odyssey* was based on it. Welcoming *Phaeacians* were the opposite of the *Laestrygonians* or the *Cyclops*, and any action that was disrespectful towards hospitality was punished by the gods. Not hosting people also meant an act of cannibalism, such as the case of *Polyphemus* in *The Odyssey*. The Judeo-Christian tradition dictated the need to provide cities of refuge (see Houston's chapter) and terrible punishments for those who did not respect hospitality duties. Interestingly, squatting of churches develops in different ways, according to the contexts. For example, in the occupation of the Caserta's main church, squatting is used as an alternative to property speculation (Filhol's chapter). In other cases, it is unclear how much autonomy migrants have acquired through their experiences of sanctuary in the US and how religious groups have supported "sacred squatting" (Houston's chapter). Migrants pose a challenge to the credibility of the Church in countries where the political role of religion is still heavy, for example in the Italian case. The reputation of spaces of "universal welcome" is also challenged by state repression politics. This makes relevant the question of who offers asylum when the national authorities deny it. In fact, we face ambiguous and supplemental welfare functions by solidarity groups and NGOs (Lippert and Rehaag 2012). People participating in Social Center activities are not alone in directly and actively resisting state repression, but they are "unique" as they simultaneously and holistically address a set of other oppressive apparatuses that are operating not only against migrants but also against society at large (see Table I.1).

The repertoire of collective actions is the set of means that are effectively available to a given set of people in order to make claims on individuals and groups (Tilly 1978). The repertoire of contention in resisting racist practices against migrants involves several actors and social movements and is very broad. In fact it encompasses a wide variety of actions, ranging from "conventional" political strategies to cultural expressions, from confrontational tactics to violent acts (see Table I.1). Direct action protests include self-immolation or hunger strikes or self-harm, neither to be considered acts of desperation or a willful heroic agential subject (Nyers 2015). The repertoire of actions adopted by migrants, as a social movement that anticipates, circumvents and fights oppressive and regulatory regimes of control, puts their agency capable of indeed transforming political space-time. Migrants have mostly the chance to engage in "unconventional" forms of political participation, analyzed in-depth in this book. The asymmetry of the meeting between migrants, squatters and different organizations is embedded in a complex interaction of solidarity, engagement and empowerment (Martínez's chapter). Solidarity operates within strategies of juridical support and public campaigns, engagement within material support and direct actions, and empowerment in self-managed experiments. Different tactics, see the case of the *lotta-vertenza* (struggle-dispute) in Caserta (Filhol's chapter) or the case of Calais (Calais Migrant Solidarity Group chapter) or the analysis of Lampedusa in Hamburg (Borgstede's chapter), provide the impossibility of a single formula or recipe to lead this kind of struggle. Furthermore, tactics and strategies reveal the contradictions regarding the cooperation between squatters, activists and migrants. In fact, while squatters and activists fight against governments, politicians, and immigration laws, illegalized migrants need to negotiate and compromise to get a legal status (Tyler and Marciniak 2013). Illegalized migrants have a political objective of "Papers for all", against "No papers at all" of squatters (Dadusc's chapter). "No papers at all" risks being a perspective only for those who have the privilege to refuse documents and rights, leaving migrants with the feeling of being instrumentally "manipulated" by activists to fight, by a privileged position, their own struggles against the governments (Dadusc's chapter).

How do we work out the reproduction of these contradictions without provoking the deadlock of social movements and struggles? How do we exit the dynamics by which governments use squatters and squatters use migrants? How can

Table I.1 Repertoire of Contention in Resisting Racist Practices against Migrants

Tools	*Actions*	*Goal*	*Participation*	*Examples*
Public campaigns	Press releases; publishing reports; discussion forums; demonstrations; petitions, civil disobedience; canvassing; marches against detention/ deportation; art, food and music festivals; workshops; revival of sancturay cities and other right-based struggles.	Recognition and extension of rights and welfare benefits; political empowerment; regularization, freedom of movement, fight against abuses, domestic violence, racism and xenophobia; awareness raising campaigns towards migration/ migrants; resisting racist regulations.	Migrant individuals and families/communities; activists; journalists; artists/graphic designers; NGOs; political parties; squatters; virtual involvement of aforementioned activists.	Campaigns against detention centres, No Borders (Europe); We are here to Stay (Netherlands and Germany); Sans Papiers, Fondation Abbé Pierre, ATD Quart Monde, Emmaüs, Médecins du Monde, l'Armée du Salut, les Enfants de Don Quichotte (France); Refugees Welcome campaign (Germany); Sin papeles, SOS Racismo (Spain); Info Immigranti (Italy); DRUM, New York (US), New Sanctuary Movement (NSM) (US and Canada). No one is Illegal (NOII), Toronto, Vancouver, Montreal (Canada);
Legal support	Legal aid; advice and advocacy in negotiating civic rights (such as marriages, divorce, separation and alike); legal procedures towards abusive authorities, employers and traffickers; negotiating anti-deportation legal procedures and prison systems; legal action against rape, abuse and domestic violence.	Knowledge and exercise of rights, access to legal resources; logistical support on immigration law and welfare system.	Migrants, Immigrants lawyers, prefecture and the judicial system, other professionals, supportive community and activists.	Specific services within Social Centers. Services provided by civic organizations, churches, progressive law firms, radical non-governmental organizations.

Sanctuary provision and personal support	Providing clandestine or disruptive housing for migrants.	Combination of support to individuals / families and overt civil disobedience.	Individuals and families, church groups.	New Sanctuary Movement (US and Canada).
Political infrastructure and capacity-building for migrant-led initiatives	Technical assistance in e.g. setting up Internet presence or registering associations, providing space for groups to meet or run offices.	Development of migrant-led organizations for mutual support and political expression.	Migrants, NGOs, Social Centres, individual volunteers.	Trampoline House, Copenaghen (Denmark); Metropoliz, Rome (Italy).
Material, Cultural Functional and Emotional support	Temporary or stable housing (makeshift camps, squatted buildings, legalized squats); decent and affordable housing; food; welfare services (child care, social health education); counseling; language training; spatial orientation; collective conversation; education and other provisions for children.	Provision of shelter and material subsistence; access to available resources; local cultural assimilation;, social network, personal wellbeing.	Migrants and migrant communities, activists, neighbors, staff in public services, social workers/ NGOs/Grassroots and philanthropic organizations, housing organizations and squatters; impromptu networks and diaspora, solidarity groups.	COSTI Immigrant Services (Canada); Food No Bombs, Doctors without Borders (MSF), Partners of Health, Yo Sí Sanidad Universal, Brigadas Vecinales de Observación de Derechos Humanos (Watch the Cops); State Watch, Amnesty International (Global).
Direct action (aside from Squatting)	Protests against the policing of borders, internment camps and forced confinements and detentions, strikes, blockades, protest marches/demonstrations, protest camps, occupation of vacant properties; interference in deportations; interruption of anti-immigrant official meeting;	Stop deportations and detentions, xenophobia and racism; abolition of internment camps for migrants; fight police control, harassment and abuse, to get legal documents of stay and citizenship.	Migrants and migrant communities, activists, professionals, civic organizations.	Lasciate CIEntrare (Italy); Calais Migrant Solidarity, Calais (France); Borderlands Autonomous Collective, No More Deaths, Samaritans, Humane Borders (US).

(Continued)

Table I.1 Repertoire of Contention in Resisting Racist Practices against Migrants (Continued)

Tools	*Actions*	*Goal*	*Participation*	*Examples*
	cultural events; open gates, fences and walls in deportation and detention centres and borders; protests against deportation of migrants (sea, land and air transits); self-immolation/self-harm of detained migrants.			
Squatting	Planning and execution of the occupation. Provide legal help; language lessons; housing needs; other inter-cultural/material needs; migrant-squatter protest marches, demonstrations, blockades; publishing reports, Creation of spaces to live, recycle, grow fruits and vegetables, workshops (sewing, bike) self-management of the premises and activities; Self-organization - running their own activities inside squats, Collective/solidarity activities of domestic work; defense of the squat; Hosting cultural events.	Development of self-managed practices; satisfaction of housing needs; equality in daily life; promotion of social and cultural events; build on social networks of mutual support, tackle gender issues.	Migrants and squatters, participants and visitors of squats, local native residents.	International Women's Space, Berlin; Rote Flora, Hamburg (Germany); Prosfigika, Athens (Greece); ex Canapificio, Caserta (Italy); Scalo Internazionale Migranti, Bologna (Italy). Seco and Eskalera Karakola, Madrid (Spain).

becoming aware of privileges condition our struggles? Some genealogies are needed (van der Steen *et al.* 2014).

Most studies ignore the relationships across various forms of protests and their connections with migrants within Social Centers or among Social Center activists. Immigration policies promote spatial segregation and political disconnection from demonstrations and protest actions. The moment in which migrants are treated within security and criminal legislation, an entire set of apparatuses is devoted to migrants' classification and control; when public housing projects disappear and policies are just a ping-pong between neoliberal right and nationalistic right, then the possibility of participating in any democratic process vanishes. Urban degradation, privatization and gentrification provide the tools to segregate populations.

Segregation is not only a physical pattern but also a mechanism that impedes building a critique of the living situation, a culture to oppose the colonization of everyday life by narratives, music, spectacles and symbols created by alien big corporations. Here, the involvement of migrants in squatted Social Center activities represents an important theme. The intersection between squatting and migration has a long history, as happened in the Netherlands in the 1970s or in Germany in the 1980s (Seibert 2008). Generally, in contrast to the UK where migrants from former colonies were enjoying citizenship rights, in France, Germany and Italy, the organizations of the working-classes included migrant workers, mostly factory workers, in their projects and efforts to transform society. This inclusion had important limitations: assimilationism, paternalism, electoral instrumentalism, and economic reductionism (Però and Solomos 2010). As a matter of fact, the majority of migrants in Europe could only engage in alternative forms of political participation (Però and Solomos 2010). The involvement of migrants within the radical squatting projects is relatively new, although it can be traced back to the beginning of Social Centers' existence (Mometti and Ricciardi 2011). This involvement and resulting relations can be documented (see Figure I.1, the case of Bologna in Italy).

The intersections between migrants and radical squatters are driven by two disparate motivations. Firstly, the asymmetrical antagonistic-malicious discourse constructed on migrants within power relations of the host societies informs the justification of illegitimate policies and state repression through capital punishment, discipline and surveillance. Secondly, a general discourse on solidarity and "fraternité" is carried out in squatted spaces. In fact, solidarity with those in need and the oppressed, which include many migrants, is a founding principle of autonomists, anarchists and squatters.

On the first issue, it has to be recognized that the radical left was the first and most serious collective actor in denouncing and opposing the instrumental creation of camps (Temporary Detention Centers) to "host" illegalized migrants, ironically called "hospitality centers" or *filoxenia* in Greece (Makrygianni's chapter). Their belief was that these detention centers constituted a new form of concentration camps and, at some point, they would be extended to other "clandestine"/"illegal" actors. The emergence and increasing construction of of self-management being corrupted, that is, no "demands" to the council, but the imposing of our conditions through concrete actions. This goes on since our first demonstration, with the banner "we don't wanna live in animal conditions" and 5000 participants (which has been hampered by the cgil), until the recent struggle of the immigrants living the cpa in via Arcoveggio and via Guelfa.

WHY SENZA FRONTIERE?

Because a no borders world is the world we want! It must be free from any damned border and from any government which wants us to live in a real apartheid regime. European democracies let goods be carried everywhere, while they arrange jails and deportations for us; they classify us as "regular" or "irregular", and army their borders. Only the world of trade and exploitation has no borders now; we fight to destroy all borders in our world – to make it free from any form of racism, repression and exploitation.

CAN I TAKE PART TO SENZA FRONTIERE INITIATIVES?

Senza Frontiere antiracist group is open to anybody who needs to change his/her life conditions deeply, and who wants to DESTROY THE CAGES of isolation and urban decay local institutions create for immigrants, workers, people with no job and students. We want to COME OUT FROM THE GHETTO to prevent police repression and racist laws from turning our lives into nightmares, in an atmosphere of general indifference. We want to make our wishes true, and carry out our plans!

WHERE AND WHEN CAN I MEET SENZA FRONTIERE?

We meet each Thursday from 6 p.m. on, in via Avesella 5/a (near via Indipendenza), but you can also come other days, from Monday to Friday, from 6 p.m. on.

Telephone number: 051-260556
e-mail: senzafrontiere@riseup.net

In November we'll open an independent and free internet point you can use from Monday to Friday, from 6 p.m. on.
Freedom of communication is our right, we have to save it!

WHAT IS SENZA FRONTIERE?

Senza Frontiere (which means: no borders) is a group of workers, immigrants, students and generally people with no sure income, which is autonomous and independent from any political party, trade union or institution. We fight for the right to a home, to freedom of movement and to dignity for everyone. We want the end of racism and exploitation in all its forms.

HOW LONG HAS SENZA FRONTIERE EXISTED FOR AND WHAT HAS IT DONE SO FAR?

Senza Frontiere started in 1990 when we used as "centro sociale" a squatted building in Via Serlio (then named "La Fabbrica"), where we were already living with 30 immigrants. A little while after that, it was already 200 of us there. That's why we needed a bigger place to live in, and after discussing it in several assemblies, some buildings in Via Stalingrado have been squatted, where today more than 60 families of immigrants are still living. After that, hundreds of immigrants (men, women and kids) joined the fight for home and dignity, and we took action again. Some buildings in the area "La Barca" have been squatted (by people of the Senegalese community), buildings in Via Gobetti (by more than 200 immigrants), Via Rimesse (where the Pakistani community squatted several places) and Porta Zamboni, an action joined by both Italians and immigrants. In the meanwhile, the left-wing council throws people out of the "centro sociale" "La fabbrica" and destroys it, but other squatting actions go on; in Rastignano for example 25 families get their flats, in via Saffi another 20 families squat new houses and the same also happens in Via Don Minzoni, where Senza Frontiere together with the immigrants manage to prevent the police and the fire brigades from evicting the squatters, and even initiates a negotiation with the Council which results in a victory for the immigrants. This is the time when the "Uno Bianca" (some local policemen who after their work shift went on with their dirty job) ruled the all town through theft and murder and especially took it on immigrants. In such a moment the need for self managment became bigger and bigger, and was of course strongly sup-

ported by our group. Once again we needed to take action: 50 immigrant families initially squatted some iacp (a no longer existing council body which decided who got council houses and who didn't) buildings at via Rimesse, and then 30 more joined them. The police tried to evict them after 10 days but the immigrants resisted; after the second attempt we decided to demonstrate on the streets and to begin a day of action, which finished with the immigrants squatting the basilica S. Petronio 2 days and 2 nights long. After that we entered into negotiations with the council; in the meantime some buildings in via del Pallone were squatted, and after a month the police tried to throw us away. The immigrants imposed themselves and gained settlement and were awarded houses, thanks to their determination and to many demonstrations and street blocks.

During these 13 years of movement struggling for the right to a home and to dignity and against any kind of racism we managed to squat houses and to keep them through demonstrations, street blocks and various initiatives. We have always acted consistently with our ideas and with determination, and never let our antiracist policy

Figure I.1 Bologna: Senza Frontiere since 1990 from the Social Center "La Fabbrica"

internment Centers has been identified as the realization of a continuous "state of exception" (Arendt 1973; Agamben 2003). If exception is the way neoliberal states function, it needs to be critically addressed to better understand its formation and the possibilities to deconstruct and tackle it. The state of exception cannot be considered as an absolute, total condition because struggles involving migrants and the ways in which they organize and act indicate how resistance is possible, despite their lack of access to the rights and protections of citizenship. Struggles through squatting allow us to move beyond the dichotomies of camp and city (Sanyal 2010). And even those under extreme conditions of subjugation are able to act in ways that allow them to constitute themselves, although problematically, as political subjects (Tyler and Marciniak 2013).

The relationship between migrants and squatting is problematic because squatting involves a risk of confrontation with police that is higher than other political activities. Yet squatting, in particular, and Social Center activities towards self-liberation, self-determination and autonomy of migrants and their solidarity networks and alliances have developed into an autonomous anti-racist/sexist/classist movement to support migrants' rights-based struggles (in the EU and North America). In Europe, an important radical activity that has evolved, since 1999, is the "no border camp" held in different parts of Central and Southern Europe (Walters 2006).[5] After the no border camp in 2009 in Calais, the Calais Migrant Solidarity Group was formed (see Calais Migrant Solidarity Group's chapter). The Calais Migrant Solidarity Group decided to not campaign on behalf of the Calais migrants, but to take direct action in the struggles against borders with the migrants themselves. This has allowed disarticulating the implicit border (and hierarchy) between 'protest' and 'movement' and inventing a new politics beyond citizenship (Rigby and Schlembach 2013). In Germany, similar to other European Social Centers, direct support has been provided to immigrants and asylum seekers. In many Autonomous Zentrums, there are many collectives dedicated to migrant issues and foreigners.[6] In 2012, several demonstrations for "freedom of movement for all" have been organized and went halfway through Germany; for example, on October 6 of that year, a "Refugee Strike Support" was organized in Cologne, a series of events against racism were also simultaneously held in Hamburg on October 12 through 14, 2012 (see Borgstede's chapter), and the "Refugee Protest March" ended on October 13, 2012, in Berlin (see azozomox and IWS chapter).[7] Migrants' involvement in squatting struggles is also remarkable in France (Bouillon and Müller 2009). The Italian case is quite rich of chronicles and records of the meeting of migrants with the radical squatter movement. For example, in 1990 for eight months in Rome, the Pantanella, a former pasta factory close to the center, was squatted. The Pantanella was the first mass squatting by migrants since the heydays of squatting by southern Italian migrants that characterized and regulated the development of Rome and Northern Italian cities in the 1950s and 1960s. Ten years later in the 2000s, migrants represented the largest share of people squatting within the "Movimenti per il diritto all'abitare" (Right to Inhabit movement). The "right to the city" movement transformed the 1960s rhetoric towards a radical reconsideration of the urban life. This capacity of the Right to Inhabit movements to plan a new urban model has produced an intersection, reversing a development almost in parallel for a couple of decades between housing movements and Social Centers. This intersection is difficult, and it has been played within a non-circumscribable area of reclamation of denied needs and the refusal of precarious labor exploitation. Rediscussing attitude toward working conditions could not be avoided, and new experiments, such as Officine Zero in Rome, RiMaflow in Milan, Fralib in Marseille, Kazova in Istambul or Vio.Me factory in Thessaloniki, arose in Europe to organize squatting around new working models.

The refusal of work, as it is conceived by capitalists and countered by autonomists and anarchists, does not fit migration trajectories easily. People participating in squatting produce distinct economies in various ways, which include the fundamental practice of self-management. Participating in a self-managed experience means sharing responsibilities with other people and breaking the path of being "normalized" by and within the mainstream "hosting" society. This means also building a pattern that opposes the construction of the "model minority" or capitalist immigrant entrepreneurs' patterns (see Cattaneo's chapter on the practices of migrants who work as waste collectors, in the urban economy, in precarious forms of employment, often negotiated daily).[8] In fact, spaces invented through the struggles of migrant-squatters and squatters are not the classic spaces of migrants: enclaves, ghettos, and suburban areas. Their actions materialize spaces and times autonomous from oppression and reveal the simultaneity of openended multiplicities of contested territories (Massey 2005).

The difficulties of defining and arranging diversity among heterogeneous subjects

As long as we think of social issues solely in terms of binary distinctions like "us and them", there is no way towards social justice and no other way to read the conflicts produced by migrations if not through charitable practices, always very limited and ineffective, carried out by a front of anti-racist organizations, from secular to Catholic groups. Social Centers denounce the vision of a society based on the difference between "us" and "them" (Anderson 2013). In many contexts for many years, squatters have been composed of different kinds of people in terms of social background, roles and identifications. For example, in the US, squatters of diverse ages and genders have mostly been "white" people, such as in "Homes Not Jail" in San Francisco (Corr 1999). A counter-perception is not at all unproblematic, and, in fact, within Social Centers the positions of migrants have not always been linear. "[…] the literature on squatting in Amsterdam wholly ignores the consequences of the radically changing face of the city's population. By only focusing on a particular profile of white squatter activist, the historical texts present a misleading and distorted view" (Kadir 2014: 32). Squatters have to build a good reputation in the neighborhood where they live, which is quite a difficult task in the face of racism, xenophobia and neo-fascism (Antonelli and Perrotta, Frazzetta chapters). Apart from this, within different contexts squatters have been accused of being white and privileged (see Pruijt's chapter). The composition of squatters both for housing and Social Centers and its evolution in space and time is an important factor to consider. Many cases show that migrants have taken a central role in squatting, such as the housing movements in Italy (Nur and Sethman, and Antonelli and Perrotta chapters) and France (Bouillon chapter). The cases of Rome and Bologna show prominent involvement of migrants but also a misunderstood field of divergence between activists who see migrants as the "subject" of revolution and those who analyze things in a more complex way (Antonelli and Perrotta's chapter). More recently, campaigns against the "precaritization of labor" built common interests between students, precarious workers and migrant laborers (see Di Feliciantonio in this volume), like the Mayday demonstrations (Euskirchen *et al.* 2009) and labor mobilizations (Pulido 2007). The heterogeneous experience of squatting also includes examples of great difficulties in self-managing and being able to repudiate mastery and rejection ideologies of rejection and practices between local activists and migrants. Cultural production in a broad sense has often opened the door to horizontal cooperation able to overcome "us and them" divisions. Classical artistic production, theatre, music, painting and new forms of art that allow people to express themselves are often features of squatted spaces. Culture relations have the power to bind people together in squatted spaces, but many other relations are difficult to orient. The decision to squat and build alternative spaces also has heavy gender implications that are complicated by the origins of migrants.[9] For example, in 1981 in Kottbusser Straße in Berlin, migrant women had serious problems when squatting (see azozomox and IWS chapter). These difficulties are important issues to be raised and discussed if squatters want to deeply self-reflect on self-management and decision-making. Decision-making processes in horizontal collectives are various and challenging (Mudu 2012; Piazza 2011). To make decisions, communication is fundamental, and language barriers are still problematic (see Martínez's chapter). The need for action, for direct action, cannot hide the confrontation of cultures, the asymmetry in organizing radical performances, the daily routine of running self-managed spaces, and the capacity for mobility and travel that activists have (Owens *et al.* 2013). In squatted spaces, migrants have to explain to their supporters that they are not victims that they do not need someone for help, but they need people who want to work together (see azozomox and IWS chapter). Working together means organizing activities, like who is going to clean the toilets and kitchens at the end of the day, and defining or forecasting what problems might arise and how to solve them. Briefly, this experiment attempts to set up a different economy. In the book, there is an interesting argument about using migration questions more than economic policy as the key explanatory variable to distinguish between left- and right-wing visions of economic models (chapter by Cattaneo). Proposals on self-sufficient economies and local and traditional productions are both present in radical left discourses as well as right-wing suggestions. De-growth narratives differing on the "self-sufficiency/open local economies" scenarios oppose the "closed and patrolled frontiers" and "national autarky" strategies of right-wing policies as well as differences based on "cultural mix" versus "national identity" approaches (detailed in Cattaneo's chapter). For an understanding of collective autonomy it is then important to consider the possible social intersections due to the reclaiming of commons (Caffentzis and Federici 2014). We collectively see the radical struggles described in this book as one potential way to create social structures in which it is not possible to establish who is a foreigner and who is not, or at least de-link any "foreign" status from exclusion and oppression.

Beyond citizenship and borders, integration and segregation, rights and illegality

Although it is difficult to sum up the situation in Europe (at least, in the western part) and North America, there is a common thread of action that links most of the radical movements (in particular, anarchists and post-autonomists) and migrants. This thread is represented by the refusal of the "status" of migrants and the manifestation of legality and illegality that has surfaced in North America and the EU around migrants coming from various marginalized countries affected by socio-economic, environmental and political crises (De Genova 2004).

The intersection of radical groups that originate at many Social Centers is quite irregular, and in each country and in different cities there exist extremely heterogeneous situations. Today, citizenship, legality and rights are problematic concepts that envelop the discourses on migration and migrants, usually synthesized under the patterns of "integration", "regularization", and "normalization". These three patterns converge often to "segregation", which is a complex process within the colonization of people by neoliberal policies to hinder any form of subjective and collective autonomy. In each European country and city migrants and "minorities" are considered integrated just using a few statistical indicators when in reality different degrees of welfare policies affect them. In some countries, particularly in southern Europe, migrants are excluded from access to social housing, and, in the case of Roma people, they are segregated in camps. It has to be recognized that migrants participating in squatted Social Centers or in collective squatting are, although contradictorily, one of the few examples of de-colonizing colonized spaces and opening borders. It is important to take into account all the struggles described in this book because these struggles have always had an impact on policies and laws. Moreover, they help seize and shape the global agenda to redefine such matters as human rights, gender, poverty, and environment (Appadurai 2006). Such a global agenda can only emerge by connecting these struggles to other radical struggles beyond rhetoric on general human rights, gender equality and the fight to end poverty, and draw linkages to the current ecological crises.

Notes

1 Since 2009, SqEK has held informal meetings in Social Centers all over Europe, and also in the United States (USA); its research agenda is published in four languages (see SqEK 2010).

2 All the websites quoted in the book were accessed on the 23 of November 2015.
3 Gender and race transnationally generate problems of naming, since racial terms have different meanings depending on location, context and history. In this proposal, we use "black" as the common term for women or men of African, Caribbean and Asian origins in Britain only, since "black" in North America (USA and Canada) refers only to women and men of African descent. We use "non-white women" or "non-white persons" to refer to women and men of African, Asian, Latin American and indigenous communities transnationally. We also use the terms "women of the global south", "men of the global south" or "people of the global south" since this is now widely used by activists to refer to women and men in what is often and problematically called "the Third World" (developed from Sudbury 2002). We also prefer to use terms such as "European", "West", "South", "Arab" and "Greek" to designate (contested) cultural formations, not geographical locations, or "racial types" (Castoriadis 1991).
4 For instance, follow the activities of Can Piella a r-urban squat in Barcelona, Spain (http://www.canpiella.cat/).
5 http://noborder.antira.info/.
6 See the documentary on the Swiss case, 'Zurich: learning german autonomously' (2010): a-films.
7 See the posting by refugeetentaction.net (see also Euskirchen, Lebuhn and Ray 2009; http://www.kanak-attak.de/ka/about.html).
8 On specific situations of migrant struggles in Southern Europe since the 2008 crisis, and their relationship to anti-austerity struggles, there is some literature based on national cases (for example for Greece: Mantanika and Kouki 2011 and Pistikos 2012; for Italy: Oliveri 2012; for Spain: Varela 2009).
9 In Rome in the Pantanella nearly 20,000 migrants passed through the Pantanella during the occupation, all of whom were men with the sole exception of one woman (De Angelis 2005).

References

Agamben, G. (2003) *State of Exception*. Chicago: Chicago University Press.

Anderson, B. (2013) *Us and Them? The Dangerous Politics of Immigration Control*. Oxford: Oxford University Press.

Appadurai, A. (2006) *Fear of Small Numbers*. Durham, NC: Duke University Press.

Arendt, H. (1973) [1951] *The Origins of Totalitarianism*. New York: Harcourt Publishers.

Aureli, A. and Mudu, P. (2015) 'The logic of squatting within and beyond legality', Paper presented at the RC21 conference in Urbino 28 August 2015.

Balibar, E. and Wallerstein, I. (1991) *Race, Nation, Class*. London: Verso.

Bauder, H. (2003) Equality, justice and the problem of international borders. The case of Canadian immigration regulation. *ACME* 2(2): 167–182.

Benjamin, S. (2008) 'Occupancy urbanism: radicalizing politics and economy beyond policy and programs', *International Journal of Urban and Regional Research* 32(3): 719–729.

Berardi, F. (2007) Genesi e significato del termine "Autonomia". In Bianchi S. and Caminiti L. (eds) *Gli autonomi*. Vol.II. Roma: Derive ed Approdi, pp. 40–54.

Bhabha, J. (1998) 'Get back to where you belonged: Identity, citizenship, and exclusion in Europe', *Human Rights Quarterly* 20(3): 592–627.

Bigo, D. (2006) Security, Exception, Ban, Surveillance. In Lyon, D. (ed.) *Theorizing Surveillance: The Panopticon and Beyond*. Cullompton: Willan Publishing, pp. 46–68.

Böhm, S., Dinerstein, A. C. and Spicer, A. (2010) '(Im)possibilities of autonomy: Social movements in and beyond capital, the state and development', *Social Movement Studies* 9(1): 17–32.

Bojadžijev, M. and Karakayalı, S. (2010) 'Recuperating the sideshows of capitalism: the autonomy of migration today', *e-flux* 17(6): 1–9.

Bookchin, M. (1995) *Social Anarchism or Lifestyle Anarchism: An Unbridgeable Chasm.* Edinburgh: AK Press.

Bouillon, F. and Müller, F. (2009) *Squats. Un autre point de vue sur les migrants*. Paris: Editions Alternatives.

Brown, G. (2007) 'Mutinous eruptions: Autonomous spaces of radical queer activism', *Environment and Planning A* 39(11): 2685–2698.

Caffentzis, G. and Federici, S. (2014) 'Commons against and beyond Capitalism', *Community Development Journal* 49(suppl 1): i92–i105.

Calavita, K. (2005) *Immigrants at the Margins: Law, Race, and Exclusion in Southern Europe*. Cambridge: Cambridge University Press.

Casas-Cortes, M., Cobarrubias, S., and Pickles, J. (2015) 'Riding routes and itinerant borders: Autonomy of migration and border externalization', *Antipode* 47(4): 894–914.

Castles, S. (2004) 'Why migration policies fail', *Ethnic and racial studies* 27(2): 205–227.

Castles, S. and Kosack, G. (1973) *Immigrant Workers and Class Structure in Western Europe*. London: Oxford University Press.

Castoriadis, C. (1991) *Philosophy, Politics, Autonomy*. Oxford: Oxford University Press.

Chatterton, P., Hodkinson, S. and Pickerill, J. (2010) Beyond scholar activism: Making strategic interventions inside and outside the neoliberal university', *ACME* (9)2: 245–275.

Chattopadhyay, S. (2013) Commentary. Border choreography, penal states and bare bodies. *Environment and Planning D,* Pion. Available: http://societyandspace.com/feed/.

Chattopadhyay, S. (2014) 'Post-colonial development state, primitive accumulation of Nature and social transformation of ousted Adivasis in the Narmada valley', *Capitalism Nature Socialism* 25(4): 65–84.

Cohen, S. (2003) *No One is Illegal*. Stoke on Trent: Trentham Books.

Common Place (2008) *What's This Place? Stories from Radical Social Centers in the UK and Ireland.* Leeds: University of Leeds.

Cornelius, W., Martin, P., and Hollifield, J. (1994) *Controlling Migration: A Global Perspective*. Stanford: Stanford University Press.

Corr, A. (1999) *No Trespassing*. Cambridge MA: South End Press.

De Angelis, R. (ed.) (2005) *Iperurbs-Roma. Visioni di conflitto e di mutamenti urbani*. Roma: DeriveApprodi.

De Genova, N. (2004) 'The legal production of Mexican/migrant "illegality"', *Latino Studies* 2(2): 160–185.

De Genova, N. (2008) 'Inclusion through exclusion or implosion', *Amsterdam Law Forum* 1(1): 43–52.

Derrida, J. (2005) *On Cosmopolitanism and Forgiveness*. London: Routledge.

Düvell, F. (2003) 'Some reasons and conditions for a world without immigration restrictions', *ACME* 2(2): 201–209.

Engel-Di Mauro, S. (2006) 'Reflections on "The Struggle against the Rebel Body"', *Capitalism, Nature, Socialism* 17: 66–73.

Escobar, A. (1992) 'Imagining a post-development era? Critical thought, development and social movements', *Social Text* 31/32(10): 20–56

Euskirchen, M., Lebuhn, H. and Ray, G. (2009) 'Big trouble in borderland: Immigration rights and no-border struggles in Europe', *Migration Struggles and Migratory Aesthetics*, paper presented at the Radical Art Caucus panel. College Art Association Annual Conference in Los Angeles.

Federici, S. (2004) *Caliban and the Witch*. New York: Autonomedia.

Fekete, L. (1997) 'Blackening the economy: the path to convergence', *Race and Class* 39 (1): 1–17.

Feldman, G. (2011) *The Migration Apparatus: Security, Labor, and Policymaking in the European Union*. Stanford: Stanford University Press

Fominaya, C.F. and Cox, L. (eds) (2013) *Understanding European Movements: New Social Movements, Global Justice Struggles, Anti-austerity Protest*. London: Routledge.

Foucault, M. (1979) *The History of Sexuality* (1). London: Allen Lane.

Gaillard, É. (2013) 'Habiter autrement: des squats féministes en France et en Allemagne. Une remise en question de l'ordre social', *Thèse École Doctorale Sciences de l'Homme et de la Société*. Rabelais de Tours: Université François.

Gidwani, V. K. (2006) 'Subaltern cosmopolitanism as politics', *Antipode* 38(1): 7–21.

Golash-Boza, T. (2009) 'The immigration industrial complex: Why we enforce immigration policies destined to fail', *Sociology Compass* 3(2): 295–309.

Goldman, E. (1917) *Anarchism, and Other Essays*. New York: Dover Publications.

Harding, J. (2012) *Border Vigils: Keeping Migrants Out of the Rich World*. London: Verso Books.

Harvey, D. (2009) *Cosmopolitanism and the Geographies of Freedom*. New York: Columbia University Press.

Hayter, T. (2000) *Open Borders: The Case against Immigration Control*. London: Pluto Press.

Holloway, J. (2010) *Crack Capitalism*. London: Pluto Press.

Holm, A. and Kuhn, A. (2011) 'Squatting and urban renewal: The interaction of squatter movements and strategies of urban restructuring in Berlin', *International Journal of Urban and Regional Research* 35(3), 644–658.

Isin, E. (2002) *Being Political. Genealogies of Citizenship,* Minneapolis: University of Minnesota Press.

Kadir, N. (2014) Myth and reality in the Amsterdam Squatters' Movement, 1975–2012. In van der Steen, B., Katzeff, A. and van Hoogenhuijze, L. (eds) (2014) *The City Is Ours: Squatting and Autonomous Movements in Europe from the 1970s to the present*. Oakland: PM Press: pp. 21–61.

Katsiaficas, G. (1997) *The Subversion of Politics: European Autonomous Social Movements and the Decolonization of Everyday Life*. Atlantic Highlands: Humanities Press.

Kingsley, P. and Jones, S. (2015) *EU sidestep on migrants will do nothing to curb Mediterranean death toll.* Available: http://www.theguardian.com/world/2015/jun/26/eu-sidestep-migrants-mediterranean-death-toll.

Klein, N. (2007) *The Shock Doctrine: The Rise of Disaster Capitalism*. New York: Picador.

Lefebvre, H. (1996) *Writings on Cities*. Oxford: Blackwell.

Lippert, R., and Rehaag, S. (eds) (2012) *Sanctuary Practices in International Perspectives: Migration, Citizenship and Social Movements*. London: Routledge.

Lorenzi, E. (2012) 'Alegría Entre Tus Piernas: To Conquer Madrid's Streets'. In C.C.E.L.E.E.A.C. (ed.) *Shift Happens! Critical Mass at 20*. San Francisco: Full Enjoyment Books: 59–70.

Luxemburg, R. ([1913] 1963) *The Accumulation of Capital*. London: Routledge.

Mantanika, R. and Koukib, H. (2011) 'The spatiality of a social struggle in Greece at the time of the IMF. Reflections on the 2011 mass migrant hunger strike in Athens', *City* 15(3–4): 482–490.

Martínez, M. (2012) 'The Squatters' Movement in Europe: A durable struggle for social autonomy in urban politics', *Antipode* 45(4): 866–887.

Membretti, A. (2007) 'Centro Sociale Leoncavallo Building Citizenship as an Innovative Service', *European Urban and Regional Studies* 14(3): 252–263.

Merrill, H. (2006) *An Alliance of Women: Immigration and the Politics of Race*. Minneapolis: University of Minnesota Press.

Mezzadra, S. (2011) 'The gaze of autonomy. Capitalism, migration and social struggles', In Squire, V. (ed.) *The Contested Politics of Mobility: Border zones and Irregularity*. London: Routledge: 121–143.

Mikkelsen, F. and Karpantshof, R. (2001) 'Youth as a political movement: Development o the Squatters' and Autonomous Movement in Copenhagen, 1981–95', *International Journal of Urban and Regional Research* 25(3): 609–628.

Mometti, F. and Ricciardi, M. (eds) (2011) *La normale eccezione. Lotte migranti in Italia*. Roma: Alegre.

Moore, A. and Smart, A. (eds) (2015) *Making Room: Cultural Production in Occupied Spaces*. Barcelona: Other Forms, the Journal of Aesthetics & Protest.

Massey, D. (2005) *For Space*. London: Sage.

Mudu, P. (2004) 'Resisting and challenging neo-liberalism: the development of Italian Social Centers', *Antipode* 36 (5): 917–941.

Mudu, P. (2007) 'The people's food: The ingredients of "ethnic" hierarchies and the development of Chinese Restaurants in Rome', *Geojournal* (68)2–3: 195–210.

Mudu, P. (2012) 'At the intersection of anarchists and autonomists: Autogestioni and Centri Sociali', ACME, 11(3): 413–438.

Mudu, P. (2014) 'Self-managed Social Centers and the right to urban space', in Clough Marinaro, I., Thomassen B. (eds) *Global Rome, Changing Faces of the Eternal City*. Bloomington and Indianapolis: Indiana University Press, pp. 246–264.

Nyers, P. (2003) 'Abject cosmopolitanism: The politics of protection in the anti-deportation movement', *Third world quarterly* 24(6): 1069–1093.

Nyers, P. (2015) 'Migrant citizenships and autonomous mobilities', *Migration, Mobility, & Displacement* 1(1): 23–39.

Oliveri, F. (2012) 'Migrants as activist citizens in Italy: Understanding the new cycle of struggles', *Citizenship Studies* 16(5–6): 793–806.

Owens, L., Katzeffi, A., Lorenzi, E., and Colin B. (2013) 'At home in the movement: Constructing an oppositional identity through activist travel across European squats'. In Fominaya, C. F. and Cox, L. (eds) *Understanding European Movements*. London: Routledge, pp. 172–185.

Paasi, A. (2015) "Hot Spots, Dark-Side Dots, Tin Pots": The Uneven Internationalism of the Global Academic Market. In Meusburger, P., Derek, G. and Suarsana, L. (eds) *Geographies of Knowledge and Power*. Springer Netherlands, pp. 247–262.

Palidda, S. (2013) *Racial Criminalization of Migrants in the 21st Century*. Farnham: Ashgate.

Papadopoulos, D., Stephenson, N., and Tsianos, V. (2008) *Escape Routes. Control and Subversion in the 21st Century*. Ann Arbor: Pluto Press.

Pattaroni, L. (2007) 'La ville plurielle. Quand les squatters ébranlent l'ordre urbain. Enjeux de la sociologie urbaine' In Bassand, M., Kaufman, V and Joye, D. (eds) *Enjeux de la sociologie urbain*. Lausanne: PPUR, pp. 283–314.

Però, D. and Solomos, J. (2010) *Migrant Politics and Mobilisation: Exclusion, Engagements, Incorporation*. London: Routledge

Piazza, G. (2011) 'Which models of democracy? Internal and external decision-making processes of Italian Social Centers in a comparative study', *Center of Studies on Politics and Society – Working Paper Series* 1(1): 3–54.

Picker, G. and Pasquetti, S. (2015) 'Durable camps: The state, the urban, the everyday', *City* (19)5: 681–688.

Pickerill, J. and Chatterton, P. (2006) 'Notes towards autonomous geographies: Creation, resistance and self management as survival tactics', *Progress in Human Geography* 30(6): 1–17.

Piotrowski, G. (2011) *Squatted Social Centers in Central and Eastern Europe*. Warsaw: ICRA.

Pechu, C. (2010) *Squat et squatters*. Paris: Presses de Sciences Po.

Pistikos, I. (2011) 'About the relation between theory and action: Drawing on the movement solidarity to refugees in Greece', *ACME* 11(2): 194–201.

Pred, A. (2000) *Even in Sweden: Racisms, Racialized Spaces, and the Popular Geographical Imagination*. Berkeley: University of California Press.

Pruijt, H. (2012) 'The logic of urban squatting', *International Journal of Urban and Regional Research* 37(1): 19–45.

Pulido, L. (2007) 'A day without immigrants: Racial and class politics of immigrant exclusion', *Antipode* 39(1): 1–7.

Rigby, J. and Schlembach R. (2013) 'Impossible protest: Noborders in Calais'. *Citizenship Studies* 17(2): 157–172.

Rigo, E. (2007) *Europa di confine*. Roma: Meltemi.

Ruddick, S. (2004) Activist geographies: building possible worlds. In Cloke, P., Crang, P. and Goodwin, M. (eds) *Envisioning Human Geographies*. London: Arnold, pp. 229–241.

Sanyal, R. (2010) 'Squatting in Camps: Building and Insurgency in Spaces of Refuge', *Urban Studies* 48(5): 877–890.

Seibert, N. (2008) *Vergessene Proteste. Internationalismus und Antirassismus 1964–1983*. Münster: Unrast.

Smith, N. (1996) 'Spaces of vulnerability: The space of flows and the politics of scale', *Critique of anthropology* 16(1): 63–77.

SqEK (2010) 'The SQEK: Squatting Europe Research Agenda - v. 1.0', *ACME* 9(3): 377–381.

SqEK (2013) *Squatting in Europe. Radical Spaces, Urban Struggles*. New York: Autonomedia.

SqEK, Cattaneo, C., Martínez, M. (eds) (2014) *The Squatters Movement in Europe. Everyday Communes and Alternatives to Capitalism*. London: Pluto Press.

Tilly, C. (1978) *From Mobilization to Revolution*. Reading, MA: Addison-Wesley.

Tyler, I. (2013) *Revolting Subjects: Social Abjection and Resistance in Neoliberal Britain*. London: Zed Books.

Tyler, I. and Marciniak, K. (2013) 'Immigrant protest: An introduction', *Citizenship Studies* (17)2: 143–156.

van der Steen, B., Katzeff, A. and van Hoogenhuijze, L. (eds) (2014) *The City Is Ours: Squatting and Autonomous Movements in Europe from the 1970s to the Present.* Oakland: PM Press.

van Houtum, H. (2010) 'Human blacklisting: The global apartheid of the EU's external border regime', *Environment and Planning D* 28(6): 957–976.

Varela, A. (2009) 'Residency documents for all! Notes to understand the movement of migrants in Barcelona', *Refuge: Canada's Journal on Refugees*, 26(2): 121–132.

Vasudevan, A. (2011) 'Dramaturgies of dissent: The spatial politics of squatting in Berlin, 1968', *Social and Cultural Geography* 12(3): 283–303.

Walia, H. (2013) *Undoing Border Imperialism.* Oakland: AK Press.

Walters, W. (2006) 'No border: Games with (out) frontiers', *Social Justice* 33(1): 21–39.

Websites

http://en.contrainfo.espiv.net/

Part I

Borders and frontiers

1 From the desert to the courtroom

Challenging the invisibility of the Operation Streamline dragnet and *en-masse* hearings

Andrew Burridge

On October 11, 2013, twelve individuals locked themselves to two buses operated by the private contractor *GEO Group* (formerly Wackenhut Corporation prior to 2003). At the time, the buses were transporting 70 undocumented persons who had been apprehended by United States Border Patrol (USBP; an arm of the Department of Homeland Security) in the southern Arizona desert just north of the Mexico–US border, and were *en route* to the Evo A. DeConcini federal courthouse in downtown Tucson, Arizona. Meanwhile, another group of six chained themselves to the entrance of the courthouse. Those being transported were caught up in Operation Streamline, a program started in 2005 in Texas, which expanded across the southwest of the US, including to Tucson in 2008. The protesters were successful in halting the buses and shutting down hearings at the courthouse, meaning those detained that day avoided Streamline and thus a criminal conviction (though all were still immediately 'voluntarily' deported). Importantly, the protesters drew wider attention and visibility to this largely invisible process. Those arrested for locking-on around the wheels of the buses – comprised of local Tucson activists involved in various other campaigns, organizing around border militarization and immigration policing in the region – were faced with several charges, most of which were dropped at the first hearing. Two remaining misdemeanour charges heard in a second hearing in July 2015 resulted in a ruling of time served (Ingram 2015). Those involved in blocking the courthouse entrance were found guilty earlier and also sentenced to time served (Ingram 2014).

Zero tolerance and the further criminalization of movement

Operation Streamline, a fast track and 'zero tolerance' program that removes judicial discretion, seeks to prosecute individuals apprehended crossing the Mexico–US border as a form of deterrence (Lydgate 2010; NIF 2012). Through the partnership of the Department of Justice (DOJ) and the Department of Homeland Security (DHS), Streamline operates daily to criminally prosecute individuals for the act of unauthorized border crossing, leading to a criminal conviction and a prison sentence. In 2010 the DOJ reported that Streamline costs between US$7–10 million per month to operate, while US$40 million was spent in Arizona alone in 2009 on incarcerating those prosecuted under Streamline, with US$10 million in court costs per month (NIF 2012: 2). Meanwhile, between 2005 when Streamline was first announced and 2012, the federal government spent US$5.5 billion incarcerating migrants (Grassroots Leadership, 2012: 3). In the Tucson sector of the USBP (which covers almost the entirety of the southern portion of the state of Arizona, and 262 miles of linear border with Mexico), 70 persons per day are randomly selected to be placed in Streamline, and are heard *en masse*, taking anywhere between 30 minutes and three hours (Santos 2014), all while shackled at the hands and feet on average for six hours at a time (Slack *et al.* 2015) – a startling and dismaying scene for observers entering the courtroom. The sheer size of the program, taking place daily in several locations across the US southwest, has resulted in the criminal prosecution of immigration violations outnumbering all other federal criminal cases in the US combined, yet there is little supporting evidence to suggest it has a deterrent effect on those crossing (NIF 2012; OIG 2015), many of whom have US citizen family members awaiting them (Slack *et al.* 2015). Prosecution ranges from a few weeks up to 180 days in the first instance of being caught within the Streamline dragnet, and results in a federal misdemeanour being placed on someone's record, followed by a felony for those caught a second time, impacting any future efforts to gain legal residence in or entry to the US (see Casella Colombeau for analysis on EU Frontex, Chapter 2, in this volume).

Challenging Streamline – A question of procedural justice, or a need for direct action?

Streamline has been contested on a number of grounds since its inception. Perhaps most notably in September 2013, a federal appeals court found that the procedure of mass pleas used in Arizona did not follow due process (Boyce and Launius 2013). As a result, the presiding judge must now address defendants individually. Little else has changed, however, and in 2016 Operation Streamline continues apace, linked in part to an unprecedented growth in prison and detention construction, with private prison companies such as *Corrections Corporation of America* (CCA) and *GEO Group* benefiting (Grassroots Leadership 2012).

The effects of Streamline are multiple: direct-aid humanitarian groups operating in southern Arizona, such as *No More Deaths*, have reported that many crossing will opt to continue walking when in need of hospitalization, rather than risk being placed in Streamline if USBP is contacted (Burridge 2009). Families face separation from loved ones, first incarcerated and then deported, often through a practice of "lateral repatriation" (known as the Alien Transfer and Exit Program, or ATEP) in order to frustrate attempts at reconnecting with guides to cross again, deporting individuals to remote and unfamiliar border towns where they have at times fallen prey to criminal organizations (Slack *et al.* 2013).

Those who are potentially eligible to claim asylum are unlikely to be told of this right. Thirty minutes (though typically much less) with a court-appointed legal representative and a mass hearing provides little opportunity to articulate such a request. Further, a recent inquiry by the DHS Office of the Inspector General (OIG) found that USBP agents do not have guidance on how to deal with those who express a fear of persecution on return, or claims for asylum made under Operation Streamline (2015). Meanwhile the quality of defense attorneys and their legal counsel, who represent several people per hearing, has been repeatedly questioned, as shown through a study of 1,100 persons who went through Streamline. Within this study it was found that only 40 per cent of those interviewed had been informed about their rights by their lawyer; 40 per cent stated that "their lawyer simply informed them they needed to sign their deportation and plead guilty," while 7 per cent reported that their lawyers had not informed them of anything. Perhaps of more concern, only 2 per cent of those interviewed had been informed of their right to denounce abuses, while in 1 per cent of responses it was found that a lawyer had "checked for legal migration options due to family connections, which is generally the first and most important duty of an immigration attorney" (Migrant Border Crossing Study, Slack *et al.* 2013: 29).

Activists in Tucson have also attempted to work with lawyers involved in representing those detained, in the hope of challenging the process from within the courtroom, reminding the lawyers that they were:

> the grease on the wheels of this criminalization machine. Their willingness to limit their arguments, encourage their clients to accept plea deals, and avoid making arguments to reduce detainee sentences is necessary for moving the seventy or so men and women through the system each day. (Borderlands Autonomist Collective 2012: 197)

While Operation Streamline continues to have a significant impact in Arizona, there is an older and ongoing history of border militarization, immigration policing, and prison and detention expansion within this region (Loyd 2012a; Boyce and Launius 2013). A now well-documented outcome of fencing roughly 700 miles of the international boundary between Mexico and the US, beginning in the mid-1990s and continuing until 2008, has been the massive death toll of those crossing undocumented through the Sonoran desert of southern Arizona (Goldsmith *et al.*, 2006; Nevins 2010). Around 2004, the growing presence of media-savvy citizen border patrols, such as the Minutemen (not to mention the continual presence of less publicly visible militias; Doty 2009; Shapira 2013), and the continued targeting and racial profiling of migrant communities by Pima County Sheriff Joe Arpaio, have drawn considerable attention to this region. Humanitarian aid groups providing direct aid to undocumented migrants lost, ill, or injured in the desert began to form in the late 1990s and early 2000s, including *No More Deaths*, *Samaritans*, and *Humane Borders*, who subsequently have been opposed, and at times threatened or prosecuted, by the US government, Border Patrol, and anti-immigrant groups for their often life-saving work. Recent years have seen a continuing pernicious creep of border and immigration policing into local communities, such as through Senate Bill 1070 (2010), which required police officers to take on the role of immigration policing, but have also seen a number of grassroots and community-led campaigns against such measures, including the use of sanctuary in local churches and community response networks (Loyd 2012b; Williams and Boyce 2013; see Houston Chapter 14, in this volume).

The ongoing churn of detention and deportation under streamline

Although ongoing protests in relation to Streamline – employing a diversity of tactics from legal challenges, to court watching programs, marches, vigils, and civil disobedience (Burridge 2009; Borderlands Autonomist Collective 2012: 197–198) – have drawn attention to this program of criminalization and provided much needed solidarity, those caught within Streamline's web are typically reluctant or unable to challenge its existence, as the prospect of a lengthy wait in prison to appeal a case that is unlikely to succeed is understandably unappealing, and relies in part on lawyers advocating effectively for their client (McNeil 2013). There is difficulty also for activists and supporters in providing solidarity or forming any direct relationship with those caught in Streamline, as persons apprehended while crossing the border will be detained and have their hearing within 72 hours, before being deported or incarcerated (and later deported) immediately thereafter.

Yet the ongoing challenging of this system – in part through recognition of its connections to the wider border and migration securitization infrastructure, and to practices of racial profiling, criminalization and incarceration of undocumented migrants – is essential in making the *in*visibility of Streamline visible and contested. As the Tucson-based activist group Borderlands Autonomist Collective deftly noted, reflecting on efforts to challenge the courtroom procedure of Streamline:

> while activists' efforts have by no means ended Streamline or the injustices involved, slowing down the system by insisting that defendants be granted due process is a first step toward broader legal and political strategies to fight the expansion of this program. (Borderlands Autonomist Collective 2012: 197)

References

Borderlands Autonomist Collective (2012) 'Resisting the security-industrial complex: Operation Streamline and the militarization of the Arizona–Mexico borderlands'. In Loyd, J. M., Mitchelson, M. and Burridge, A. (eds) *Beyond Walls and Cages: Prisons, Borders, and Global Crisis*. Athens: University of Georgia Press, pp. 190–208.

Boyce, G. and Launius, S. (2013) 'Warehousing the poor: How federal prosecution initiatives like "Operation Streamline" hurt immigrants, drive mass incarceration and damage U.S. communities', *DifferenTakes*, (82). Available: http://popdev.hampshire.edu/projects/dt/82.

Burridge, A. (2009) 'Differential criminalisation under Operation Streamline: Challenges to freedom of movement and humanitarian aid provision in the Mexico–US borderlands', *Refuge* 26(2): 78–91.

Doty, R. (2009) *The Law Into Their Own Hands: Immigration and the Politics of Exceptionalism*. Tucson: The University of Arizona Press.

Goldsmith, R. R., McCormick, M. M., Martinez, D. and Duarte, I. M. (2006) 'The "Funnel Effect" and Recovered Bodies of Unauthorized Migrants'. Report submitted to the Pima County Board of Supervisors. Binational Migration Institute: University of Arizona.

Grassroots Leadership (2012) *Operation Streamline: Costs and Consequences*. September 2012. Available: http://grassrootsleadership.org/categories/operation-streamline.

Ingram, P. (2015) '12 Operation Streamline protesters sentenced to time served'. Available: http://www.tucsonsentinel.com/local/report/071915_streamline_protesters/12-operation-streamline-protesters-sentenced-time-served/.

Ingram, P. (2014) 'Operation Streamline protesters found guilty', *Tucson Sentinel*. Available: http://www.tucsonsentinel.com/local/report/100314_streamline_protesters/operation-streamline-protesters-found-guilty/.

Loyd, J. M. (2012a) 'Live, Love and Work', Interview with Luis Fernandez, August 2010, In Loyd, J. M., Mitchelson, M. and Burridge, A. (eds) *Beyond Walls and Cages: Prisons, Borders and Global Crisis*. Athens: University of Georgia Press, pp. 228–238.

Loyd, J. M. (2012b) 'Human Rights Zone: Building an antiracist city in Tucson, Arizona', *ACME* 11(1): 133–144.

Lydgate, J. (2010) 'Assembly-line justice: A review of Operation Streamline', *California Law Review* (98): 481–544.

McNeil, S. (2013) 'Streamlined deportation: No one here in this room can help you', *Truthout*, Available: http://www.truth-out.org/news/item/13580-streamlined-deportation-no-one-here-in-this-room-can-help-you#.

National Immigration Forum (NIF) (2012) 'Operation Streamline: Unproven benefits outweighed by cost to taxpayers'. Available: https://immigrationforum.org/blog/operation-streamline-unproven-benefits-outweighed-by-cost-to-taxpayers/.

Nevins, J. (2010) *Operation Gatekeeper and Beyond: The War on 'Illegals' and the Remaking of the U.S.-Mexico Boundary*. Abingdon: Routledge.

Office of the Inspector General (OIG) (2015) *Streamline: Measuring its Effect on Illegal Border Crossing*. Department of Homeland Security. Available: https://www.oig.dhs.gov/assets/Mgmt/2015/OIG_15-95_May15.pdf.

Santos, F. (2014) 'Detainees sentenced in seconds in 'streamline' justice on border'. Available: http://www.nytimes.com/2014/02/12/us/split-second-justice-as-us-cracks-down-on-border-crossers.html?hp&_r=5.

Shapira, H. (2013) *Waiting for Jose: The Minutemen's Pursuit of America*. Princeton: University Press.

Slack, J., Martínez, D.E., Whiteford, S. and Peiffer, E. (2015) 'In Harm's Way: Family separation, immigration enforcement programs and security on the U.S.–Mexico border', *Journal on Migration and Human Security* 3(2): 109–128.

Slack, J., Martínez, D.E., Whiteford, S. and Peiffer, E. (2013) *In the Shadow of the Wall: Family Separation, Immigration Enforcement and Security*. The Center for Latin American Studies, University of Arizona. Available: http://las.arizona.edu/sites/las.arizona.edu/files/UA_Immigration_Report2013web.pdf.

Williams, J., and Boyce, G. (2013) 'Fear, loathing and the everyday geopolitics of encounter in the Arizona borderlands', *Geopolitics* 18(4): 895–916.

2 Frontex and its role in the European border regime

Sara Casella Colombeau

Frontex

Frontex was established as a European Agency by the Council Regulation (EC No 2007/2004) of 26 October 2004 with the general objective to coordinate the operational cooperation between member states in matters of migration control at the external borders of the European Union (EU). In recent years, Frontex has not only experienced important developments (in terms of tasks and means), but it has also been the subject of growing media coverage. In this chapter, the objective is to present the role played by the agency in the EU border regime, ten years after its creation (for an analysis of Operation Streamline, see Burridge Chapter 1, in this volume).[1]

More about Frontex

Frontex is a European agency granted with increasing staff (317 people in 2014) and budget (from 6.3 million in 2005, to 87 million in 2010, and finally 114 million euros in 2015). Its main activity is the coordination of joint operations expenditures (without joint return operations) which represented 37.6% of the total budget of the agency in 2014 (Frontex 2015). They consist of deploying materials and staff at the external borders of the member states in order to reinforce border control and surveillance. Frontex depends upon the staff and equipment that the member states are willing to provide. Some of the national corps participating in the Frontex operations are actually military corps, as is the case of the Spanish *Guardia Civil* and the Finnish border guard.

The agency organizes "joint returns" operations, which consist of chartering expulsion flights of detained migrants in different member states. In 2014, 45 flights (Frontex 2014a: 51–52) were organized by Frontex for a cost of 9,497,000 euros (9.69% of the total budget). Frontex also provides training courses to the border police officers taking part in its operations. These training sessions have strong socialization effects on officers who belong to varying police occupational cultures, providing a common standard and technical skills (Horii 2012).

Another important task relating to the production and use of data concerning border crossings is "risk analysis", which consists of gathering, synthesizing and analyzing the data produced by the border police from all over Europe and during Frontex operations. The "risk" refers here to the "migratory risk". It is presented by Frontex as a fundamental activity, as it is used as a "predictive" tool to determine the characteristics of the joint operations. Thanks to the risk analysis, the board of the agency, which is composed of the representatives of the heads of the borders authorities and two representatives of the European commission, decides upon the operations. This reflects the rather great autonomy in the decision-making that Frontex, as a European agency, enjoys (Pollak and Slominski 2009). This autonomy can be analyzed as a result of the process of the creation of the agency itself (Jorry 2007; Leonard 2009; Neal 2009; Wolff and Schout 2012; Ekelund 2013).

Frontex's regulation has been revised twice, in the years 2007 and 2011. This new regulation provided the agency with the possibility of gathering and using operational nominative data. This regulation also furthered the budget and decision-making autonomy of the agency. The agency is now able to buy or rent its own equipment. A range of measures were also adopted in order to prove the commitment of the agency to the fundamental respect of rights.

Frontex and the building of the European borders

It is quite difficult to assess the impact of the agency's operations on the migratory situation at the EU borders. Indeed, Frontex is only one of the instruments of the general tightening of border control. First, since the mid-1970s several measures have contributed to a general tightening of the border control – visa policies (Guild 2002; Infantino 2013), migrants' detentions and deportations, externalization of migrants' control either to private actors (through carrier sanctions) (Guiraudon 2002; Boswell 2003; Lavenex 2006) or to countries of origin or transit. Second, border surveillance and checks at the ports of entries remain as a national prerogative as the national border authorities watch over the EU external borders.

However, some elements can be outlined. First, Frontex's ten years of activity has not affected the number of people trying to reach the European territory. For instance, a 2014 Frontex report shows that 104,302 illegal entries in 2010 were stopped while more than 285,000 people crossed the border with permits. In some cases, they just diverted the flows. One example is the Hera-I Sea joint operation deployed from 2006 around the Canary Islands (Carrera 2007) and in the territorial waters of Senegal, Mauritania, Cape Verde and Gambia. In 2006, 31,863 migrants entered via the Canary Islands. Today very few migrants are entering the European territory through these islands, and the media attention has switched to other parts of the Mediterranean.

Second, despite the rise of the budget and duties of Frontex, the number of migrants who die in their journey to the EU has greatly increased in the last 15 years. The United Kingdom (UK) NGO United estimates that between 1993 and

2012 more than 16,000 persons died at European borders. Furthermore, in 2004, 1,497 migrants died in their attempts to reach Europe; 3,537 died in 2014.[2]

Against this background, more precisely, two consequences of Frontex's creation can be emphasized. First, the principle of Frontex's interventions produces a legal limbo, where the legal responsibilities are really difficult to impute (is it to the member state, to the individual police officer, to the agency, or to the Commission?). And Second, there is an important symbolic dimension in Frontex's activities and a production of narrative tools that contributes to the definition of the "outsiders" and to the nature of the European border regime.

Frontex's blurred legal responsibility

One of the consequences of Frontex's operations is the difficulty to assign responsibilities in case of fundamental rights violations. There has been intensive academic debate on the subject (Papastavridis 2010; Moreno-Lax 2011; Marin 2014; Fink 2015) to determine if the member state hosting the operation, the individual border officer or the agency should be held responsible. This interrogation arose mostly in relation to practices of "push-backs" reported by Human Right NGOs and other organizations[3] (UNHCR 2009, HRW 2009a, 2009b). This practice consists of intercepting and returning the migrants' boats to their point of departure without considering the presence of people looking for international protection. It violates the principle of "non-refoulement" provided in Article 33 of the 1951 Convention Relating to the Status of Refugee. However, this practice is at the core of the Frontex intervention in third countries' territorial waters (HERA II and III) and has been widely documented as a common practice at the sea border between Greece and Turkey (Migreurop FIDH, EMHRN 2014; Médecins sans Frontières 2014). The adoption in May 2014 of a regulation that prohibits push-backs represents a step forward, but doubts have been formulated with regards to the capacity of the Commission to enforce it (Peers 2014).

Frontex as narrative producer

The relatively marginal position of the agency with regards to the overall effort deployed to secure the European borders is compensated by its strong determination to publicize each of its interventions. Frontex is now a source for mainstream media and as such has important narrative production ability. The creation of Frontex has been characterized as a response to the terrorist attacks of 9/11 (Jorry 2007; Vaughan-Williams 2008) and as a consequence of the securitization or as an "instrument of securitization" (Balzacq 2008) of the migration policies at the EU level. The link with 9/11 can be nuanced; even if the security agenda had some effect on the timing of the creation of the agency, it is inscribed in the long-running dimension (Wolff and Schout 2013). Still, the communication activities of Frontex contribute to a discursive link between migrants and criminal activities, typical of the securitization process (Waever 1995). Risk analysis is one of the main communication tools for Frontex. Reports are published quarterly and annually and describe the 'routes' used by the migrants to reach the EU. Several 'routes' are identified, such as the western Mediterranean route, central Mediterranean route, and western Balkan route.[4] These routes emerge from the aggregation of individual migrant's data at a border crossing and monitoring of migrant itineraries (such as the nationality of the migrants, their point of departure, the means of transportation, and the *modus operandi* to cross). Every migrant's border crossing is systematically associated with criminal activity (migrant smugglers, use of forged documents, etc.). The image depicted here reveals a strong link between migration and criminality.

There is a second aspect of the agency's communication: the ambiguity between search and rescue duties on one hand, and security and border control objectives, on another. Frontex's communication emphasizes the humanitarian aspect of its operation (Horsti 2012). This communication could be interpreted as a smoke screen to hide less valuable activities and practices. But it can also be interpreted as the building of 'humanitarian borderlands' (Aas and Gundhus 2015), where the strong emotional dimension resulting from shipwrecks and tragedies at the border is mobilized in a utilitarian way by the agency in order to depoliticize the consequences of the border policies (Cuttitta 2014). However, the agency's regulations provide no specific mention of search or rescue activities, and these practices only rely on international law of the sea and the obligations attached to it.[5] This aspect also stands out in the statements of the Executive Director of Frontex, Fabrice Leggeri, in April 2015.[6]

Conclusion

Since its creation, Frontex, as an agency, has become more robust if the increasing budget, its autonomy in regards to joint operations, training activities, and data assimilation are considered. Despite an emphasis put on human rights apparatus, there are still growing concerns about the respect of fundamental rights. To conclude, two paradoxes can be outlined as a result of Frontex's activities. First, the emergency-driven activities are particularly at odds with the recurrence of the migrant crises at the EU borders. Second, the agency's role in the definition of the European border regime can be compared to that of border control operations at the US–Mexico border (Andreas 2001: 4): "the rising budget, the sharper tone in the policy discourse participate to the increase and professionalization of the criminality (human smugglers, document forgering) that is supposed to fight". If we consider the focus on the criminal-network dimension of the border crossings in the risk analysis produced by the agency, it seems to ignore the role played by the agency itself, among other border control measures, in the "escalation" process. In sum, Andreas (2001) defined this notion as the process of rising budget of the border agencies, increase in the surveillance technologies, and a sharper tone in the policy discourse to imply a strengthening and a professionalization of the activity of borders crossings linked to criminality.

Notes

1 To see the structure of Frontex see: http://frontex.europa.eu/about-frontex/organisation/management-board/.
2 Several organizations have tried over the past several years to secure the statistics on migrant deaths at the EU borders, such as United (from 1993 to 2012) http://www.unitedagainstracism.org/campaigns/refugee-campaign/fortress-europe/; Fortress Europe at http://fortresseurope.blogspot.no/p/la-strage.html (from 1988 to 2014); the Migrant Files http://www.themigrantsfiles.com/#/counting-the-dead (from 2000 onwards); and also International Organization of Migration http://missingmigrants.iom.int/.
3 The European Court of Human Rights has condemned Italy in the case "Hirsi Jamaa and Others" for forcibly returning Libyan migrants.
4 See for details: http://frontex.europa.eu/trends-and-routes/migratory-routes-map/
5 See the International Convention for the Safety of Life at Sea (SOLAS), http://www.imo.org/en/About/Conventions/ListOfConventions/Pages/International-Convention-for-the-Safety-of-Life-at-Sea-%28SOLAS%29,-1974.aspx.
6 In the *Guardian*, 22 April 2015, Patrick Kingsley and Ian Traynor notes that "EU borders chief says saving migrants' lives 'shouldn't be priority' for patrols", http://www.theguardian.com/world/2015/apr/22/eu-borders-chief-says-saving-migrants-lives-cannot-be-priority-for-patrols.

References

Aas, K. F. and Gundhus, H. O. I. (2015) 'Policing humanitarian borderlands: Frontex, human rights and the precariousness of life', *British Journal of Criminology* 55 (1): 1–18.

Andreas, P. (2001) *Border Games: Policing the U.S.-Mexico Divide*. Ithaca, United States: Cornell University Press.

Balzacq, T. (2008) 'The policy tools of securitization: Information exchange, EU foreign and interior policies', *JCMS: Journal of Common Market Studies*, 46 (1): 75–100.

Boswell, C. (2003) 'The "external dimension" of EU immigration and asylum policy', *International Affairs* 79(3): 619–638.

Carrera, S. (2007) 'The EU border management strategy: Frontex and the challenges of irregular immigration in the Canary Islands', *CEPS Working Document* (261), March 2007: 1–35.

Cuttitta, P. (2014) '"Borderizing" the island setting and narratives of the Lampedusa "border play" ', *ACME* 13(2): 196–219

Ekelund, H. (2013) 'The establishment of FRONTEX: A new institutionalist approach', *Journal of European Integration* 36(2): 99–116.

Fink, M. (2015) 'A "blind spot" in the framework of international responsibility? Third party responsibility for human rights violations: The case of Frontex', in Gammeltoft-Hansen, T and Vedsted-Hansen, J. (eds), *Human Rights and the Dark Side of Globalisation: Transnational Law Enforcement* (forthcoming).

Frontex (2014a) *Frontex General Report 2014*. Available: http://frontex.europa.eu/assets/About_Frontex/Governance_documents/Annual_report/2014/General_Report_2014.pdf.

Frontex (2014b) *Frontex Annual Risk Analysis 2014*. Available: http://frontex.europa.eu/assets/Publications/Risk_Analysis/Annual_Risk_Analysis_2015.pdf.

Frontex (2015) Frontex budget, 07.01.2015. Available: http://frontex.europa.eu/assets/About_Frontex/Governance_documents/Budget/Budget_2015.pdf.

Guild, E. (2002) 'The border abroad – visas and border controls'. In Groenendijk, C (ed.), *In Search of Europe's Borders*. The Hague: Kluwer International Law, pp. 87–104.

Guiraudon, V. (2002) 'Before the EU border: remote control of the "huddled masses"', in Groenendijk, C (ed.), *In Search of Europe's Borders*. The Hague: Kluwer International Law, pp. 191–214.

Horii, S. (2012) 'It is about more than just training: The effect of Frontex border guard training', *Refugee Survey Quarterly* 31(4): 158–177.

Horsti, K. (2012) 'Humanitarian discourse legitimating migration control: FRONTEX public communication', in Messer, M., Schroeder, R., and Wodak, R. (ed.), *Migrations: Interdisciplinary Perspectives*. Vienna: Springer, pp. 297–308.

Human Rights Watch (2009a) Report *Pushed Back, Pushed Around*. Available: http://www.hrw.org/en/reports/2009/09/21/pushed-back-pushed-around-0.

Human Rights Watch (2009b) *No Refuge: Migrants in Greece*. Available: http://www.hrw.org/sites/default/files/reports/greece1009.pdf.

Infantino, F. (2013) 'Bordering at the window: Schengen Visas policies and allocation practices at the Italian Embassy and Consulate in Morocco'. In Carrera, S., Guild, E. and Bigo, D. (eds), *Foreigners, Refugees or Minorities? Rethinking People in the Context of Border Controls and Visas*. London: Ashgate.

Jorry, H. (2007) 'Construction of a European institutional model for managing operational cooperation at the EU's external borders: Is the FRONTEX Agency a decisive step forward?' *Research Paper Challenge - Liberty and Security* (6): 1–32.

Lavenex, S. (2006) 'Shifting up and out: The foreign policy of European immigration control', *West European Politics* 29(2): 329–350.

Leonard, S. (2009) 'The Creation of FRONTEX and the politics of institutionalisation in the EU external borders policy', *Journal of Contemporary European Research*, 5(3): 371–388.

Marin, L. (2014) 'Policing the EU's external borders: A challenge for the rule of law and fundamental rights in the area of freedom, security and justice? An analysis of Frontex joint operations at the Southern Maritime Border', *Journal of Contemporary European Research*, 7(4): 468–487.

Médecins sans Frontières (2014) *EU and Greece turn their backs on refugees arriving at Greek island*, Available: http://www.msf.org/article/eu-and-greece-turn-their-backs-refugees-arriving-greek-islands.

Migreurop, FIDH, EMHRN (2014) *Frontex Between Greece and Turkey, the Border of Denial*, 2014, Available: http://www.frontexit.org/en/docs/49-frontexbetween-greece-and-turkey-the-border-of-denial/file.

Moreno-Lax, V. (2011) 'Seeking asylum in the Mediterranean: Against a fragmentary reading of EU member states' obligations accruing at sea', *International Journal of Refugee Law* 23(2): 174–220.

Neal, A. W. (2009) 'Securitization and risk at the EU border: The origins of FRONTEX', *Journal of Common Market Studies* 47(2): 333–356.

Papastavridis, E. (2010) 'Fortress Europe and FRONTEX: Within or without international law', *Nordic Journal of International Law*, 79(1): 75–111.

Pollak, J. and Slominski, P. (2009) 'Experimentalist but not accountable governance? The role of Frontex in managing the EU's external borders', *West European Politics*, 32(5): 904–924.

Peers, S. (2014) *Analysis of the New EU Rules on Maritime Surveillance: Will they stop the Deaths and Push-backs in the Mediterranean?* Available: http://www.statewatch.org/analyses/no-237-maritime-surveillance.pdf.

UNHCR (2009) 'UNHRC deeply concerned over returns from Italy to Libya', Available: http://www.unhcr.org/4a02d4546.html.

Vaughan-Williams, N. (2008) 'Borderwork beyond inside/outside? Frontex, the citizen–detective and the war on terror', *Space and Polity* 12(1): 63–79.
Waever, O. (1995) 'Securitization and Desecuritization'. In Lipschutz, R. (ed.), *On Security*. New York: Columbia University Press.
Wolff, S. and Schout, A. (2013) 'Frontex as agency: More of the same?', *Perspectives on European Politics and Society* 14(3): 305–324.

3 Undocumented territories

Strategies of spatializations by undocumented migrants

Henk van Houtum and Kolar Aparna

Introduction

Squatting is often seen as an "illegal" occupation from a state-centric perspective in terms of not abiding to formal procedures and laws of claiming space. This often leads to a classic purification state policy, in which the (undocumented) inhabitant-migrant is unilaterally accused, criminalized and made to feel "out of place", without looking beyond the binary of state versus migrant. And also without critically looking at the state's own practices in systematically reproducing the legal/illegal divide as fixed and static. In this chapter we attempt to deconstruct this postulated illegality of squatting places in a state. We will do so by turning the gaze back to the state. We ask: to what extent may the state lay claim to be, in today's globalized and interconnected world, the only legitimate "occupier" of space. What does legality morally mean if this implies a misrecognition, criminalization, dehumanization, and even "collateral-izing" the deaths of people who travel without the documents requested by the state, thereby discriminating between human beings and going against international human rights? What is more, this illegalization by the legal entity of the state may come across as uncontestable, yet, in reality the difference between legal and illegality is obscured daily and shifted by various practices on the ground. For in daily urban practice the illegalized migrants are, at the same time, actively engaged in various transnational social movements for the right to "be here". They are also passively allowed (*gedoogd*) to make or squat places and seek shelters in cities in which they are often helped and supported by actions of various citizens and municipalities of that very same state. Despite also being confronted with exclusion, hostility, and rejection, they also form and create lived relations with documented inhabitants based on friendship, solidarity or love. We will zoom in on and investigate what we call "strategies of spatializations" emerging from such everyday relational practices of undocumented migrants along with fellow inhabitants that break or go beyond the legal/illegal, inclusion/exclusion binary frameworks of spatial relations. These strategies, we argue, together make up the "Undocumented Territories", the territories created by and for human beings that are illegalized by states, not on the state map, yet real.

Seeing States as Illegitimate Squatters

> They ask me for documents, but where can I find the documents they ask for? Can I buy them from the shop? (Inhabitant of Nijmegen identified as "undocumented" by the Dutch state, Interview with authors, 11 September 2014)
>
> [...] a letter from Afghanistan was rejected as documentation for an application for asylum by the immigration office because it did not have a house number. But there are no house numbers in that region and they are not aware of this. (Interview with volunteer at local migrant support organization Stichting Gast, Nijmegen, 15 August 2014)

States have borders and these borders are in essence of paper. Modern nation-states require passports and visas to allow human travel to and across their borders – this phenomenon is historically relatively recent (see e.g. Torpey 2000). The carefully crafted, and minutely detailed paper control mechanisms, that are full of biometric checks these days, have become the global legal standard for identification. This means that states monopolize the legality of international movement. This creates the tautology that one can only be "welcomed" or "unwelcomed" by states on the basis of the documents issued and identified by states themselves (Torpey 1998). It is with paper documents that states define who are their citizens and who are not, as well as which foreign nationals require visas and which of them can travel visa-free across their borders. States control the selection mechanisms that grant citizenship rights to some, while someone without a passport is principally "rendered stateless", and reduced to the bare life of being "only" a human being, in this state-ruled world. For those rendered as "only" human beings, "only" human rights apply and not citizenship rights, and without the sanctionary power of an international organization to legitimize them as citizens (Arendt, 1951; Agamben, 2000).

In a romantic wish to naturalize their governance, nation-states create their own reality or a simulacrum in a Baudrillardian (1994) sense, of the family of subjects born in (naissance) as members of the same family, the nation. Others not born in the nation have to go through a process of naturalization. Illustrative for this b/ordering process are the terms natives, fatherland and motherland, patria and patriots (van Houtum and van Naerssen 2002). In the modernistic era, as Foucault in his writings has made clear (1975- 76; 1977; 1991), the romantic paternal notion has been coupled by the "all-compassing" view of a panopticon. In a way, states have become the Father and the Big Brother at the same time. These modern modes of identification based on detailed and rational mapping and measuring of bodies attached to documented identities (fingerprinting, biometric scanning, etc.) are seen as essential in the desired governing of state subjects (van der Ploeg 2006), were developed notably in the specific contexts of World Wars and European colonial expansions of the late nineteenth and early twentieth centuries. The techniques of ruling through a way of "seeing" colonial subjects on the basis of measurable, rationalized "subject data" emerging from anthropometric experiments in the colonies that were used to identify "dangerously mobile subjects" were exported to the metropoles and today underpin global (im)mobility regimes (Cole 2001; RAQS 2004). Today, this discrimination on race, this apartheid has taken the shape of discrimination based on the grounds of birth, in the privileging of people born in certain states over others for issuing of visas (van Houtum 2010; also see Anderson's foreword, in this volume). More than the passport, it is the paper border of the visa, the paper that a priori irregularises people who cross the border of another state without a visa, while allowing some others to travel visa-free, that is the most powerful control mechanism of the state today (Bo 1998). Hence, it is such enframing (Mitchell 1988) techniques, that not only render people as "subjects" of their own or another state, whose movements of possible entrance need surveillance, but also render people invisible, rejectable and in the end sacrifisable as "homines sacri" (Agamben 1998) because those that enter/inhabit only as "human beings", not as a priori approved citizens.

At the same time, given that it is a human right to flee without proper documents, how can such state-centric imaginaries, and practices be considered legitimate if they continue to be blind to the lived realities of peoples inhabiting and traversing these spaces on the ground? What rights do states have to monopolize the authority under the header of legality to restrict movements of peoples in an increasingly interdependent world? When seen from the perspectives of transnational spheres of migrants who are unwelcomed by states, and the affective and the solidarity-based movements and networks they become part of on the ground, states emerge as illegitimate squatters. In an increasingly transnational sphere and interdependent world, claiming sole authority over human movements and identities by demanding fixed identifications on the basis of pre-determined documents, reproduce divisive rather than inter-relational spatial imaginaries.

Strategies of Spatializations

> When you are not welcome in the streets and also the asylum center puts you out, it feels better to know that here someone says you are welcome. (A migrant on being welcomed by a local support organization, Conversation dated 23 July 2014)

Linking the debates on hospitality to immigration the philosopher Jacques Derrida argues for a hospitality that is not limited by conditions of identity, such as documents, or interrogation for political and legal origin, among others. For Derrida, the non-citizenship of people such as asylum-seekers and undocumented migrants, urges us to rethink democratic relationships beyond the borders of nation-states (Derrida 2000). Not only do spatialities of undocumented migrants and actors working alongside and on behalf of them, serve as a critical vantage point to question the authority of states in regulating (im)mobilities, but also allow for rethinking the harsh borders of legality applied to human relations and identities.

If states grant status to individuals based on legal borders that "dehumanize" and claim ownership of human identities on the basis of specifically required procedures and documents, then those who are unable to identify themselves by the state (so-called "undocumented migrants") often see no other option than to claim a space of existence in ways that "rehumanize" their identity. As a migrant whose asylum application has been rejected explained to us, her asylum interview with the immigration officer was a boring four-hour long interview in which she was denied the possibility of speaking in a common language known to both her and the immigration officer – English, but instead was required to speak in either Dutch (national language of country), or her mother tongue Amharic which would be interpreted, thereby denying her full control over her story. However, her everyday interactions cannot be controlled and regulated by the immigration officer. Her networks with people on the ground have been built on affective and solidarity based relations, rather than the paper technocracy that governs the hospitality of the Dutch state. Meeting her own husband who is a legal citizen did not happen via formal procedures despite her own movements being highly regulated by the state. While she was on her way to the immigration center for one of her many interviews regarding her asylum procedure, she and her current husband first met on the streets as "familiar strangers" because of their shared language of Amharic. While the Dutch state on paper assumes a homogenous "Us" (Dutch) versus "Them" (migrants), on the ground however these boundaries are rather blurred between legal citizens and so-called undocumented migrants, thereby making it much easier to "come-together" and build affective and human social ties. Despite of or precisely because of being criminalized for their very existence and their bodies rendered as invisible and illegal, undocumented migrants rely on forging human relationships, and are actively producing citizenship based on everyday negotiations in ways that precisely challenge this invisibility and illegality.

Apart from everyday relationships that are forged on affective and emotional ties, undocumented migrants also claim space by linking to social movements and local hospitality networks led by legal citizens and local municipalities. For instance, the social movement "We are Here!" led in practice by legal citizens and undocumented migrants (see chapter by Dadusc). As the phrase already suggests the movement claims a "We" (all those unwelcomed by the state as) as opposed to the rejection of individual bodies by immigration procedures. The movement claims political agency and citizenship based on collective practice (Balibar 1997) with an "Are", as active beings rather than the passivity imposed onto them by the state. And finally, a claim is made to the relation between space and human identities with a "Here!" – a here that is claimed based on embodied inhabitance regardless of possession of documents issued by the state. Each of these words challenge the state gaze towards their bodies and existence by precisely inverting the state-gaze that separates documented from undocumented inhabitants as incapable of sharing space and identities.

Local support groups, be it of religious networks of churches and mosques or activist and solidarity networks, reclaim spaces of hospitality with alternative visions of human relations based on trust, and co-habitance, however complex that might be in practice given the inequalities separating undocumented migrants and legal inhabitants. The persistence of such movements, and the diverse range of local migrant support organizations across the world that claim indeed direct relational collaborative practices between legal citizens and undocumented migrants and refugees underline the urgency for, and the feasibility of, a new kind of hospitality beyond that offered by states. But also, even local municipalities that are legal bodies of the state can and do challenge the state gaze precisely because they are faced more closely with the embodied and intertwined struggles of undocumented migrants and legal citizens inhabiting everyday life. As a volunteer from a local migrant support organization in Nijmegen said: "Despite knowing that what we do is not legal according to the state the municipality supports our activities a lot and we have accumulated a lot of goodwill for our work" (interview with authors, 23 June 2014). Such spaces of welcome, and everyday relational practices forged by legal citizens, local civil society, activist groups and local municipalities challenge and transform Agamben's notion of "state of exception" beyond the exclusionary politics experienced by migrants with state immigration systems.

Each of the above relational practices, we argue, stand for the strategies of spatialization inevitably emerging from the lived embodied dimension of borders as spaces of inter-human action that are hard to bind within pre-determined categories of inclusion/exclusion, legal/illegal, citizen/migrant, and unidirectional relations of us/them. As seen in the above examples, ranging from the affective-linguistic dimension to more embodied struggles of everyday life, or to more solidarity and goodwill-based relations, inter-human actions and relations producing space, already confront and de-legitimate the functioning of state (paper) borders and territories as a "line in the sand". Border space then emerges as a space of active reclaiming, constant negotiating, mediating and rehumanizing relations of hospitality based on more reciprocal and embodied practices by inhabitants that are not bounded within fixed and divisive containers but always hospitable to a continuous becoming of relational practices.

This territory is not on the state map

> Yes, I am illegal, but you know I also have a life, and not everyone knows I am illegal … and I want to keep it this way. (Interview with authors, 22 June 2014)

The struggle for space and identity is constantly negotiated on the ground, of which states are but one among the many actors. Hospitality relations forged by and for undocumented Others are not only negotiated around borders of legality/illegality of citizenship produced by states, but more fundamentally come to be intertwined to notions of an "ideal citizen", what Anderson calls "community of value" (2013), that is negotiated at the everyday level. However, the sole authority over space claimed by states, and state-centric notions of communities of value as bounded and inward looking needs to be questioned given the transnationalities of human interactions and relations in everyday life. Rather than looking at "undocumented migrants" as illegal squatters, we need to invert the gaze and question the illegitimate squatting practices of states that are unable to fully see human identities as mobile and inter-relational. Rather than accusing migrants of being "undocumented" we need to question the exclusionary and bare life producing documenting practices imposed by states onto human bodies and relations. In times of globalization, human practices, be it of "staying in place" or "moving/living across places", come to acquire inter-relational and transnational dimensions that go beyond the binaries of state versus migrant, Us versus Them, insiders versus outsiders and natives versus non-natives. State maps are caught in a territorial trap of boundedness and divisive spatial imaginaries. What we call 'strategies of spatialisations' of undocumented migrants and actors working along with and on behalf of them challenge the squatting practices of states that impose a priori visions onto human identities and relations. As opposed to the dehumanising tendencies of state bureaucracies, such strategies are based on rehumanizing spatial relations. Rather than seeing territories of communities as walled and separated, such strategies are led by affective and practical ties between people regardless of legal categories imposed by states. Rather than seeing inhabitants as legal versus illegal, such practices are based on an Arendtian inter-*esse*, the giving meaning to the space "in-between people" (*inter homines esse*) (Arendt 1958). Such *inter-esse* asks for "making space" for continuous negotiation, mediation, as well as embodied relationalities and interactions between people understood as politically charged, open-ended and never fully finished or bounded. So, rather than exceptionalizing, illegalizing and marginalizing such strategies of spatializations of undocumented migrants and local actors, we argue that it is these in-between spaces that become central to reimagining and co-producing new territorial relationalities with each other that are now simply missing from state and world maps. In other words, we make a plea for the recognition and acceptance of pluralities of human relations that cannot be ironed out to fit within the perfect divisive lines of state borders. We make a plea for a world that does not morally divide human beings between documented and undocumented. Human relations can not be bound by a world divided by paper.

References

Agamben, G. (1998) *Homo Sacer: Sovereign Power and Bare Life*. Stanford: Stanford University Press.

Agamben, G. (2000) *Means without End. Notes on Politics*. Minnesota: University of Minnesota Press.

Anderson, B. (2013) *Us and Them? The Dangerous Politics of Immigration Control*. Oxford: Oxford University Press.

Arendt, H. (1951) 'The decline of the nation-state and the end of the rights of man', in *Imperialism, Part Two of The Origins of Totalitarianism*. New York: Harcourt Brace.

Arendt, H. (1958) *The Human Condition* (second edition). Chicago: University of Chicago Press.

Balibar, E. ([1997] 2013) 'What we owe to the Sans-papiers'. *Transversal, 02*. (Retrieved from http://eipcp.net/transversal/0313/balibar/en)

Baudrillard, J. (1994) *Simulacra and Simulation*. Ann Arbor: University of Michigan Press.

Bo, B. (1998). 'The use of visa requirements as a regulatory instrument for the restriction of migration'. In Böcker, A., Groenendijk, K., Havinga, T and Minderhoud, P (eds.) *Regulation of Migration: International Experiences*. Amsterdam: Het Spinhuis, pp. 191–202.

Cole, S. A. (2001) *Suspect Identities: A History of Fingerprinting and Criminal Identification*. Harvard: Harvard University Press.

RAQS Media Collective (2004) 'Machines made to measure: On the technology of Identity and the manufacture of difference'. In Nold, C. (ed.) *Emotional Cartography: Technologies of the Self.* Welcome Trust and Space, pp. 15–26.

Derrida, J. (2000) *Of Hospitality: Anne Dufourmantelle Invites Jacques Derrida to Respond*. Stanford: Stanford University Press.

Foucault, M. (1975-76) *Society must be Defended, Lectures at the Collège De France*. New York: Picador.

Foucault, M. (1977) 'Discipline and Punish, Panopticism'. In Sheridan, A. (ed.), *Discipline & Punish: The Birth of the Prison*. New York: Vintage Books, pp. 195–228.

Foucault, M. (1991) 'Governmentality'. In *The Foucault effect: Studies in Governmentality*. Chicago: University of Chicago Press: 87–104.

Mitchell, T. (1988) 'Enframing'. In *Colonizing Egypt*. Cambridge: Cambridge University Press.

Torpey, J. (1998) 'Coming and going: On the state monopolization of the legitimate "means of movement"', *Sociological Theory* 16(3), 239–259.

Torpey, J. (2000) *The Invention of the Passport: Surveillance, Citizenship and the State*. Cambridge: Cambridge University Press.

Van der Ploeg, I. (2006) 'Borderline identities: The enrollment of bodies in the technological reconstruction of borders'. In Monaham, T. (ed.) *Surveillance and Security: Technological Politics and Power in Everyday Life*. New York: Routledge.

van Houtum, H and van Naerssen, T. (2002) 'Bordering, ordering and othering', *Tijdschrift voor Economische en Sociale Geografie (TESG)* 93(2), 125–136.

van Houtum, H. (2010). 'Human Blacklisting: The global apartheid of the EU's external border regime', *Environment and Planning D: Society and Space* 28(6), 957–976.

4 Trapped on the border

A brief history of solidarity squatting practices in Calais

Calais Migrant Solidarity[1]

Introduction

The illegalized migrants in Calais have no stable place to live. They are constantly denied shelter as a result of French and British immigration and housing policies and are often quickly and repeatedly expelled if they settle anywhere. In this context, when migrant communities squat and carve out a space for themselves, it is both a militant and resistant practice. Squatting not only reasserts people's rights to an autonomous and dignified life while trapped in France, but also actively subverts British border controls by supporting those who attempt to cross the border clandestinely. In this chapter, we first give a brief history and explanation of the border in Calais and the living situation of clandestine travelers trapped there as background to their squatting practices. Then we present a general description of the different ways in which people squat in Calais and their motivations for using these spaces. After that, we discuss two different squatted spaces, their histories, and our personal reflection on how these spaces function. Finally, we end with an account of the current situation as the state begins to change its strategy to deal with migrant squats and their supporters in the city.

Brief background

In 2003 the British and French governments signed the *Le Touquet Treaty* in which they agreed to establish juxtaposed immigration controls on cross-Channel ferry routes. This meant that all travelers between the two countries would have to clear immigration in the country of departure rather than on arrival. This externalized the entire UK border to France in order to keep would-be illegalized entrants off the British mainland, as they would be in France if they were caught rather than on British soil.

The juxtaposed controls were only one part of a broader strategy to prevent people moving to England; the other half of which was the closure of the Red Cross managed refugee center in Sangatte, existing since 1999. In the eyes of the governments, the center had a "magnet effect" and attracted "illegalized migrants", supposedly creating the problem rather than responding to it. To justify their erasure, the governments equated the minimum humanitarian standards for sustaining life with "pull factors" that threatened the sovereignty of United Kingdom's (UK) border with undesirable migrants. This is an excuse that has been repeated over and over again for the refusal of any sort of meaningful humanitarian support for migrants in Calais at a state level. Today politicians still cite fears that a "New Sangatte" would attract more migrants than those already present, and this is their primary reason for denying the provision of any housing arrangements for migrants in Calais.

The year 2003 marks the beginning of the modern deterrent policies and practices in Calais, which consist of the simultaneous tightening of security measures at the ports and attacks on the clandestine travelers' living spaces. This is an ongoing pattern that has had much iteration. For example, in 2009 there was the eviction and destruction of the "jungle" (see p. xx), where approximately 2,000 people were living, while at the same time the British made a £15 million investment in new technology to search goods and vehicles at the port. Again in 2014, £15 million were pledged by the British to increase funding for border controls and border police forces in the city amidst the eviction and destruction of the jungles and squats where migrants were dwelling. Despite the increasing difficulty of the border crossing, the number of people coming to Calais has been rising over time along with the urgency of the housing crisis that faces them here.

Squats on the border

Most of the people who are squatting in Calais do not want to become squatters. They are trapped in the city without the papers needed to continue their journeys to the UK and are forced to accept whatever form of shelter they can while they try to cross the border – a task which takes most migrants many months. Most of them have not squatted before and do not plan to continue squatting once they regularize their status and are able to work and access social services. Even those who choose to stay and ask for asylum in France often remain without accommodation and in the squats for years while their cases are considered. For all of them, squatting is not a choice but rather a necessity for survival.

There are also handfuls of native Europeans, or people otherwise having legal status, who squat in the city along with the migrants in Calais. These people often have previous squatting experience and a political analysis around housing and migration issues in Europe, and often are associated with Calais Migrant Solidarity (CMS). For them, to squat and share skills and knowledge about squatting with the migrants there is an act of solidarity stemming from political convictions (see Buillion Chapter 5, this volume). They have varied involvement in the different squats in Calais, either by just visiting, spotting buildings for potential squats, opening buildings with/for migrants to take shelter, trying to establish legalized squats, or occupying squats along with other residents. This is done in combination with other forms of support work such as documenting police violence; collecting food, clothing or blankets to distribute; and sharing information on the asylum process in England to migrant groups (see Azozomox and Gürsel Chapter 9, this volume). We write here as members of this group of squatters in Calais and from the collective perspective of CMS.

Types of squats in Calais

Squats in Calais take the form of a multitude of abandoned and non-abandoned spaces such as houses, factories, vacant land, holes that are dug under roads and sidewalks, spaces behind stacks of washing machines in laundromats, park benches, or abandoned World War II pillboxes. Despite the extreme variety of squats, there are three broad types of occupations popularly known as "jungles", *kharabas*, and "legal squats". The people who create and inhabit these squats often depends on their legal status in France (European Union (EU) citizen, asylum-seeker, or clandestine not wanting to regularize their status until reaching the UK), their relationship to Calais (whether staying long term or trying to cross as quickly as possible), and their previous experiences of squatting in Europe.

Most common in Calais are the "jungles", a term taken from the Pashto word *dzhangal*, which is used to refer to squatted camps. Jungles are highly improvised and autonomous squats, the purpose of which is to provide temporary shelter for migrants to rest while they traverse the French/UK border. These encampments are usually comprised of tents and makeshift structures made of pallets and tarpaulin, and are set up in open spaces either in the city itself or slightly outside its urban center. Jungles can be quite small, consisting of no more than a few tents, or large, with hundreds of individual living arrangements. Some of Calais' larger jungles have even had impressive mosques, churches, restaurants, and stores in addition to sleeping spaces. The jungles are situated close to amenities necessary for the survival of the squatters living there, such as water standpipes, food distribution centers, or lorry parks to travel to England. While there are covert jungles where people live unobtrusively to avoid contact with police, most jungles in Calais are obvious and have the maximum number of people possible living in them. Usually a jungle is squatted by a small group of people and, as time goes by, more people join the occupation. In this way the jungles grow until the point where the state gathers enough political support and police for their eviction and destruction.

Another articulation of squatting in Calais is that of the *kharaba* – an Arabic word meaning an empty or dilapidated house. This term is used to refer to the occupation of abandoned and often heavily damaged buildings. These squats seldom have basic utilities, lockable doors, or complete shelter from weather conditions but are not as exposed as the jungles. However, just like the jungles, they are not secure spaces, and residents face frequent ID controls by police and the constant threat of eviction. Calais is a city of run-down and abandoned buildings, and often kharabas are located in the heart of the city. This gives the occupants more access to the city's resources, but creates a greater distance to travel to its periphery where they can try to cross to the UK. Kharabas are often squatted by a mix of people including native Europeans, migrants without status, and people that are claiming asylum in France or who have been deported to France from UK, under the Dublin Regulation. Generally the occupants are long-term squatters who have been living in Calais for many months or years, or have had previous experience squatting in other European countries. Kharabas are often used by people who desire to live in secluded and private spaces. In addition, Kharabas have marginally more comforts than the camp-like living arrangement in the jungles.

The final type of squatted space for migrants in Calais is the "legal squat", which is also the main type of squat that we will be discussing in the following sections. This is a squat that has a formal legal complaint taken against its occupation. French law stated, until recently, that if a squatted building is the primary residence of the occupants and if they can show they have been there for a significant amount of time (in practice this worked out to be 48 hours although this was not written in the law), then the police cannot evict them until the case is resolved in a civil court. However, this law was recently changed in the French Senate through a process initiated by Calais' mayor to make it easier to legally evict migrant squats, an issue that will be returned to later. In securing this type of squat, citizenship status came into play. People seeking asylum and native European citizens were needed to declare the squat as their primary residence and defend the squat in the first 48 hours of occupation in order to enforce the legal process. After the legal process was started, a squat became normalized and secure from police, so those without papers could join the occupation without fear of a sudden eviction. Legal squats are some of the best-resourced spaces as the security they provide allows infrastructure to be built within them. They also usually have utilities and are some of the only places in the city where migrants can access toilets, running water, or electricity. They are frequented by all types of people for many different reasons; including getting food and tea, charging telephones, meeting up with friends, permanently residing there, or just occasionally resting.

"Legal squatting" in Calais: history and practice

In this section, we will discuss the situations of a few legal squats we were involved in occupying, managing, and supporting from winter 2012 to summer 2014. Squatting by illegalized people in the city goes back much further than 2009. However, this is a well-documented period due to the constant presence of CMS-affiliated people in Calais who concerned themselves heavily with documenting and intervening in police raids and evictions of squatted spaces. During this period, although some unsuccessful attempts were made at legal squatting and there were certainly people squatting smaller places clandestinely in the city, the main articulations of squatting were those of the jungle and large kharabas. Life for illegalized people in these spaces involved a constant game of cat and mouse with the state authorities. An area of land or an industrial space would be occupied and encampments built to which the police would then regularly come to harass people living there by checking papers, pepper spraying belongings, and arresting people. This pattern would continue until police had gathered sufficient information and mustered large enough forces for mass evictions. They would then clear the occupation, destroy the encampment, brick the building up, or raze it to the ground. Following this clearance a new jungle or kharaba would appear in the same place or another abandoned space in the town, and the cycle would continue. Over the years, this process of occupation, harassment, and eviction has become a constant part of life in Calais for its illegalized migrants.

Rue Caillette

In response to this pattern of violence, a small group of people, in the winter between 2012 and 2013 occupied a house owned by the Town Hall, in the center of Calais, on a street named Rue Caillette. They were, for the most part, native Europeans affiliated with CMS who had previous experience squatting in other cities in Europe and were aware of the legal rights of squatters. This group explicitly took this house in order to resist the police when they arrived, inform them of the law, and force them to make a legal action against the occupation in the civil courts. The tactic behind this occupation was to set a new precedent in Calais for dealing with squats where police did not enter the space until the courts had ruled. It must be stressed again that this is how squats should be dealt with as per French law; however, in Calais, because squatters for the most part have no access to the law as they are undocumented, squat evictions are conducted there illegally and with impunity.

The space existed for quite some time without any contact with the authorities. However, in early February, almost a month after it was first opened, an employee of the municipality tried to enter the house but could not as the locks had been changed. He called the police. Then, for the rest of the day the authorities tried to enter with a locksmith but were unable to due to the barricades. On the door of the occupied building, a legal notice was posted that explained the rights of the occupants, and that a bailiff was needed to begin the legal process required for an expulsion. There were also masked people in the windows of the occupied house explaining the law to the police and dogs barking from behind the door. This further deterred them from trying to enter, and eventually the police returned with a bailiff who took the name written on the front door and went off to file the paperwork. This was the beginning of the first successful "legal squat" in Calais.

During the next couple of months, this squat was primarily used as a work and sleeping space for people involved with CMS. At this time, a previous office rented by CMS was being used as a sleeping space by almost forty men who were trying to cross the border. For this reason, many people connected with CMS wanted a separate space where they could rest, work, and feel safe, and so advance their presence in Calais. It was also unknown how the police would react to the new legal squat if they knew a large number of people without papers were residing in it. However, at the end of March 2013, the contract on that rented office was terminated, and the people who had been staying there were now without accommodation. Another legal squat was targeted for occupation to house those migrants; however, it was immediately and illegally evicted. Thereafter, the squat at Rue Caillette was opened to shelter those migrants now living on the street.

The first weeks with everyone were difficult as attempts were made to try to divide the spaces and assert control over them. In particular there were a few problems when people started fights inside the house using extremely racist and sexist insults. However, gradually the situation stabilized and the cohabitation process continued smoothly. But problems persisted, like migrants charging others rent in order for them to stay in the squat or claiming that it was a space only for one specific ethnic group. The CMS squatters had to spend a lot of time combating this and insisting that it was a free and open space. As counter-examples, CMS detailed other squats in the city that were controlled by specific migrant communities or smuggling networks, and were unwelcoming to outsiders or those without the money to pay for a place to sleep. The police also respected the fact that they had no right to enter this squat and did not try to make any early morning raids there as they had previously done in the kharabas or as they were continuing to do in the jungles.

During that time, from when papers were served to the date for the court hearing, Rue Caillette was an illustration of how a very diverse group of people sharing extraordinary circumstances could live together in solidarity. Although not without occasional problems, it was remarkable that thirty to sixty people shared a single house with one toilet and seven rooms, in spite of all individually experiencing the mental agony of being far from home, hunted by police in the street, and having to find a way to cross one of the most highly secured borders in the world. More than that, Rue Caillette became people's home and provided them a place they felt was their own in such a hostile city.

This positivity continued and in some ways was strengthened by the intense pressure on the squat from outside. The city tried to cut the water and electricity four times; however, with the skills and knowledge of the people living inside the buildings, utilities were restored. Native French racists also shouted offensive slogans and threw objects at the building. Patrols of CRS (French riot police) also regularly stopped and checked people's papers down the street from the squat as an intimidation tactic. Additionally, in late May, after the eviction and destruction of a squat on the east side of town, all the inhabitants of that building were directed to Rue Caillette by police. This was particularly ironic because the state authorities were proposing an illegally occupied building, which they were also looking to evict as soon as possible, as the accommodation solution for those who they were now evicting. A local humanitarian association also thought this was the best option for the recently evicted and helped the migrants transport themselves and their belongings from the evicted building to the door of the squat on Rue Caillette. This doubled the population of the squat overnight and was obviously an attempt by the authorities to increase the pressure on the people inside, hoping that they would begin to fight more and give the state a pretext for their own eviction. However, this did not happen, and the inhabitants welcomed and gave the new arrivals space in the house that was already far too crowded. There was recognition from those already staying in Rue Caillette that the existence of their squat was also fleeting and that any shelter that they as illegalized migrants had access to in Calais needed to be shared with the entire community.

The city received another defeat in their strategy to evict Rue Caillette when in July the court in Calais declared the municipality's request for eviction "unreceivable". This meant the city had made procedural errors in submitting their case, and they had to resubmit it before the case could be heard by the court. This failure further highlighted that the municipality was unaware of the procedures to evict a squat through a legal process. This response from the courts also gave the occupants more time, stability, and security inside the squat. However, not long after this victory, the squat unfortunately was brought to a very sad and abrupt end by a small group of people that did not respect the space or even live in it themselves.

Members of a smuggling network brought one person who used to be involved with them to sleep in the squat one night as he was just released from prison and had nowhere else to go. In the meantime, they organized a revenge attack from a previous dispute with him for the morning. Although "only wanting to teach the guy a lesson", and despite the efforts of the paramedics who responded to the incident, the person whom they attacked died from his wounds inside the building. The police came immediately forcing everyone outside and closed the squat as a crime scene. After the initial investigation, they bricked over all entrances, doors, and windows, not allowing anyone back in, even to retrieve their belongings. This was an extremely frustrating end for all the occupants as these actions did not reflect the spirit of the place at all and provoked the eviction that everyone had been fighting so hard against. Furthermore, this tragedy provided legitimacy to the authorities in calling for the evictions of future squats, and further served to demonize migrants in the local media. However, despite this terrible ending, many took their experiences from Rue Caillette as a positive example of how a very diverse group of people can live together in solidarity in Calais if given a small amount of security and autonomous control over their living space. It was also a very strong motivator for people to try and create similar spaces in the future.[2]

Struggling to secure more legalized squats

After the closure of Rue Caillette, Calais' authorities began to implement a no tolerance policy to the squatting of new buildings in the city. While the *Mairie* (Town Hall) appeared to tolerate the various jungles spread around the city, any attempts at making another legal squat were swiftly, and often illegally, crushed by state forces. Eight attempts were made between autumn and winter of 2013, but every time the municipality discovered the squats, they were immediately evicted and the residents arrested, no matter the period of time they had been squatting for. The pattern was always similar. Soon after the squatters announced their occupation and provided the evidence to prove that they had been living there already for many days, the police would arrive and collect a testimony from a neighbor or the owner who would say they had not seen us there before. The police would then declare that the occupation only had begun on the day we made it public and so they still had 48 hours to act. Ignoring the squatters' evidence, they would then come back to break down the doors and arrest the squatters. During this time, CMS affiliates ended up pursuing a court case against the *Prefecture of Nord – Pas de Calais* for their illegal eviction of squats. However, parallel to this legal action,

CMS also decided to take direct action in its bi-annual network meeting, which happened in December 2013. It was decided there to organize a mass squatting action in which squatters from all across Europe would be invited to come and support the opening of multiple legal squats in one coordinated action. The idea behind this was that if squatters opened enough buildings in one day, state authorities would be overwhelmed and unable to evict all of the squats, only sending police to one or two of the new squats before the others would have been in public occupation for the minimum 48 hours needed to start a legal process.

Beginning on 25 February 2014, nearly a hundred squatters, arriving in groups of five to ten, occupied empty buildings around the city. The occupants tried their utmost to stay unnoticed, taking their supplies in with them and limiting their movements in or out of the squatted buildings. Furthermore, a team of "decoy squatters" was tasked with making unoccupied abandoned buildings appear like they were squatted by posting banners, turning on lights or making noise inside the buildings. One factory this team pretended to occupy was raided twice during the action week by police and was in fact just down the road from one of the real targets. It was a funny moment when the CRS vans tore down the street to evict an unoccupied building driving right past the squat that was actually being opened.

In a single night, a total of five buildings (one large factory and four smaller town houses) were occupied across the city. During this period, one of the buildings was discovered and evicted prior to the 48 hours required to begin a legal process. However, on 28 February 2014, CMS announced the occupation of a number of buildings across the city. After this announcement, one of the squats at Impasse Leclercq (which happened to be the only one owned by the Mairie) was evicted, re-squatted and then evicted again by police. This left a total of three squats in the city after 48 hours had passed – a large factory on *Rue Massena*, and two town houses at *Rue Auber* and *Rue de Vic*.

Now that we had some assurance that these spaces could be securely inhabited by migrants for at least a couple of months, decisions had to be made about how these spaces could be utilized. After extensive debate for many days between CMS affiliates, asylum seekers in Calais, and people trying to cross about how to use and maintain the new legal squats, decisions were made based upon the form and the utility of the buildings. For instance, Rue Auber was occupied by a collective of long-term residents in Calais and was chosen as a private and closed residence, with limited access. Rue Massena was initially chosen to be a non-residential Social Center that would offer bike repair, phone charging, a multilingual library, language classes, and a boxing gym, but eventually became a sleeping space. Finally the third house, Rue de Vic, became a private residence for a group of Eritreans and Ethiopians who were also trying to cross the border and whose previous house had recently been burnt and destroyed. Each of these squats offers interesting insights to the different needs of people squatting in Calais and how those needs are negotiated in the midst of the humanitarian crisis found there. However, only the occupation of Rue Massena will be discussed in depth in the following section because it represents the most successful Social Center and housing solution for migrants and non-migrants in Calais that came out of this action.

Rue Massena

Rue Massena was an old factory that had been abandoned for many years and was in a state of extreme disrepair when it was occupied in February 2014. The space began its existence as a Social Center shortly after its occupation with little or no sleeping space except for a few residents who were working on repairing and organizing the space. Within a few weeks, rotten floors had been rebuilt, holes in the roof and the ceiling were repaired, a kitchen was built, a running water system was established, and a bike workshop with dozens of frames and components was set up. The space began to attract more and more visitors, and people started to take help from the workshops and other utilities offered by the center. New English language classes were started and so were the boxing and fitness activities. The Social Center was a residence for some people but simultaneously developed as a center for various skills training for all, free of cost.

After the eviction of the "Sudanese Jungle" in the North of Calais on 11 April 2014, Rue Massena became an emergency sleeping space and then eventually transformed into a permanent home for sixty Sudanese people, most of whom were traveling through Calais on their way to England. These transformations in Rue Massena offered an interesting and vibrant dynamic to the squat. The space was symbolic as a huge collective social space, where many people of different cultures, backgrounds and aims could meet, organize, and manage the space together. On 15 April 2014, the squat hosted a trans-European gathering of antifascists who were responding to the threat of a fascist demonstration and the anticipation of attacks against other squatted spaces in the city. The gathering attracted around one hundred people, many of who stayed at the Rue Massena squat for a week. This strengthened the connections between the residents of Rue Massena, local urban inhabitants, and international antifascist groups. The squat also became well known in the European squatting community and was visited on many occasions by traveling musicians, people who provided information tours about autonomous struggles in other cities, and by friends of the squatters.

This open form of squatting and the formation of a Social Center run collectively by people with European passports, people claiming asylum, local Calaisiens, and people attempting their journey forward to England stood as a fantastic example of community self-organization – indicative of how squatting can create bonds of solidarity, friendship, autonomy, and self-management across cultures, race, citizenship statuses, skills, age, and gender. The Massena squat was a creative resource for all of these different communities, not only offering material support in the form of food, shelter, and free bicycles, but also emotional and political support. The links that were created through active solidarity at the Social Center had a larger message than simple mutual aid. What was happening at Rue Massena was emblematic of collective strength as people struggled together against police brutality and repressive border control mechanisms.

The Massena squat also played a fundamental part in supporting and organizing the waves of hunger strikes in Calais in June 2014. Many squatters from this building were involved in the occupation and defense of the *Salam* food distribution center following the evictions of the jungle just outside it that month, as well as the later occupation of the Galloo squat. Additionally it functioned as a radicalizing space for some local Calaisiens, who were exposed to the situations, ideas, and analysis of migrant and radical communities they would otherwise not have known. Furthermore, the space took an active role in resisting the British immigration system, by offering skill shares and workshops on what to expect from the asylum interview process once people arrived in England, and formulated an important and cohesive element to the "no borders" struggle.

Fort Galloo and the repression of squatters' rights in Calais

There has also been another major legal squat in addition to those mentioned here and which lasted for 11 months before it was evicted on 2 June 2015. The ground and buildings on it were taken as part of a large demonstration of five hundred people supported by local associations. In this respect, Galloo was important as it was the first squat that was occupied and then openly supported with the help of local *associations* (NGOs and voluntary groups). Before this, CMS had organized squatting actions autonomously or with small migrant groups, but working together with NGOs for this occupation not only reduced the legitimacy of any state repression against it but also involved many people not experienced in direct action. The site was an abandoned metal recycling workshop located in an industrial area that backed onto railway lines. Surrounded by a high wall, the 12,000 square meter complex included a house, a hanger space, and a large courtyard. It was open to absolutely everyone, except the police, and at its high point around 400 people lived there. The residents included migrants of different genders, backgrounds, and with various legal statuses, as well as native Europeans and activists associated with CMS. With so many people staying in one place, it was sometimes difficult for the residents of what became known as Fort Galloo to manage their living space collectively. In this regard, Galloo was both a success and a failure; it was a nightmare to organize collectively but provided a space safe from police harassment. In fact, though Fort Galloo was deemed an illegal occupation after only four months, it was occupied for almost one year.

The current situation

The eviction of Galloo and the other large jungles existing in Calais during 2015 only took place after the opening of a "day center" by the state where people could access basic services such as showers and medicine but under no condition sleep there (remember the paranoia of creating a "New Sangatte"). However, the state did openly declare that it would tolerate the establishment of a jungle on the outskirts of Calais close to this center, and even specified the land that could be used for this purpose. This was an acknowledgment that their previous strategy of evicting squats without any alternative accommodation solutions did not in fact make the people living in those places disappear, and so it had to change. Although it may be tempting to consider this as a victory and positive response to the high profile legalized squats and political struggles that had taken place around migrant accommodation over the years, this would be a naïve assumption. This day center and tolerated jungle are in fact part of a larger strategy to finally defeat the autonomous living spaces created by migrants and their supporters, and fulfill the mayor's declared goal of "zero squats" in Calais. They coincided with the adoption of her new anti-squatting law through which she is seeking to remove the previous legal protections for squatters in all of France. They also serve as a ready-made political justification for the eviction of any future squats by being the alternative accommodation solution that never existed before. By concentrating the migrants into such a small area, they also increase the tensions between communities and make the population as a whole easier to police. In fact, the recent changes have not had much to do at all with seriously tackling the problem of accommodating migrants in Calais, but rather represent an effort to ghettoize the migrant community and transform their situation from a political struggle for dignity at Europe's borders into a neutral humanitarian crisis to be managed by the state and large NGOs.

Given these recent changes in terms of the legal landscape of squatting and the creation of the Jules Ferry ghetto for migrants in Calais, the future of squatting and particularly of legal squatting in this city is quite uncertain. As the new generations of migrants passing through the city are taken in by this cynically "tolerated jungle", they do not get exposed to the squatting tactics or the radical solidarity networks that work to make them happen. Unless we can find ways of transforming how we get into contact and build relationships with those who continue to get stuck on this border, it will become more difficult to work together in the future on confrontationally occupying new buildings to highlight and protest the conditions migrants are forced into at Calais.

Notes

1 Calais Migrant Solidarity (CMS) is an international network of autonomous people involved in practical solidarity work with the migrant communities of Calais. First getting involved in 2009, the group has since been busy with monitoring police activity, squatting, supporting migrant's political protests and direct actions, distributing food, clothing, water, blankets, and tents, doing outreach and publicizing the situation in Calais, and providing migrants with asylum and immigration information for the UK and other European countries. [see https://calaismigrantsolidarity.wordpress.com/]

2 In the time between the Rue Caillette occupation and the mass squatting action that took place in 2015, there was another legal squat that was taken in the East of the city on a street called Boulevard Victor Hugo. It was set up to be a safer space for women, children, and particularly vulnerable people rather than a general shelter for all of those migrants homeless and on Calais' streets like Caillette had been. Unfortunately, space in this text does not permit us to discuss it as its history is very complicated and deserves a full length analysis, but this was a very important action in the history of squatting in Calais and directly led to the French state setting up a shelter for migrant women and children in the city.

Part II

Squatting for housing

5 Why migrants' squats are a political issue

A few thoughts about the situation in France

Florence Bouillon

Introduction[1]

In France as elsewhere, research workers' and activists'[2] publications about squats have generally included typologies of squats designed to show the complexity of the problems involved, while suggesting some criteria for tidying up the teeming picture of reality. These systems of classification can be placed in turn under various headings, depending on whether they are based on "configurations" (Pruijt 2013), squatters' residential trajectories (Bouillon 2009), the objectives they pursue and the resources at their disposal (Aguilera 2013), the ideological background involved and the range of possible actions available (Péchu 2010), to mention just a few examples. These authors have often made the distinction between squatting for political ends, which is a form of militant action and commitment, and squatting simply as a means of obtaining a roof over one's head. This distinction is often made implicitly: many studies have focused entirely on "Social Centers", where people occupy empty buildings as part of an overall move to protest against capitalistic, patriarchal, elitist, non-egalitarian systems. But it is important to note that this definition is far too restrictive because there are so many kinds of "political squats" and objectives, depending on the historical, geographical and political context involved (Péchu 2010; SqEK, Cattaneo and Martínez 2014). Unlike many European countries, in France the term "Social Center" denotes publicly funded centers providing a specific locality with social outreach services. To avoid confusion, some activists occupying squats talk about "self-managed Social Centers" or more frequently about "political squats".

Several questions therefore arise in the first place as to why relatively little attention has been paid so far to the issue of squatting for housing purposes in countries such as France, where squatters of this kind are nevertheless a large majority. Among the various explanations for this situation are the discretion and invisibility of those squats. For instance, there is the possibility that those who publish information about squats may have some ideological, social, or class affinities and potentially subversive intentions in common with the squatters. The aim of this chapter, in line with other contributions to this volume, is to describe more closely the complex, ever-changing world in which squats are used as a means of accommodation by migrants, while at the same time underlining the essentially political aspects of this form of squatting (see Calais Migrant Solidarity Chapter 4, in this volume). It is worth pointing out, in view of the increasingly complex migratory patterns observed nowadays, that the term "migrant" is now being used in French in the field of social science. The term "immigrant" is mentioned either to evoke public statistics, or the migration movements, which took place within the framework of the "foreign workers using policies" which were effective until the 1970s (Mazzella 2014).

There certainly exist many good reasons for distinguishing between squats of various kinds, but any attempt to draw up an ethnographic picture of squats in terms of rigid classifications is bound to fail. The fundamental starting point of any analysis on the logic of squatting should originate from the squatters' point of view. In my field studies, which have mostly been conducted in France,[3] I have observed several empirical reasons for this: many squats tend to change with time, and the way they function also frequently changes as the weeks go by. This was the case of one squat in the center of Marseille, whose occupants originally claimed to be "artists", but which gradually began to open its doors to North-African migrants and migrants with no papers. Similarly, in a political squat, a change of ideological background occurred when its original occupants were replaced by people who were simply looking for a place to live, whose aim was to negotiate extra time with the authorities before being evicted. Besides these examples, there are other squats of mixed characteristics, in terms of their occupants' history and their social origins as well as the activities which take place therein through their everyday life, workshops, meetings, celebrations, debates, and the like. Lastly, the criterion of choice often used to distinguish between squats occupied by poverty-stricken and migrant populations who have no other housing options and hence squats are deliberately chosen as a way of life and a way to commit (by artists or militants, for example) does not stand up to scrutiny, since a much wider range of possibilities and constraints than those suggested by this sole distinction have been found to exist. The main aim of the chapter is to present the idea that squats used for housing purposes, which are mainly inhabited by migrants in France, pose a number of essential political questions to the societies in which they are located.

Squats occupied by migrants result from political decisions

As other western countries, but maybe with a specific violence and strength because of its colonial history, France is the theater of multiple forms of racism and discriminations towards "foreigners", real or supposed. Unequal treatments, contempt, and hostility often characterize ordinary relations toward those who are perceived as non-natives, whether it is because of the color of their skin, their accent, their name, or their nationality. Unequal treatments are also the fact of the State, as foreigners do not have the same rights on several issues as national residents. The current migratory policies join this long history of domination, as well as a European context characterized by border-closing and internal limitation policies. Border control, policies against unlawful immigration, quota-based immigration policies, voluntary immigration: these are the keywords in current French migration policies and those adopted all over the European Union.[4] Since the French frontiers were closed to immigrant workers in 1974, a series of laws have made the conditions governing migrants' access to the country and their residential rights increasingly strict. In 1991, for instance, asylum seekers' right to work in France was curtailed and in 1993, access to social and medical services (apart from emergency medical aid) was withdrawn from those with no papers under the "Pasqua laws". New migration laws have been passed continuously for the last ten years: restrictions on family reunification, tighter checks on mixed marriages, shorter deadlines for appealing against the decisions of the OFPRA, longer penal sentences for infringing administrative rules, and larger numbers of illegalized migrants have been expelled (amounting to approximately 30,000 per year).[5] For the sake of safety and efficiency, regardless of the political leanings of the successive governments, the fundamental rights of all the foreigners present on the French soil have been eroded, their freedom of movement has decreased, and their chances of integration have been eliminated.

In this context, and because of a huge poor housing problem (specialized associations agree on 3.5 million people badly accommodated in France), squats are one of the many signs of vulnerability shown by migrant men, women and children. These people deprived of access to work and social protection, some of whom have no proper legal status and are obliged to work under precarious conditions or live in derelict slums, foreigners from the Global South and migrants suffering from discriminative practices[6] are also often a million miles away from being entitled to decent housing. In France, especially in the Ile-de-France region around Paris, a system of "social hostels" has developed considerably during the last decade. These special hostels have been set up by State- and Region-assisted child protection associations to help homeless people and families, many of who are migrants (Le Méner 2013). Then, the questions remain: how could they possibly escape having to adopt the most unstable, uncomfortable places of abode when they cannot afford even the cheapest lodgings, or when they can no longer resort to relatives' homes? Under these circumstances, they are propelled to gather the courage to break into empty buildings to obtain a roof over their heads.

The various occupants of squats in France cover the whole range precarious squatters who are challenged by present-day poverty such as run-away teenagers, homeless people, single parent families with no resources, and penniless students. Currently, like the slums in the 1960s, migrants mainly inhabit squats. Unsurprisingly, many of them have no papers, which adds to the difficulty in finding a place to live: no residence permits, no legal access to work or to council housing. The asylum seekers are hardly any better off. However, thousands of individuals and families inhabiting these squats are actually normal, law-abiding citizens who have been unable to find proper lodgings. In Marseille, the squats in the city center are mostly occupied by North African migrants and isolated foreign youths. As in Lyon and Paris, Romas originating from Central Europe and the Balkans have been occupying large squats on the outskirts of the city for several years: these squats are not only the largest but also the most insalubrious and the least well tolerated by the neighborhood.[7] The squats in the Ile-de-France region around Paris are occupied by many families originating from sub-Saharan Africa, who are known to undergo the most discrimination in terms of access to housing (Simon 2001).

When I first met Bakari from Senegal in 2008, he was living with his wife and their 3 year old daughter in a 20 square meter studio in a building entirely occupied by squatters in the 20th arrondissement of Paris. After arriving in France in 2001, Bakari lived at first with his elder brother, who had come to France in 1994. Although this brother had a 10-year residence permit and a long-term work contract with a cleaning firm, he had been unable to find a flat to rent and had been squatting a small one-bedroom flat for 5 years with his wife and four children. Despite the shortage of living space, Bakari spent 3 years with this family, sleeping on a sofa in the living room beside the first two children, who slept on mattresses on the floor. But when the third and fourth children arrived, the situation became impossible. Bakari was working long hours at the restaurant without papers, he did his best to find lodgings elsewhere. He eventually heard that a squatted studio flat in the next-door building was about to become vacant. After paying the previous occupants the fee of 3000 euros they demanded, Bakari moved into this flat in 2005 with two colleagues who were also looking for accommodation. But when his wife came to join him, his flat-mates had to leave his squat and find lodgings elsewhere.

This short account of events shows how various threads in the institutional production of squats come together: people with no regular status and low incomes who are not entitled by law to proper housing qualify only for living in squats. To complete the picture, there are more than 2 million vacant homes in France. According to INSEE, the number of vacant homes increased from 1.9 million in 2004 to 2.64 million in 2014. The vacancy rate has therefore shot up from 6.3% to 7.8%. All these ingredients have resulted in migrants squatting empty buildings and offices. When they are evicted, they have no other alternative but to create new squats.[8] Although there is no reliable available data about squats in France, because of their transient and often hidden nature, presumably the squats occupied by migrants will continue to increase in number, and the quality of the living conditions they provide will continue to decrease as it becomes always harder for disadvantaged people to obtain decent lodgings.

We must make no mistake about the publicity sometimes given to some "alternative" squats: in France, these places are just the tip of the grim underlying iceberg. As my field data have shown, the majority of the occupants of squats are migrants who have recently arrived from the Maghreb countries, Central Europe and sub-Saharan Africa. These findings have been confirmed by other field studies. Research carried out between 2004 and 2008 focused on the most insalubrious squats in the Ile-de-France region, showed that less than 5% of among the 84 occupants of squats were born in France, and 83% originated from a sub-Saharan African country (Bouillon and Dietrich-Ragon 2012). Bachelors, couples and families awaiting the outcome of their applications for regular papers and those whose applications have been refused grasp squatting as the last opportunity of obtaining a roof over their heads. Even foreigners whose papers are in order are often also obliged to move from one temporary home to another without any hope of ever obtaining stable lodgings. The various squats they inhabit form migrants' trajectories, as they go through a whole series of temporary abodes. For instance, in the case of one Algerian family, who was housed at first in a center for homeless persons before moving to a small flat in the center of Marseille: when their application for asylum was refused, they had to look for another accommodation. In another case, a young man from Casablanca lived at his aunt's flat for a while before moving to a squat in the city center because he was not in very good terms with her. Likewise, a Roma family was forced to leave their caravan park and spent several nights in an abandoned vehicle before being told by other members of their community about a vacant flat in a dilapidated tenement building. Another example is that of a family from Senegal who, after being evicted from the squat that they were sharing with other occupants, moved to one squat after another without ever being offered a chance of proper housing or any alternative solution. In all these cases, the links with public housing policies and migration policies are obvious: squats for housing are indeed truly political places. Far from being homes for outsiders and misfits as they are often thought to be, squats reflect and materialize decisions deliberately taken by the wider society, based on the idea that the right to ownership should have priority over the right to proper housing for all and the idea that strict policies should be applied to deter migrants from settling in France (see Calais Migrant Solidarity Chapter 4, in this volume). These conditions actually force some members of the population to adopt an irregular way of life although they aspire to lead a normal existence, and as the French administration puts it, "to become properly integrated".

When associations become involved: squats for migrants and people's right to proper housing

The exclusion of migrants from public housing policies is not a recent practice: one only needs to think of the slums, brothels and shabby homes for workers that existed during the nineteenth century. Migrants have not always passively accepted the unworthy housing conditions imposed on them, however. During the 1960s, many Committees consisting of North African immigrants were set up to do away with dilapidated housing and the exploitation of the poor by the landlords, which resulted in some huge strikes at workers' hostels. In these places, the living conditions were particularly intolerable as immigrant workers were crammed together in tiny and uncomfortable rooms. They were not allowed to have visitors and had to keep strict hours and obey the overseers, many of who were brutal racists (Bernardot 2008). In addition to these deplorable living conditions, the landlord's decision to increase the rent taken by the landlords (the public companies SONACOTRAL and AFTAM[9]) was the last straw. The residents' committees therefore founded by the first immigrant workers' movements were endowed with a real political power which was subsequently transmitted to future generations. The various movements created by workers, revolutionaries and Christian believers thus strove to defend the rights to proper housing. The know-how thus acquired about squats, strikes, negotiations and the protection of the homeless was put to good use during the 1990s, when some new battles had to be won. In 1987, after a series of tragic fires that occurred in Paris, the *Comité des Mal Logés* (the committee for the poorly housed) was created. Three years later, fifty or so families originating from Mali occupied the Place de la Réunion after being evicted from their lodgings. This event led to the creation of one of the main present-day associations defending the rights of the homeless, *Droit Au Logement* (the right to lodgings) acronymously DAL. One of the branches of DAL called *Comité des Sans-Logis* (the committee for the homeless), or CdSL, was particularly active during the 1990s.

For more than 20 years now, DAL and other associations such as *le Comité Action Logement, Droits Devant!!, AC, Jeudi Noir*, and others have organized sit-ins, requisitioning vacant buildings and occupying public spaces in order to make the French government and the public aware of the fact that the housing crisis has reached such serious proportions. During the 1990s, some memorable large-scale protests took place at Château de Vincennes (1992), rue du Dragon (1994), Saint-Bernard's Church (1996), the Saint-Denis Basilica (2002), and rue de la Banque (2007), for example. The fact that these protests were given considerable coverage by the media certainly contributed to a series of important laws on housing for the underprivileged being adopted during the 1990s. The lobbying carried out along with other movements defending the rights of the homeless (la Fondation Abbé Pierre, ATD Quart Monde, Emmaüs, Médecins du Monde, l'Armée du Salut, les Enfants de Don Quichotte, etc.) eventually led to the Law on the Right to Proper Housing being passed early in 2007. The aims of these "self-requisitions", as some associations have called them, were three-fold: giving homeless people, including many migrants, proper homes; drawing public attention to the problem of derelict housing; and putting these issues on the political agenda by urging local and national decision-makers to take steps to promote housing for the underprivileged (Aguilera 2012). "Self-requisitions" was used in order to avoid the stigmatizing connotations of the term "squatters" and to remind people of the requisition laws of 1945, which authorize the public authorities, under specific conditions, to requisition vacant premises in order to house homeless people. The associations defending people's right to proper housing have been constantly demanding the application of this law, of which little use has been made since it was passed.

When squats, sit-ins and protests become headline news, they *make visible what was invisible*. They give the "undesirables" (Agier 2008) a place in a system from which they were completely excluded. Squatting is a way of proclaiming one's very existence directly, physically, and materially in order to become visible and gain a hearing, i.e., to take part in the life of the city. Associations which take advantage of this fact have found a direct means of acting and awakening public opinion, speaking in public and subverting law and order: these constitute alternative ways of constructing political discourses.

Squats give rise to political clashes

The third political aspect of migrants' squats focuses on the people squatter's meetings and the effects of these meetings in terms of "reciprocal politicization". In this case, squats are not just a space but also a context for an experience. There has been a long-standing quarrel in France between so-called "political squatters" and those pursuing artistic goals. But rather than the activities they pursue, the distinction between the two groups is rather that the latter tend to negotiate with the authorities the right to occupy the premises they have squatted, whereas the former refuse to enter into discussions with the authorities at all and object that giving activists the formal right to occupy squats might facilitate the development of criminal activities. Although squats do not constitute a homogeneous social world, their inhabitants have opportunities of creating occasional as well as more durable social links with each other. Some squats of the political or artistic kind occasionally take in single migrants or families originating from various parts of the world, for instance. These migrants can also be artists or militants, of course, but they are often worlds apart from the activists living in the squat, who play the role of hosts to the newcomers. Squats occupied by activists and/or artists are thus frequently infiltrated by individuals who are fairly indifferent to the initial goals of their occupants. These individuals can be migrants in difficulty, youths with no resources, former prisoners whose social links have been severed, vulnerable people with psychiatric disorders or addictions, penniless women with or without children, and travelers of all kinds who need to recover their strength before moving on.

Although many different configurations exist involving activists' and artists' squats, ranging from those which accept only people who share their "antisocial" values and ideals to those which are so wide open to outsiders that their functional patterns are constantly changing, a certain amount of friction can nevertheless generally be observed between the two attitudes, which is never completely resolved. Clashes tend to occur between the principle of openness and the need for squatters to protect themselves, the feeling of solidarity and the need for privacy. Various arrangements can therefore be made, such as allocating a few rooms in the squat to passing travelers and other needy candidates or directing them to other squats. Activists' squats often include what they call "sleepings" or doss-houses in France. Some individuals frequenting "sleepings" are eventually co-opted and allowed to graduate to squats, where they acquire the status of residents (which is endorsed by their being given a set of keys). In the large French cities, there are people who know exactly where to find vacant places, thanks to which hundreds of destitute people find places of shelter.

It has been known to occur on the contrary that some squatters who claim struggle for freedom refuse to give temporary lodgings to Algerian families in dire straits on the grounds that they do not share their political opinions. Should obeying political rules mean welcoming only those who toe the party line, or should it mean keeping an open mind and accepting a mixture of people with various ideas/practices/backgrounds? The responses of political squatters tend to oscillate between these two extremes, which are difficult to reconcile. In order to explain why many political squats are closed to outsiders, it is necessary to take the problems which arise into account, and the fact that cohabiting too closely with other people can be extremely wearing. The problem is obviously not due to militants and migrants having difficulty in living together, but it often rises from the differences between in their lifestyles and their ideas about what is public and what is private (Breviglieri 2009; Pattaroni 2013). However, whether or not the members of the two groups share the same living space, it is worth noting that the occupants of political squats have made many outreach efforts to help needy migrants by dispensing French lessons, running activities for children, collecting and distributing food, helping to apply for residence papers, and providing those about to be evicted with support.[10]

Since meeting up with and/or cohabiting with political activists means to participate in discussions, migrants who know little about the French political context, and even less about the practices of autonomy/liberation/direct democracy, often undergo politically socializing processes in joining meetings, debates and discussions. On similar lines, feminist ideas have been known to circulate among migrants who could never have imagined the need for such discussions in their country of origin. But a process of political indoctrination sometimes also occurs in the opposite direction. These encounters bring militants face to face with harsh realities which they have rarely experienced themselves: the extreme poverty and exile which are the constant lot of these migrants, many of whom have recently fled the horrors of armed violence in Sudan, Syria and Eritrea. It is one thing to have an intellectual knowledge about this violence, but it is very different when these places of refuge and these moments of sharing, although with conflicts, make people's pathetic condition infinitely more real and tangible, expressed in words and emotions and not just in terms of facts and figures and warning messages. Since the residential statuses of activists inhabiting the squats with the migrant families are similar and because they socialize on a daily basis, activists are in an excellent position to comprehend the pain and challenges for migrant men, women and children from other countries along with their efforts to survive. And in this point of time, often depressing, when tens of thousands of shipwrecked migrants fleeing the North-African coast are drowning in the Mediterranean Sea, squats can stand as an option to provide migrants with some solace and solidarity, thanks to the efforts of activists in supporting the migrants.

Conclusion

Squats occupied by migrants can be said to be truly political issues for at least three reasons. Firstly, they result from the political mechanisms involved in governments' migration and housing policies (not to mention questions such as labor policies, racial discrimination, urban management policies and the criminalization of squats, all of which at least indirectly impact squatters' situations). Secondly, they epitomize the need for everyone to have a proper roof over their head, not to say the need for a more equitable society. Thirdly, squats constitute political arenas because they are meeting-places where migrants come into contact with people from other social environments. Migrant squatters have contacts not only with political activists, in and around their squats, but also form networks with people who do not reject the precarious and, often, stigmatized migrants. Neighbors, social workers, journalists, students, and union members can also enter this social arena as a place of action and communication. Politicization is a two-way process, which works like contagion, promoting the circulation of ideas. It can therefore be said to involve a kind of trajectory from one point of view to another, corresponding to a deep belief in social justice and the hope of building a better and more equitable society.

Notes

1 Part of the contents of the first two sections were previously published in French (Bouillon and Muller 2009). I want to thank Baptiste Colin for his very relevant remarks on this chapter.
2 The term "activists" is taken here to mean people who practice squatting as a form of political commitment, mainly for anarchist or anti-capitalistic reasons and/or to support national independence/devolution. The recent development in various European countries of squats whose occupants support extreme right policies will not be addressed here.
3 My field studies took place mostly in Marseille but also in Paris from 2000 to 2010, at thirty squats of various kinds (ranging from a disused factory occupied by several dozen squatters to a tiny flat occupied by a single family and from a squat in the city center occupied by a group of militant freedom-fighters to a group of Bosnic gypsies squatting premises in a large block of suburban flats, etc.).
4 According to the official statistics, 5.8 million immigrants (people of other nationalities born in other countries), amounting to 8.8% of the country's population, were living in France in 2013. Contrary to popular opinion and the statements made on the media and by politicians, an increasingly large proportion of those who migrate to France are of European origin: almost one out of every two immigrants who arrived in France in 2012 was born in another European country, whereas only three out of ten were of African origin. The great increase in the number of female immigrants is another point worth noting.
5 OFPRA (Office français de protection des réfugiés et apatrides) is a public institution in charge of the application of French and European laws as well as international conventions regarding a refugee's status.
6 French research has lagged far behind other countries (the United States in particular) in terms of studies and analyses of discrimination. The debate in France has focused mainly on the relevance of producing ethnic statistics, which some authors have claimed to be a useful means of measuring racial discrimination, whereas others fear that these statistics may have stigmatising effects. In any case, the data of this kind which exist are usually confidential. On the discriminating processes affecting people's access to housing, see the study by Valérie Sala Pala, who analyzed the social and ethnic "sorting mechanisms" at work in the allocation of council housing in France and Britain (Sala Pala 2013).
7 Regarding the repression of populations of Romas and their expulsion from squats and slums see Olivier Legros and Tommaso Vitale (2011), and Martin Olivera (2011), as well as the fine collective work coordinated by Sébastien Thiéry (2014), in which artists, architects and philosophers have described the violence and the absurdity of the events which occurred when the "Nationale 7" slums in Ris Orangis near Paris were evacuated in 2013. See also Aguilera (this volume, Chapter 12) and Antonelli and Perrotta (this volume, Chapter 13).

8 In France, the decision to evict people from squats has to be made by a judge and the intervention of the police for this purpose has to be authorised by the Préfet (a state representative). Although this lengthy procedure often gives squatters a little extra time, the judges nearly always agree to the eviction being carried out. The judicial procedure is sometimes bypassed, as in those cases where squatters are caught in the act of unlawfully entering premises and illegal cases involving Romanies. It is worth mentioning that even the flimsy protection from which squatters benefit at present is being frequently threatened by new bills of law whereby, if they are passed, the *Préfets* will be able to simply demand the eviction of squatters without further formality on grounds of home invasion. Four bills on these lines have been presented to the French Parliament during the last 20 years, the last one as recently as June 2015. Although these bills have never been passed so far, there is a risk that squats may become assimilated in the future to criminal activities in France, as in many other European countries. Check also: http://sanspapiers.internetdown.org/.

9 SONACOTRAL stands for "Société nationale de construction de logements pour les travailleurs algériens" and AFTAM for "Association pour la formation des travailleurs africains et malgaches". SONACOTRAL became ADOMA in 2007 and AFTAM became COALLIA in 2012. Both structures have widened their clientele to all kinds of "precarious people"since the 1990s.

10 Information about actions and companionship of this kind at the Attiéké squat in Saint-Denis near Paris can be found on the website giving information about French political activists' squats: https://fr.squat.net/tag/attieke/.

References

Agier, M. (2008) *Gérer les indésirables. Des camps de réfugiés au gouvernement humanitaire*. Paris: Flammarion.

Aguilera, T. (2012) 'DAL et Jeudi noir: deux usages du squat dans la lutte contre le mal-logement'. *Métropolitiques*, Available: http://www.metropolitiques.eu/DAL-et-Jeudi-noir-deux-usages-du.html.

Aguilera, T. (2013) 'Configurations of squats in Paris and the Ile-de-France region'. In Squatting Europe Kollective (ed.) *Squatting in Europe. Radical spaces, Urban struggles*. New York: Autonomedia, pp. 209–230.

Aguilera, T. (2016) *Racialization of Informal Settlements, De-Politicization of Squatting and Everyday Resistances in French Slums*. London: Routledge.

Bernardot, M. (2008) *Loger les immigrés. La Sonacotra 1956–2006*. Bellecombe-en-Bauges: Éditions du Croquant.

Bouillon, F. (2009) *Les mondes du squat. Anthropologie d'un habitat précaire*. Paris: PUF.

Bouillon, F. and Muller, F. (2009) *Squats. Un autre point de vue sur les migrants*. Paris: Editions Alternatives.

Bouillon, F. and Dietrich-Ragon P. (2012) 'Behind the façades. Ethnographies of Parisian squats', *Ethnologie française* 42: 429–440.

Breviglieri, M, (2009) 'L'insupportable. L'excès de proximité, l'atteinte à l'autonomie et le sentiment de violation du privé'. In Breviglieri M., Lafaye C. and Trom D. (eds), *Compétences critiques et sens de la justice*. Paris: Economica.

Legros, O. and Vitale, T. (2011) 'Les migrants roms dans les villes françaises et italiennes: mobilités, régulations et marginalités', *Géocarrefour*, 86/1. Available: http://geocarrefour.revues.org/8220.

Le Méner, E. (2013) 'Quel toit pour les familles à la rue?' *Métropolitiques*. Available: http://www.metropolitiques.eu/Quel-toit-pour-les-familles-a-la.html.

Mazzella, S. (2014) *Sociologie des migrations*. Paris: PUF.

Olivera, M. (2011) *Roms en (bidon)villes*. Paris: Editions Rue d'Ulm.

Pattaroni, L. (2013) 'Politiques de la différence. Critique et ouverture des possibles dans l'ordre de la ville'. In Cogato Lanza, E., Pattaroni, L., Piraud, M. and Tirone, B (eds) *De la différence urbaine. Le quartier des Grottes/Genève*. Genève: MétisPresses.

Péchu, C. (2010) *Les squats*. Paris: Presses de Sciences po.

Pruijt, H. (2013) 'Squatting in Europe'. In Squatting Europe Kollective (ed.) *Squatting in Europe. Radical spaces, Urban struggles*. Brooklyn: Minor compositions/Autonomedia, 17–51.

Sala Pala, V. (2013) *Discriminations ethniques: les politiques du logement social en France et au Royaume-Uni*. Rennes: Presses universitaires de Rennes.

Simon, P. (ed.) (2001) 'Les discriminations raciales et ethniques dans l'accès au logement social', Note de synthèse N°3 du GIP GELD. Available: https://hlm.coop/IMG/pdf/rapport_geld.pdf.

SqEK, Cattaneo C. and Martínez M. (eds) (2014) *The Squatters Movement in Europe. Everyday Communes and Alternatives to Capitalism*. London: Pluto Press.

Thiéry, S. (ed.) (2014) *Considérant qu'il est plausible que de tels événements puissent à nouveau survenir. Sur l'art municipal de détruire un bidonville*. Paris: Post-Editions.

6 Migration and mobilization for the right to housing in Rome

New urban frontiers?

Nadia Nur and Alejandro Sethman

Introduction

Migrants entering from other countries are a relatively new phenomenon in Italy. Its boom has coincided with that of a housing crisis. Indeed, because of their often lower incomes and fragile legal situations, migrants form an important weak portion of the increasing demand for housing. Faced with a lack of public provisions for social housing, migrants often undertake alternative housing strategies, but they have also undertaken collective actions to claim their right to housing. During the 1900s and 2000s, the city of Rome became one of the main Italian destinations for migrants. By the turn of the century, Rome was also the scene of the emergence and development of a new right to housing movement that became a relevant actor in the city's housing policy arena. These two processes have intertwined, resulting in the migrants playing a relevant role in the mobilization for the right to housing. Through their involvement in Rome's *Diritto all'abitare* (*Right to Inhabit*) movement, migrants have combined the expression of their demand for housing with direct access to it through squatting. But their action goes well beyond their residential issues: it unsettles national citizenship. Therefore, while the frontiers of the national political space remain closed by anti-migration laws and racist discourses, the political activity of migrants can open up space for the imagination, demand and practice of new kinds of citizenship at the urban level.

The purpose of this chapter is to describe the participation of migrants within Rome's Right to Inhabit movement and to analyze its implications for the expansion of rights for non-native Italians. The struggle for the right to housing, as a concrete expression of the struggle for the right to the city, has provided migrants access to the political space they lack at the national level. The chapter is organized in three sections. First we analyze the relation between migration and the new housing question, focusing on the emergence of right to housing movements. Then we discuss the housing patterns of migrants and their participation in squats in relation to the recent policies on migration and housing. Finally we examine the political dimension of migrant housing activism, particularly in the way in which it manifests a concrete formulation of an emergent *urban citizenship*.

Migration and housing issues

The recent growth of a foreign migrant population in Rome was simultaneous with the emergence of a "new housing question" (Cremaschi *et al.* 2007; Tosi 2006). During the post-war period, Italian housing welfare was centered on facilitating access to housing through property, leaving a marginal role to social rental housing. During the 1990s, social housing practically disappeared, and real estate prices soared while the private rental market was deregulated. Due to their average lower income and fewer assets, migrants especially suffered from this situation. In many cases, they were constrained to shelter in de facto social housing (i.e., precarious, low-cost units), squeeze into overpopulated rooms, or, as we will see in the following section, squat vacant buildings. The objective of this section is to briefly describe the housing condition of migrants in Rome in recent years and to trace the origins of the Rights to Inhabit movement.

An overview on migrants' presence, residential patterns and housing conditions in Rome

In various European countries, migrants can count on a strong social housing sector inspired by a universalist model (Scanlon and Whitehead 2010). But, as in other southern European countries, contemporary Italy lacks a significant housing welfare system. This is the result of a combination of long- and short-term trends. On the one side, the housing policy path undertaken since the post-war period has financially and legally privileged proprietary tenure, keeping the public housing stock at quantitatively marginal levels. On the other side, the regulation of the housing rental market through the *Equo Canone* (introduced by Law 392 on 27 July 1978) was abolished in 1998.

During the 1990s and the first half of the 2000s, home buying prices soared. Due to the economic situation, household incomes did not follow the same trend. The gap was partially filled with a financialization of access to housing through mortgage loans. The deregulated housing rental prices also increased during the same period. Moreover, as the labor market evolved towards a decrease of contracts of indefinite duration, rental arrears grew together with eviction orders. As home buying and rental prices went up, creating increasing social need for aid, more pressure was put on a public housing system. With no turnover and no new public residences being built, the demand for housing by the most vulnerable groups was caught between an unaffordable market and the absence of supply provided or subsidized by public institutions. Besides this, evictions of tenants from their homes have reached 30,000 cases in the last five years and are likely to reach 150,000 in the next five years. Gradually this unmet demand produced a situation that was defined in the public agenda as a "housing emergency".

The city of Rome has historically been the Italian capital of migration. However, the economic crisis and the spatial complexity of the city have caused a shift in the territorial distribution of migrants to the former peripheries (and now semi-central areas). More recently this centrifugal force has reached the small municipalities in the surrounding areas with a better offer of low-cost rental housing. Analyzing settlement models, we can observe the difference in housing patterns between the behavior of migrants and the native population. In fact, housing patterns are more influenced by territorial structural factors and other variables such as transnational networks or concentration of nationals that allows for the maintaining of cultural behaviors. In addition, the geographies of migration show a pattern of conurbation which develops in relation to the availability of the transport network (roads and railroads) and extends into the suburbs and hinterland (Centro Studi e Ricerche IDOS 2014).

Among other Italian provinces, Rome is still the most attractive for newcomers, although its migrants' integration index places it at the bottom of the national chart (83rd place within the 103 Italian provinces). According to Centro Studi e Ricerche Idos (2014), foreigners residing in the whole Province represent 8.7% of the total number of the foreigners in Italy. However, 53% of migrants residing in the Province are living in the Municipality of Rome. Official data show that foreigners who are residents of Rome are over 380,000, although this figure is underestimated since it does not include those without a regular residence permit. In 2014, they represented nearly 13.1% of the city's total population (IDOS 2014), an 8.2% increase since 2012. The increasing feminization of migration is relevant, since 52.4% of migrants are women (data regarding Rome are similar to national data). Moreover, the number of unaccompanied minors is also growing, especially in recent times.

Furthermore, according to the United Nations High Commission for Refugees (UNHCR), arrivals of asylum seekers, refugees and beneficiaries of international protection have increased by 87% in the first semester of 2015; nearly 70,000 have arrived in Italy since then. However, it is almost impossible to give concrete figures on asylum seekers since thousands of migrants are just "passing through" on their route to other northern European countries where the reception system offers more opportunities. In the city of Rome, it is estimated that there are more than 20,000 refugees. A quarter of them decided not to apply for asylum in Italy. Reception centers only have room for around 2,500 asylum seekers. Associations, cooperatives and NGOs manage most of the reception centers, as the inadequate public system is unable to provide comprehensive assistance. In addition, a huge number of other people, whose applications for asylum are being processed, are not entitled to being hosted and thus have chosen to partake in squatting vacant buildings along with those who are voluntarily "invisible".

The economic crisis increased in 2008 and represented a turning point for the nexus between migrants and housing issue. Since then, migrants seemed to have turned into dynamic agents of the real estate market. But the years following 2008 exacerbated the difficulties of migrants in meeting their housing needs. Greater rigidity of banks in granting mortgage loans, combined with the weakening of economic and employment conditions for the migrants, led to an increase in demand for rental housing. The main obstacles faced by migrants when they try to enter the real estate market are several: insufficient availability of housing, higher rents, irregular or non-existent contracts, poor quality of the property, and request of additional guarantees for the execution of contracts, such as an Italian guarantor or the activation of a bank guarantee. Moreover, prejudices and a growing racist behavior inhibit most low-income migrants from entering the market and also from the possibility of full integration.

The reduction of income, along with the increase in the availability of irregular jobs, caused an increase in the number of migrant tenants evicted due to unaffordable rentals (Coin 2004). According to Sunia-Cgil, 26% of the cases of the so-called "*morosità incolpevole*" (being unwillingly in arrears) are migrants. The typical migrant family under eviction is composed of 3 or 4 people, with at least one minor (present in 60% of the families) and with an annual income of less than 15,000 euros. If we combine the data collected by *Sindacato Nazionale Unitario Inquilini ed Assegnatari* (SUNIA) in Rome with the data of the Court of Rome, we observe that nearly 80% of the eviction procedures are due to non-payment by the tenants, and 5% to the discontinuation of "free loans" (this type of contract conceals the massive use of informal renting). Compared to the Italians, migrants are also disadvantaged when they apply for access to public housing. On the one hand, migrants are increasingly mentioned in the announcements for the assignment of public housing; on the other hand, the dedicated number of apartments available is not proportional to the number of migrants present in the population. Only 1.5% of apartments owned by the ATER (the regional public housing management agency) are assigned to foreign migrants (Righetti 2010).

Unlike in the 1950s, when internal migrants from southern Italy started to move to Rome, the question of housing for migrants is now entirely absent from the political agenda. In this frame, the housing situation of migrants is particularly fragile. They are indeed fully invested in the new housing question conformed by a social demand by vulnerable groups in a context of rising prices. Moreover, many migrant households lack the formal requisites to access mortgage loans or even home rental, thus cohabitation of more than one family in the same apartment is very common. As they are experiencing the worst side of the housing situation, migrant households develop alternative strategies in which access to housing is achieved at the cost of reduced housing quality (Fioretti 2011).

The emergence of a new cycle of struggles for housing

Despite its marginality to the political arena, during the 1980s and early 1990s, different grassroots organizations, such as the *Coordinamento Cittadino di Lotta per il Diritto alla Casa* (CLDC) and *Lista di Lotta* (LdL), began to occupy public buildings with the objective of obtaining inclusion in social housing allocations. But the struggle for housing was no longer at the center of the agenda as it had been in the 1960s and 1970s. After the resurgence of leftist activism in the early 1980s, two organizational experiences focused on re-elaborating leftist political culture – the Squatted Social Centers, *Centri Sociali Occupati Autogestiti* (CSOAs) and the *Radio Libere* (Mudu 2004, 2012). This political mixture gave birth to a series of innovative initiatives like the *Tute Bianche* (White Overall) and the *Disobeddienti* (Disobedients), focused on drawing attention to the negative social and economic effects of neoliberal globalization, particularly on the youthful population (Fumagalli and Lazzarato 1999).

In 1998, a group of CSOAs activists created an organization called *Diritto alla Casa* (DAC) that, together with the CLDC, squatted an abandoned administrative building for residential use in the Quarticciolo, a poor district built during the fascist era. The objective was to shelter people evicted from the city center, to make way for access to housing, an issue that was emerging as a concrete challenge for low- or no-income households who were forced into precarious employment by neoliberal policies and the development of the housing market (Mudu 2014). The Quarticciolo squat preceded the configuration of the *occupazioni* that would be at the center of DAC/ACTION from then on. Almost all the inhabitants of the squat were members of the CSOAs who had difficulties accessing housing. In this sense the squat served not only as a protest action and as an open space for establishing links with the locality, but as a "practice of the objective", that is, a practice that in itself allowed the production of what was being claimed. After this first occupazione in partnership with another organization, the DAC continued the strategic squatting of abandoned public buildings, such as schools and kindergartens, on its own.

By the end of 2002, due to pressure from the judiciary DAC activists engaged in a series of debates. The result was to combine the right to housing approach with a more general view on the city and the political role of citizenship. This change gave birth to a new organization called ACTION. As a new version of the DAC, ACTION developed a clear understanding of the housing question by linking it both to changes taking place in policy (i.e., the retrenchment of housing welfare), in the social arena (migration and individualization of households) and in the economic arena (real estate valorization of capital and higher unemployment). From these diagnoses, ACTION developed a new political horizon – the Right to Inhabit. This right links access to housing with direct political participation at the local level in order to counter exclusionary economic policies and urban dynamics (Sethman 2016).

By putting this new formulation of the right to housing into practice, ACTION was able to end the double isolation that the struggle for housing had suffered for 20 years. By using the local deployment of the CSOAs and the new squats to interact with the rising unsatisfied demand for housing (mainly among migrants and young workers and students), the organization successfully established connections with large sectors of Roman society (see Di Feliciantonio Chapter 7, in this volume). As public opinion and political leaders began to recognize the new challenges of accessing housing, ACTION's contentious methods became more acceptable for both new activists and households in need of a roof over their heads. This phenomenon reinforced the political alliance between the left and center-left parties at the local level, giving more leverage to the organization inside the political coalition (Sethman 2015).

The space opened up by the Right to Inhabit as a catalyst of contentious politics for housing rights was later populated by other organizations such as the *Acrobax* Social Center and the *Blocchi Precari Metropolitani* (BPM), giving birth to a highly volatile network whose ties would strengthen or weaken depending on many factors, mostly related to opposing approaches to representative politics. To sum up, during the last decade, the various movements for the right to housing have acquired increasing importance in bridging the shortcomings in public policies. In particular, during the last two years, the increase in occupations of abandoned properties, both public and private, has represented the only political response "from below" to the lack of housing policies. In the complex situation we outlined, the Right to Inhabit movement acts as a vehicle of expression for their demand for adequate housing, but it also provides them with an immediate solution to their housing problems: the squatting of vacant buildings.

Migrants and the struggle for the right to housing in Rome

In this section, we describe the relevant role of migrants in the housing movement that emerged in the city of Rome. We also discuss squats that are organized around the issue of migration. The participation of migrants is not simply a result of their housing grievances, but it is also the result of their involvement in a process of mobilization that allowed them not only to access housing but also to exercise a citizenship that was denied to them in the national political space.

Home struggled home

Apart from traditionally being a city of migrants, Rome is also the city with the highest number of occupations for housing purposes, mostly led by the various constellations of the Right to Inhabit movements (ACTION, BPM, and CLDC, *Comitato Popolare di lotta per la casa, Esc Infomigrante* among the most active) that oppose the latest policies toward housing and speculation in the real estate market. The geography of squats is varied, and a survey of all existing realities is a demanding task due to the many recent occupations, the existence of more stable squats, and the continuous evictions and displacements.

Thousands of Italian and foreign households are living in abandoned schools or properties, buildings owned by ATER or by other governmental agencies, gyms and sports centers, occupied with the support and coordination of Right to Inhabit movements or Social Centers that play a crucial role in developing a dialogue with the governmental administration. Although housing should be a key factor in the integration process, in Italy, it is increasingly becoming a critical factor that is related also to national policies toward migration. The informal way of living as a housing solution is adopted by a growing number of migrants and could be interpreted as the result of the ineffectiveness of both housing and migration policies. The large number of occupations for residential purposes by migrants, asylum seekers and refugees show that "squatting" is still regarded as an effective solution to address the inconsistency of the anti-migration policies and recent government measures concerning the housing crisis. It also highlights that we cannot debate anymore about a state of emergency, but we should recognize the structural inadequacies of policies with respect to the transformations of society (Nur 2014).

The high rate of migrant participation in the occupations led by the Right to Inhabit movements suggests that a new powerful political and social subject is emerging, in which the issue of inhabitance is associated with that of migration development (Mudu 2014). In fact, in the city of Rome, migrants play a leading political role, both as constituents and as leaders, in the various fringes of the Right to Inhabit movements. In doing so, they have turned their experience of urban inhabitance into a field of struggle for the rights that they have been denied at the nation-state level. Urban citizenship appears as both the horizon and the point of departure for migrants participating in housing activism.

A report by the *Commissione Sicurezza di Roma Capitale* (2010) (Security Commission the Council of Rome) draws a map in which occupations and squats are concentrated mostly in the first five boroughs (Municipi), while they are rare in the suburbs. The Commission's report does not quantify the dwellers nor the share of migrants and does not consider the small occupations and informal settlements built by refugees and asylum seekers, some of which are characterized by slightly homogenous ethnic background and by the absence of external coordination.

Recent years have been characterized by the growing number of migrants, new occupations raids (the so-called *Tsunami Tour* of 2013) and new evictions. In 2013, approximately 2,500 family units (more than 6,000 people) were living in 60 squatted buildings. In this group, 70% were foreigners, and many of them were unemployed, or unauthorized or unable to work. According to the most updated estimates, there are now likely 90 squatted buildings, excluding the informal settlements.

We can sketch out different types of occupations (Nur 2014). The first type can be defined as "organized" or coordinated by the Right to Inhabit movements and/or other organizations. Usually the dwellers of this kind of squat are mixed, that is to say that Italian families live together with foreigner families, and residents of the squat belong to various ethnic groups. An example of this kind of squat is the one in viale Castrense, where more than 60 households live – Eritreans, Latin Americans and Italians. The building is a former school located in the central area of San Giovanni that was vacant for many years until it was occupied by ACTION in 2003. After the occupation, ACTION signed an agreement with the owner. The internal organization is typical of squats; decisions are made through periodic assemblies and participation in political activities and rallies led by the movement is almost mandatory. The squat's proximity to *Scup* (Sport and popular culture), a popular Social Center (evicted and occupied in a different place in May 2015), and to *Sans Papier* facilitated the dwellers' integration into the neighborhood, as they were perceived as part of the same movement.

In 2013, the wave of new occupations in Rome, the so-called Tsunami Tour, led to the squatting of a vacant building in via Curtatone, in the very center of Rome, near Termini Station. The former public institute, occupied by CDLC, hosts more than 600 refugees and asylum seekers, mostly Eritreans and Ethiopians. Many of them survived the Lampedusa tragedy (see Borgstede Chapter 13, in this volume). The location of the squat, in a neighborhood traditionally popular for its Eritrean restaurants, is raising new tensions but also increasing awareness on the evident gap between the living conditions of the first wave of migrants and those of the recent wave.

In the context of the same political action, nearly 100 people occupied an abandoned public building in via Santa Croce in Gerusalemme, in the multi-ethnic Esquilino neighborhood, with the help of ACTION activists. The number of squatters grew rapidly, reaching 500 individuals and following the scheme of the Castrense squat, where Senegalese, Italian, and Romanian dwellers represent a model of an ethnically diverse micro-society. In the past year, the building has hosted cultural and political initiatives, exhibitions, gathering artists, activists and professionals willing to share experiences and raise awareness on the housing emergency related to migration.

The squat mentioned belongs to a model of occupation that, in addition to solving the concrete problem of the housing crisis, poses an explicit goal of integration, through workshops and cultural and artistic initiatives. In this category, we can also mention *Metropoliz*, a former sausage factory in via Prenestina in the neighborhood of Tor Sapienza, occupied in 2009 by BPM in collaboration with the NGO *Popica Onlus*. The building was abandoned for 20 years, and when the squatters moved in to occupy, it was extremely dilapidated. Dwellers lived in open air for six months before dividing the area into small plots that were assigned to each family unit. After a few months, a group of 100 Roma people, who had been evicted from the abusive camp of *via di Centocelle*, occupied a warehouse nearby the main building. At the moment, Metropoliz is inhabited by nearly 200 migrants coming from different regions (South America, North Africa, the Horn of Africa, Eastern Europe) and a group of Roma people. This kind of squat enabled the revitalization of the territory through the innovative re-use of the occupied space. It is a multi-ethnic laboratory, livened up by the *Laboratorio di Arti Civiche*, artists and film makers or film producers.

The peculiarity of the type of squat mentioned above is the activation of synergies around migration issues across migrants, associations, NGOs, local administration and local citizens. It is worth mentioning *ZaLab*, an association of film makers and social workers who advocate for the spreading of democracy and for minority and migrants rights. *Habeshia*, an NGO, stands as a reference point for all migrants from the Horn of Africa, for all kinds of support. Moreover, a network between different types of squats has been built, aimed at mutual legal and health assistance.

Also occupations that are independent of squatter movements belong to this feature of occupation. Similar to the self-managed and self-organized squats that we will describe further, this type of occupation is generally an ethnic enclave, inhabited primarily by refugees and asylum seekers, mainly coming from former Italian colonies, Eritrea, Ethiopia, Somalia, and Sudan. Examples include *Salam* (or *Selam Palace*), the former headquarters of Tor Vergata University in the Romanina area, born in 2006 as a "new emergency solution" after the clearance of *Hotel Africa* (another ex-squat self-managed by migrants), which caused the splitting of dwellers into two different occupied buildings. Depending on the number of new arrivals, there are an average 1,000 residents of Selam, all of who hold of the status of refugee or have been granted subsidiary protected status. The length of stay depends on the outcome of migration routes. For several years, the *Cittadini nel Mondo* association has run a help desk offering medical treatment and advice on access to health services. Although the inhabitants of the building worked together with Cittadini del Mondo and with the support of Open Society foundation in order to legalize the squat, Selam Palace is still illegal and recently became a symbol of refugee crisis.

Naznet, a "historical" self-managed occupation in via Collatina in the outskirts of Rome, is giving shelter to about 700 Eritreans and Ethiopians. Some of them have occupied the building since 2004, when the building was occupied with the help of ACTION, and they hold legal residence permits for asylum, while others are "in transit" on their way to northern Europe. The latter are still illegalized, trying to escape the trap of the Dublin Regulation.

A completely different case is one of the informal settlements of Ponte Mammolo, which hosted about 200 people of 11 different nationalities (mainly Eritreans and Ethiopians, along with Ukrainians, Romanians, Bangladeshis and other minorities). The original tent city established in 2006 was transformed over time, becoming a small slum that remained disconnected from the reality of other occupations. In May 2015, the settlement was evicted, and women and children were moved to a reception center while some men are still sleeping on the street.

The estimates show that there are nearly 2,000 refugees and a greater number of asylum seekers living in occupations or informal settlements in the capital,[1] most of them are people who are waiting for international protection, people who have already acquired the status of refugee, or people temporarily residing in Rome on their way to other European countries. The absence of policy initiatives (Balbo 2005) gives rise to the emergence of new forms of housing and housing struggles that give rise to new forms of citizenship.

No room for legality: the effect of exclusionary housing policies on migrants

It is not easy to define the bond between migrants and radical movements claiming the right to inhabit, although it is clear that recent repressive migration policies, combined with the economic crisis and housing emergency, constitute a plot on which of the most vulnerable social groups are trying to build a common platform to claim a series of basic rights.

Due to the increasing number of migrants, we can say that Rome is undergoing a process of "globalization from within" (Clough Marinaro and Thomassen 2014) even though local and national policies seem unable to manage the transformations occurring in the social structure. As we described in the previous section, squats, as well as migration issues, are managed through a perpetual state of exception (Agamben 1998). Perceiving global Rome through the experiences of migrant squatters reveals a city in which public apparatuses continue to exert fundamental power over spaces and people – a power which limits their right to the city (Clough Marinaro and Daniele 2014).

While supra-national policies strengthen the barriers of Fortress Europe, denying the principles of freedom of movement on which its own principles are based, nation-states, Italy in particular, reshape anti-migration policies, tightening access to citizenship, housing and work. By inhibiting the integration process, the State is *de facto* violating its own constitutional principles. From the establishment of Identification and Expulsion Centers (CIE) and their evolutions, the abolition of Mare Nostrum and the starting of new initiatives to ward off migrants, to the recent debate on European Union's (EU) refugee quota program, the overall measures towards migrants are aimed at the restriction of the right to migrate. At a local level, the process of integration of asylum seekers goes no further than to be hosted for a brief period in a reception center. Once forced migrants have arrived on Italian territory and applied for asylum, they are taken to the reception centers (CARA) where they remain until their applications are processed by the area committee. Although the "*Sistema di protezione e accoglienza dei rifugiati*" (SPRAR) program has been amplified, there is still no satisfactory system of protection, no measures for integrating political refugees, as well as other migrants, into Italian society. Once entitlement to international protection has been granted, there is no provision for resettlement, and many of them end up living in informal or precarious conditions.

In the outlined context, it is not difficult to understand how the migrants stand at the intersection between political action of No Borders or No One is Illegal movements, and the right to housing movements such as International Alliance of Inhabitance and Habitat International Coalition, among others. At a more local level, migrant issues related to housing and residence highlight the contradictions of the whole regulatory apparatus, thus bringing to the fore the concept of right to the city. If migration, although governed by national and supra-national policies, impacts the social and spatial structure of the city, the local level should be the ground on which fair access to rights is granted and implemented. However, this appears to be an unfinished process, although many municipalities have established advisory committees aimed at encouraging the political participation of migrants.

In the twentieth century, granting the right to housing meant that integration was achieved. In the global cities of the twenty-first century, this right is becoming a tool for the exclusion of vulnerable groups such as migrants. The dynamic of inclusion/exclusion – through which global Rome is continuously keeping migrants and squatters in a permanent condition of legality/illegality, tolerance/clearance, and protection/abandonment – highlights the existence of a dramatic institutional vacuum. The recent authoritarian response to the occupations for housing purposes in the city of Rome, characterized by a series of evictions of both squats and Social Centers, shows that the struggle for the right to inhabit is a fundamental democratic right.

The already precarious situation of migrants and refugees have recently worsened due to the application of Article 5 of the Housing Plan (*Piano Casa*, also called *Piano Lupi*), a national law that prevents people from registering a formal residence in an "illegally" occupied public building. The constitutional doctrine shows that the right to housing is not restricted by the protection of property. Instead, it is related to a broader concept of rights connected to equality, both in material conditions of life and individual dignity. Thus the right to housing is a precondition for the enjoyment of fundamental freedoms and social rights, such as the right to health, education and employment. The Housing Plan applies the rule of law only for the protection of the possession, thus denying the fundamental rights granted by the Constitution. In addition, the Lupi law denies people the right to participate in the selection procedures for the assignment of public houses for a period of five years, prevents squatters from signing up for electricity, gas and water contracts, and excludes the occupants from the access to the rights connected to residency (education, health, political participation, etc.), the first steps toward citizenship recognition.

Moreover, the Lupi law, along with the entire apparatus regulating migration and housing, is in complete violation of the Constitution:

- Article 3 proposes that all citizens have equal social status and are equal in front of the law, without distinction of sex, race, language, religion, political opinion, or personal and social conditions. It is the duty of the Republic to remove those obstacles of an economic and social nature which constrain the freedom and equality of citizens and prevent the full development of the human being and the effective participation of all workers in the political, economic and social country.
- Article 16 establishes the right of every citizen to move and reside freely anywhere in the national territory.
- Article 30 claims that it is an obligation of parents to educate their children. In fact, it is not possible to attend school without providing an official residence.

The Lupi law also violates a rule of the *Consiglio di Stato*, according to which it is compulsory that citizens have a formal residence, regardless of their social and economic status. Finally, Act no. 286 of 1998 (governing migration and the status of foreigners) states that regular migrants have the right to access housing on equal terms with Italian citizens. The first applications of the law in Rome have affected Salam and another squat located in via Pecile whose residents have been denied residence registration. Currently, the *Uffici Territoriali di Governo* (UTG) (former *Prefetture*) are considering not applying the Lupi law and are inviting Municipalities to include all citizens, including squatters, in the registry.

By obstructing the possibility of migrants to access housing, the Lupi law projects the occlusion of the national political space into the experience of urban residence. As it happened with arrivals, the repressive approach has not restricted the phenomena of irregular housing. It has encouraged the development of alternative forms of access to urban space as expressed by the persistence of residential squats with a high migrant participation. However, as we have seen, the squatting activity is not just a response to migrants' housing grievances; it is part of a struggle for a right to inhabit the city that encompasses access to housing and participation in the urban political arena.

Migrants, activists, citizens

Migrants were particularly hit by the housing emergency in the city of Rome towards the beginning of the 1990s. The new cycle of protests that started in these years critically re-conceptualized the idea of housing in a way that goes well beyond the idea of a home and includes the notion of inhabiting. This allowed for the formation of a cognitive framework that turned housing grievances into a politically productive social demand for the right to inhabit.

As this demand is related to the direct participation in the set of political and economic relations that determine the means of production, circulation and consumption of land, it has opened up a concrete space of political activity for migrants in the city of Rome. As it is structured around squatting, this participation has been mostly non-formal and conflictive. But however marginal to the mainstream workings of parliamentary democracy the urban action of migrants has disrupted the exclusionary conception of citizenship defined by the political order and by the laws of the nation-state (Purcell 2002).

Different authors have conceptualized the development of political activities of marginalized subjects at the urban (rather than at the national) space. Saskia Sassen has studied the "denationalization" of citizenship and its repositioning at the urban scale, particularly through the presence of migrant subjects (Sassen 2002, 2008). James Holston (2008), in turn, has defined urban citizenship in terms of the agenda and scale of social mobilization. For him, globalization has turned cities into the "site and substance" of emergent forms of citizenship. The case of Rome shows that the urban space is indeed a contemporary battleground for marginalized populations.

Squatting an empty building is a visible public action and exhibits a new set of power relations. It simultaneously and contentiously demands and informally provides for a right recognized by national law and international treaties but not guaranteed by the authorities. For this reason, the intersection of the Right to Inhabit movement and migrants is so productive. Squatting is a political tool developed at the local level and an instrument for allowing migrants to concretely access the city as a residential space. While the civil, political, and social rights of national citizenship remain forbidden for most migrants, the city has provided them with an emergent urban citizenship.

As with national citizenship, urban citizenship is about rights. In this sense, the intertwinement between the Right to Inhabit movement and migrants raises new questions around the conception of the "right to the city" (Lefebvre 1968, 1996). How can those who are denied (national) citizenship have such a right? The rhetoric of occupation represents a right to be present, before any specific political demands are made (Mitchell 2012). The act of occupation by migrants is, in fact, the "right to have rights" (Butler 2011). The dialectic of legality and illegality of both the migrants and their forms of access to housing has opened up a breach into the economy of the native–foreigner divide. In the urban space, the status of migrant is reabsorbed into the status of urban citizen that puts national origin in the background. The act of occupying highlights the contradiction between the constitutional norms that sanction the right to adequate housing without distinction of origin, and the denial of the right to the city to both natives and aliens. Thus, as the issue of migration is rescaled, also the political claims of natives and foreigners overlap.

Rome's status is a paradoxical balance between being the capital of one of the world's economic powers and simultaneously an informal self-made city (Cellamare 2014). The institutionalization of "gray spaces" (Yiftachel 2009), semi-permanent space, suspended between legal and illegal, formal and informal, where a different citizenship level is exercised, is emblematic of this condition. Squats, "unauthorized" and "illegal" but tolerated spaces of inhabitance, are not born in conformity to law, but indeed "in the light of rules", thus they are nomotropic spaces (Conte 2011). Positioned between legality/approval and eviction/destruction, squats call into question the dominant public discourse on migration as a border or a national-security problem.

The practice of residential squatting carried on by migrants in the frame of their struggle for housing rights is a way of responding to the inadequacy of public administration. And it is also a way of formulating a new citizenship even when the classic national one is denied. This emergent citizenship is composed of rights relating the experience of urban residence and is practiced at the local level. The city appears then as a new frontier, one that lies beyond the limits imposed by national borders and makes the redefinition of the lines separating inclusion and exclusion more accessible to migrants.

Note

1 These estimates, based on Caritas statistics, reports of associations and information collected through interviews, are likely to be constantly increasing, due to the volatile situation of arrivals by boat through the Mediterranean. Some sources estimate that the number of asylum seekers living in Roman shanty towns and squats has reached a peak of 7,000. According to UNCHR (2015), refugee and migrant arrivals to Italy across the Mediterranean from January to July 2015 have been in the order of 67,000. A part of them are likely to have found a shelter in Rome.

References

Agamben, G. (1998) *Homo Sacer: Sovereign Power and Bare Life*. Stanford: Stanford University Press.

Balbo, M. (ed.) (2005) 'International migrants and the City', Venice: UNESCO UN-Habitat and Dipartimento di Pianificazione Università IUAV di Venezia.

Butler, J. (2011) 'Bodies in Alliance and the Politics of the Street, European institute for progressive cultural policies'. Available: http://www.eipcp.net/transversal/1011/butler/en.

Cellamare, C. (2014) 'The self-made city'. In Clough Marinaro, I. and Thomassen, B. (eds), *Global Rome: Changing Faces of the Eternal City*. Bloomington and Indianapolis: Indiana University Press, pp. 143–155.

Centro Studi e Ricerche IDOS (ed.) (2014) *Osservatorio Romano sulle Migrazioni. Decimo Rapporto*. Roma: Edizioni IDOS.

Clough Marinaro, I. and Daniele, U. (2014) 'Evicting Rome's "undesirables": Two short tales'. In Clough Marinaro, I. and Thomassen, B. (eds), *Global Rome: Changing Faces of the Eternal City*. Bloomington and Indianapolis: Indiana University Press, 114–127.

Clough Marinaro, I and Thomassen, B. (2014) *Global Rome: Changing Faces of the Eternal City*. Bloomington & Indianapolis: Indiana University Press.

Coin, F. (2004) *Gli immigrati, il lavoro, la casa: tra segregazione e mobilitazione*. Milano: FrancoAngeli.

Commissione Sicurezza di Roma Capitale (2010) *Mappatura degli stabili di proprietà pubblica e private occupati abusivamente*. Available: http://www.affaritaliani.it/static/upll/mapp/mappatura_stabili.pdf.

Conte, A. G. (2011) 'Nomotropismo', *Sociologia del diritto* 27(1): 1–27.

Cremaschi, M., Di Risio, A. P., Longo, G. and Lucciarini, S. (2007) 'Dinamiche dei territori e questione abitativa'. In Clementi, A. (ed.) *Reti e territori al futuro*. Roma: Società italiana degli urbanisti.

Fioretti, C. (2011) 'Do-it-yourself housing for immigrants in Rome: Simple reaction or possible way out'. In Eckardt, F. and Eade, J. (eds) *The Ethnically Diverse City*. Berlin: BWV Verlag.

Fumagalli, A. and Lazzarato, M. (1999) *Tute bianche: disoccupazione di massa e reddito di cittadinanza*, Roma: Derive e Approdi.

Holston J. (2008) *Insurgent citizenship: disjunctions of democracy and modernity in Brazil*. Princeton: Princeton University Press

Lefebvre H. (1968) *Le droit à la ville*. Paris: Anthopos.

Lefebvre H. (1996) *Writings on Cities*. Cambridge, MA: Blackwell.

Mitchell, W. J. T. (2012) 'Image, space, revolution: The arts of occupation', *Critical Inquiry*, 39(1): 8–32.

Mudu, P. (2004) 'Resisting and challenging neoliberalism: The development of Italian social centers', *Antipode*, 36(5): 917–941

Mudu, P. (2012) 'I Centri Sociali italiani: verso tre decadi di occupazioni e di spazi autogestiti', *Partecipazione e Conflitto*, (4)1: 69–92.

Mudu, P. (2014) 'Ogni sfratto sarà una barricata: squatting for housing and social conflict in Rome'. In Cattaneo C. and Martinez M. (eds) *The Squatters Movement in Europe. Everyday Communes and Alternatives to Capitalism*. London: Pluto Press, pp. 136–163.

Nur, N. (2014) 'Dalle politiche antisociali al diritto alla città Quando i migranti vogliono abitare'. *Urbanistica Informazioni* 257(100): 35–8.

Purcell, M. (2002) 'Excavating Lefebvre: The right to the city and its urban politics of inhabitant', *Geojournal*, (58)2: 99–108.

Righetti, C. (2010) 'La casa popolare resta un sogno'. *La Repubblica*, 24 February 2010.

Sassen, S. (2002) The repositioning of citizenship: Emergent subjects and spaces for politics. *Berkeley Journal of Sociology* 46: 4–26

Sassen, S. (2008) *Territory, Authority, Rights: From Medieval to Global Assemblages*. Princeton: Princeton University Press

Scanlon, K and Whitehead, C. (2010) 'Le logement social en Europe: Tendances communes et diversités persistantes'. In Lévy-Vroelant, C. and Tutin, C. (eds) *Le logement social en Europe au début du XXIe siècle: la révision générale*. Rennes: PU Rennes.

Sethman, A. (2015) 'Housing (the) crisis: urban social movements in the post-welfare western European city. A study on the cases of Paris and Rome', unpublished doctoral dissertation. Universidad Nacional de San Martin, San Martin (Argentina)/Università degli Studi "La Sapienza", Rome (Italy).

Sethman, A. (2016) "From contention to co-governance: the case of the right to inhabit movement in Rome (2001–2013)" *Antipode*.

Tosi, A. (2006) 'Povertà e domanda sociale di casa: la nuova questione abitativa e le categorie delle politiche', *La Rivista delle Politiche Sociali*, 3: 61–78.

UNCHR (2015) La via del mare verso l'Europa. Il passaggio del Mediterraneo nell'era dei rifugiati. Available: http://www.unhcr.it/sites/53a161110b80eeaac7000002/assets/5594f5c90b80eefd95005817/La_via_del_mare_verso_l_Europa.pdf.

Yiftachel, O. (2009) 'Critical theory and "gray space": Mobilization of the colonized', *City*, 13 (2–3): 246–263.

7 Student migrants and squatting in Rome at times of austerity

Cesare Di Feliciantonio

The development and diffusion of squatting initiatives both in the form of housing projects and Social Centers in Western European countries have acknowledged that migration is a fundamental right that should be defended and supported. Indeed, international migrants have been traditionally the most excluded from the benefits of the welfare state, depicted as a "threat" and danger" to the sustainability of welfare regimes (Bommes and Geddes 2000; Schierup *et al.* 2006). However, the relevance of squatting for internal migrants has not received much attention, notably in Southern European countries characterized by a severe lack of social housing in the welfare system (Castles and Ferrera 1996; Allen *et al.* 2004). This appears to be particularly relevant in a city like Rome where big waves of internal migrants arrived during the twentieth century and provided a self-organized and autonomous solution to the housing question (Martinelli 1985; Berdini 2010).

Against this background, I focus on the relation between squatting and a particular type of internal migration, i.e. young people from other Italian regions who move to Rome for their university studies. The focus concerns the emergence of several squatting initiatives aimed at giving an autonomous response to the lack of public welfare services and rights for students, these initiatives being undertaken since the worsening of the current debt and financial crisis and the consequent adoption of austerity policy measures by both national and local institutions. The social status of students has been completely reshaped in the last thirty years: far from guaranteeing access to well-paid and professional jobs, Italian universities create a new low-income and unemployed mass of people. According to 2011 data, roughly, 42% students holding a Masters degree were unable to find a job within the first year after their graduation or after the completion of their academic course (Almalaurea 2012). Under these circumstances, the options left for students are to squat/ to self-manage to continue their existence in a big city or to return to their hometown. In this context, autonomous and self-organized spaces represent a collective response towards the deteriorating material condition of young people; when deciding to remain in Rome despite material constraints, the imagery and expectations for their future (*albeit* precarious) life play a crucial role in people's choices.

When analyzing the relation between internal student migrants and squatting as an urban social movement (Martinez 2013), Rome represents an important case. In fact, Rome has a massive student population, including the largest European university, *La Sapienza*, with more than a hundred thousand students. The traditional weakness of the Italian welfare system, with a chronic lack of public accommodation, has also concerned students who mostly rely on the rental black market. At the same time, university students have proved to be among the most active resistors to neoliberal market reforms and policies. Students organized relevant movements, such as occupying the streets or blocking university activities. For example, in the fall of 2008 the *Onda* (wave) movement led to the occupation of all the main campuses of Italy and stalled regular activities (Bernardi and Ghelfi 2010). In a city where squatting and autonomous politics have a long-standing history (Mudu 2004, 2012), the connections between the students' movement and squatting initiatives have proved to be very intense. During the fieldwork I carried in Rome in 2013 and 2014 to analyze the new wave of squatting initiatives that emerged since the worsening of the current economic situation, the tight relation between the squatting and the students' movements was directly expressed by my research partners on several occasions. The *Onda* was often depicted as a crucial moment of politicization leading to a succeeding involvement in squatting initiatives. As pointed by S5T,[1] an internal migrant aged between 25 and 30 who started to get involved with the *Onda*:

> For me it was the beginning of everything, not just protesting against the cuts to university, I still see it as the moment in which I started to imagine a new world together with the others and fight for it. [...] That experience then brought me to engage with grassroots politics, constructing new spaces to imagine a new sociability – an alternative to the world of competition and individualism we live in [...]. At the same time, for me and many people around it was also a turning point concerning material conditions, such as those my family was experiencing due to financial distress and their inabilities to support my education and other financial needs. The housing prices remaining sky-high, it was impossible for me to find a decent job. (...) Squatting for us was a response to a materially constrained situation as well as a political project to claim housing rights for students [...]. we do not want to be exploited [at work] ten hours per day just to pay the rent! (personal interview, June 2013, author's translation).

This brief account reveals the complex material everyday scenario faced by students who were internal migrants – all those who wanted to live in Rome at the current moment of economic turmoil, austerity politics and cumulative poverty. The historic changes in the economic state of several European countries have deep implications on the everyday life of students in Rome in the following ways.

First, I detail on the increase of poverty as highlighted by a recent study of the Bank of Italy (2014): more than 21% of people aged between 19 and 34 years live under the poverty line in Italy. In 2012 the same index was around 18% for people aged between 35–44 years and 45–54 years. These data register a fast-rising increase of poverty especially for people aged 45–54 years, indeed those living under the poverty line within this group were only 13% in 2008. The only group for which the ratio of people living under the poverty line is decreasing despite crisis and austerity is that of people aged more than 64 years. If we consider the trend of the equivalent income between 1991 and 2012, then the study reveals a decrease of the equivalent income for all age groups, except that of people aged more than 64 years. People aged 19–34 years have been the most affected by this decrease, their equivalent income having decreased approximately by 15% in this period. In a country where family is the main pillar of the welfare system (Poggio 2008), the impoverishment of families considerably reduces the material possibilities of young generations.

Secondly, we should consider the cuts to the (already weak) welfare system provision, especially in terms of housing. So, the students' accommodation public system is able to respond only to the housing needs of 10% of the student population for affordable housing. Moreover, the severe cuts introduced by austerity politics concern several domains of student life; for instance the libraries of the Faculty of *Lettere* (Humanities) of Sapienza are now open only four hours per day because they do not have enough employees to operate for additional hours. The combination of these effects has led to a rapid decrease in the number of students enrolling to the university. For instance, in the case of Sapienza, in the last five years the number of new students has decreased more than 10%.[2]

As a response to this situation that has worsened over the last few years, squatting remains the main option and squatting initiatives have emerged in Rome to claim for a larger and more inclusive welfare system for students. Squatting initiatives can be regrouped in two typologies. The first one is that of squatted housing initiatives for students (*studentati occupati*), the main ones being *Alexis*, *Degage*, *Mushrooms*, *Point Break* and *Puzzle*. They are located in five different neighborhoods: Alexis in San Paolo, Degage in Regina Margherita/Policlinico, Mushrooms in Casal Bertone, Point Break in Pigneto, and Puzzle in Tufello. All these places are in the neighborhood of the university campuses, except for Puzzle. The links with the main metropolitan squatting networks are really strong, most of them having been squatted during the "Tsunami Tour", a big campaign launched by the three main metropolitan squatting networks; *ACTION-Diritti in Movimento, Blocchi Precari Metropolitani, Coordinamento di Lotta per la Casa* between 2012 and 2013. Moreover, they are connected also with other struggles and social movements. For instance, Mushroom is located within the property of *Officine Zero* (OZ), a squatted and self-managed former train wagons-factory.

The second typology is that of several Social Centers providing self-managed services to students, especially libraries and rooms to study. In San Lorenzo neighborhood, the closest to the main university campus of Sapienza, both *Communia* and *Nuovo Cinema Palazzo* have self-managed libraries and study-rooms open until late night. In the case of *Communia*, as well as other Social

Centers, university students are the core militants and attendants. In 2013, I submitted a questionnaire to *Communia* militants aiming to understand their socio-economic backgrounds: almost 80% of the respondents identified themselves as undergraduate or postgraduate students. In fact, at the time of its conception, *Communia* aimed at being a newly squatted students' house.

When interviewed about the reasons leading to the decision to squat, many student squatters refer to the worsening economic and financial situation of their families as a turning point – "you squat or you *leave* Rome". To better understand this process, we can refer to the words of one of my research partners (MF2) aged 20 to 25:

> The last couple of years have been really tough for me, I had a circle of close friends I met in my first two years here but then most of them had to leave: no more money to study in Rome! (*discontented chuckle*) I have been in a similar situation. For my parents it has become difficult to sustain me, so at one point I really had to decide: going back to my hometown or stay in Rome? How to deal with the lack of money? [...] I started working four days per week and it was alright, but then for my sister too it was time to start the university, so my parents could only support me marginally, and the money I made was not enough to sustain. Again I was faced with the same question whether to go back or stay? And again I took the decision of staying [...] I decided to enter this housing project as I already knew some of the people involved through university activism [...]. I am still working for four days a week and I am somehow managing without asking for more money from my parents. I understand that financial situation is constrained back home. (personal interview, author's translation, June 2013)

The narrative of MF2 reveals the strong tensions generated by the current situation in which more and more families are living under economic constraints, thus making more difficult to support their young members, a traditional characteristic of the Italian welfare system. However another question arises: what leads these young students to choose to remain and squat over going back to their hometowns?

From my research two main factors appear to explain this decision – imagery and the expectations for future life (although precarious) and a strong collective life. Concerning imagery, several research participants stressed how important moving to Rome was for them, something they have aspired to for long time. As emphasized by Weston (1995), this seems to be the case especially for queer people not conforming to indices that section our societies along gender, sexuality, and corporeality among the others. For these subjectivities, moving to a (big) city offers the possibility to "finally express what they want or to experiment, transgress, taste, learn" (VH5, personal interview, author's translation, May 2013). This relates to what Larry Knopp has defined as "the queer quest for identity" (2004). In the words of JHF, aged between 25 and 30:

> I think I have always imagined myself escaping from my hometown, from the control of people who I have known for all my life. It's an escape from

> the expectations, from family normativities. [...] I first moved to Naples and then to Rome for my Masters, it \has been a sort of continuous discovery, creating new families and communities, living the life I wanted, learning to say 'no' to someone else's expectations and peer pressures [...] I do not think I will spend my whole life in Rome, but now my affects and my community are *here*. (personal interview, author's translation, emphasis added by author, June 2013)

The words of JHF express the importance of imagery, expectations in determining the choice to remain in Rome even in a situation of economic distress. The interrelation of these factors calls into question the role of future self-expectations and projects that are inserted into people's lives in precarious times. Indeed living in Rome and being active within collective squatting initiatives seem to strengthen young people's will to continue to live in such a collective and politicized environment, even if the city does not sanction many possibilities in terms of employment and better quality of life (see, in this volume, Nur and Sethman Chapter 6). Here we can refer to the self-narrative of CWE, aged between 25 and 30:

> Squatting and constructing these kinds of political projects are not simply a form of political engagement, you build your own community, personal ties, friendships, maybe you even find a partner, or more than one, within it! [...] I cannot imagine myself without these people anymore, they are part of my life, [...] I am aware that in Rome I will probably remain hyper-precarious forever, maybe I won't even find a decent job, [...] but here I learned to imagine and live a new life, a collective life, something I do not want to renounce. [...] It gives me so much energy to imagine new struggles, new forms, new projects [...], we cannot stop thinking that we are creating new forms of relations and sociability! (personal interview, author's translation, March 2014)

To conclude, squatting represents a concrete, self-organized alternative for student internal migrants in Rome who experience a situation of increasing material constraints; the involvement in these initiatives makes people experience a new collective and politicized everyday life, leading them to remain in Rome instead of going back to their hometowns where they could get some welfare support from their families. Despite precarious living conditions and expectations for the future, squatting continues to exercise for the people involved its character of "world-making", as recently pointed out by Alexander Vasudevan (2015) borrowing from Muñoz (2009).

Notes

1 The vagueness of data about research partners responds to the ethical commitment of guaranteeing full anonymity to the squatters involved in the research project.
2 Source of data on students: http://www2.uniroma1.it/infostat/facolta.php?aa=2015&lk=1.

References

Allen, J., Barlow, J., Leal, J., Maloutas, T. and Padovani, L. (2004) *Housing and Welfare in Southern Europe*. Chichester: Wiley-Blackwell.

Almalaurea (2012) *Rapporto sulla condizione dei neolaureati*. Available: www.almalaurea.it.

Berdini, P. (2010) *Breve storia dell'abuso edilizio in Italia*. Rome: Donzelli Editore.

Bernardi, C. and Ghelfi, A. (2010) 'We won't pay for your crisis, we create institutions of the common!', *EduFactory webjournal*: 108–118.

Bommes, M. and Geddes, A. (eds) (2000) *Immigration and Welfare. Challenging the borders of the welfare state*. New York: Routledge.

Castles, F. G. and Ferrera, M. (1996) 'Homeownership and the welfare state: Is Southern Europe different?' *South European Society and Politics* 1(2): 163–184.

Knopp, L. (2004) 'Ontologies of place, placelessness, and movement: Queer quests for identity and their impacts on contemporary geographic thought', *Gender, Place & Culture* 11(1): 121–134.

Martinelli, F. (1985) *Borgate romane. Dalla marginalità alla domanda di servizi*. Milan: FrancoAngeli.

Martinez, M.A. (2013) 'The squatters' movement in Europe: A durable struggle for social autonomy in urban politics', *Antipode* 45(4): 866–887.

Mudu, P. (2004) 'Resisting and challenging neo-liberalism: The development of Italian Social Centers', *Antipode* 36(5): 917–941.

Mudu, P. (2012) 'At the intersection of anarchist and autonomists: Autogestioni and Centri Sociali', *ACME* 11 (3): 413–438.

Muñoz, J. E. (2009) *Cruising Utopia: The Then and There of Queer Futurity*. New York: NYU Press.

Poggio, T. (2008) 'The intergenerational transmission of home ownership and the reproduction of the familialistic welfare regime', In Saraceno, C. (ed.) *Families, Ageing and Social Policy. Intergenerational Solidarity in European Welfare States*. Cheltenham: Edward Elgar Publishing, 59–87.

Schierup, C., Hansen, P. and Castles, S. (2006) *Migration, Citizenship and the European Welfare State: A European Dilemma*. Oxford: Oxford University Press.

Vasudevan, A. (2015) The autonomous city: Towards a critical geography of occupation, *Progress in Human Geography* 39(3): 316–337.

Weston, K. (1995) 'Get thee to a big city: Sexual imaginary and the great gay migration', *GLQ* 2(3): 253–277.

8 Palazzo Bernini

An experience of a multicultural squatted house in Catania

Federica Frazzetta

Palazzo Bernini is a huge palace of about 2,500 square meters. It constitutes four blocks of flats, located in a "boundary zone" of Catania, about 3 kilometers from the city center, between two different neighborhoods. One neighborhood is exposed to an area, known as Borgo-Sanzio, that is mostly populated by middle-class bourgeoisie. The other neighborhood is adjacent to a historically popular area called Picanello. In 1999, Catania Municipality bought this space, spending a sum of 3.5 million euros, to allocate space in the palace to some branches of Municipality offices, but the space remained unused. In the following years, homeless people sporadically occupied this space, and a group of activists squatted it after the Genoa G8 in 2001. In this way, the palace has been intermittently squatted and evicted. In this chapter, I detail some incidents surrounding roughly 150 squatters of Palazzo Bernini that consisted of Roma and Bulgarian migrant communities and a radical collective called *Aleph.*[1]

A multicultural palace

In December 2012, one of the apartment blocks of Palazzo Bernini was squatted by some extended Roma families who previously lived in a Roma camp near Catania's airport. This group decided to relocate because of the sordid living conditions at the camp and the fear of imminent eviction.[2] Along with the Roma communities, a small group of North African migrants and natives Catanese squatted another block of apartments at the Palazzo. Then, between April and May 2012, a new wave of squatters, composed of Bulgarians and Romas from the *shacks of Corso dei Martiri*[3] (Martiri's Street) occupied some portions of the Palazzo. This way, the occupants of Palazzo Bernini multiplied to 150 additional squatters, including 60 minors. The Roma and Bulgarian families were nucleated or extended. Each family had two rooms, at their disposal, with electricity but no running water. Outside the Palazzo, there were four little common squares and a long porticus shared by all the occupants. None of the Roma women were employed, the Roma men engaged in petty daily jobs, such as street peddling or performing. Contrarily, the Bulgarian women worked as cleaners or caregivers, while Bulgarian men collected and traded scrap iron. None of the young people of these communities attended school. Keeping in view cultural differences and prejudices, the families who were dependent on begging as a form of livelihood were not open to socialization; even women, who were used to spending more time together than men, did not share anything about their lifestyle or daily challenges. Moreover, spending time together and sharing the same common spaces did not lead them to share common needs or difficulties. Among the families that coexisted at the Palazzo, there was no collaborative initiative to maintain the place or aid each other in times of need. However, in February 2012 the situation started to change through the intervention of the activists of the Aleph collective.

Aleph is an antagonist radical left collective born in January 2012, following the occupation of a self-managed Social Center called *Centro Sociale Occupato* (CSO) Ziqqurat. The occupation lasted just one day because of an immediate eviction. After this experience, the collective continued its political activity, focusing mostly on the *No Muos* (a large military radar system) struggle in Niscemi, the housing emergency in Catania, anti-fascism, and the lack of social spaces in Catania. The collective is mostly composed of students (both from high schools and the university) and young unemployed people and precarious workers. They all come from different political experiences, such as university collectives or other Social Centers. Since October 2014 the Aleph collective has been squatting the CSO Liotru, which is a self-managed Social Center located in a popular neighborhood in Catania. The collective manages an anti-eviction help desk, after school activities for children, and a gymnasium. They also organize different events on various themes.

Activists and Roma: unconventional fellows

In February 2012, the Aleph collective and other squatters in Palazzo Bernini held a meeting when the Aleph activists decided to squat the palace to create a Social Center without any knowledge of other communities squatting there. Aleph chose to squat Palazzo Bernini because it was a symbol of the Municipality's abandonment and waste, and it could be reused to organize social and political activities. When the activists arrived and found Roma families already residing at the palace, the immediate collaboration with the migrant squatters was impassable. Thereafter, for several weeks following, no collaboration between the activists and the migrant communities happened. But as the number of squatters increased, the situation started to change.

"Commonplaces" and eviction: how squatters reacted

With the increase of squatters, the middle-class bourgeois neighborhood started complaining about the Roma squatter communities debasing the sanctity of the palace demanding an immediate eviction from the Municipality. What mostly disappointed some people in the neighborhood was the cacophony generated from the palace because of loud music played at odd hours, the chatter of children playing barefoot outside the palace, and the continuous oscillation of the squatters to the water fountain in the vicinity and back. All this was perceived as an adulteration of the neighborhood and quality of life of the residents. Moreover, there was the diffused opinion that with the presence of Roma and Bulgarian people, thefts and other episodes of petty crimes increased. The Municipality organized two inspections of the palace; one was assisted by the local media. Therefore, some reactionary local media started to broadcast the palace as a hub of criminal activities and as a place that could spread epidemics owing to the people's unhygienic living habits. The situation worsened quickly with additional unpleasant episodes that occurred when resident parents secluded their children from Roma children who availed the parks close to the palace.

When the negative public opinion started building against the Bernini's squatters, the Aleph activists were the voice for the Roma and Bulgarians squatters to avoid eviction of families without any operative housing alternative for them. Activists organized several meetings, trying to involve the entire Bernini's squatters to represent the increase in allegations from the neighborhood and its possible outcomes. Owing to poor knowledge of Italian, most squatters were not aware of the local television and news journal broadcast of the neighborhood accusations. Even convincing the inhabitants of the importance of organizing and holding meetings were challenging because the inhabitants of the palace had not experienced anything similar in the past. Earnest efforts, time and endurance went into bringing the people on board. Due to internal prejudices among the Roma and Bulgarian communities, it was arduous to persuade them to collaborate, collectively manage the palace and support each other. In addition, communication was not feasible because all the squatters could not communicate adequately in Italian or spoke one common language. After some initial attempts, a couple of meetings were quite successful and at least one male representative from each family participated. In describing the situation, activists proposed to the squatters to organize the reconstruction and cleaning of the common spaces outside the palace, to cease playing loud music at night or odd hours, and to maintain the cleanliness in the vicinity surrounding the palace. These efforts were systematized to give a better impression to the neighborhood. In general, the Roma and Bulgarian communities were not appalled by the disgracing news probably because being evicted was commonplace for most of them and they were unsure and pessimistic about undoing the decision of the neighborhood or the municipality. During the discussions, some of the squatters mentioned the idea of moving before any eviction ensued. Nevertheless, in this discouraging and demoralizing environment, some squatters agreed to participate in the cleaning activities. On the public cleaning days, the local media and other political groups were invited. Very few supportive people participated along with local media. The inhabitants of Palazzo Bernini diligently participated and continued to maintain cleanliness of the common spaces. These collaborative cleaning interventions helped the groups to remove some internal prejudices among them.

Later on, the major local media was propelled to clarify the general misconceptions and prejudices created on squatters' daily activities and habits. So the activists generated an alternative information flow through fliers, web blogs, and video interviews to explain the meaning and utility of the actions and good intentions of the squatters in maintaining the safety and hygiene of the Palazzo and the neighborhood. The media started re-considering Palazzo Bernini squatter initiatives with seriousness and requested interviews from squatters. Most of the squatters refused media attention but wanted the activists to speak on their behalf. The Roma and Bulgarian squatters did not trust the journalists, specifically because of the previous image crafted by their previous media reports. Some groups signed a common document in support of the squatters requesting the Municipality to forgo eviction. Aleph activists also started a negotiation with the Municipality in order to find some alternative homes for the squatters. In spite of all of these efforts, the eviction could not be stalled from the 17 to the 23 July 2012. The Municipality of Catania even offered the Roma and Bulgarian squatters bus tickets to relocate to their respective countries. Roughly, 10 people accepted the offer. While some squatters decided to move, many squatters (mostly Romas) started sleeping at the portico of the palace as a sign of protest against the eviction. After weeks of protest, they gradually relocated to different places in small batches. Many of them went back to the shacks of Corso dei Martiri, while some of them squatted abandoned vehicles (such as cars or vans). Activists proposed to a group of families to squat another house, but after the Bernini experience, the squatters decided to avoid another collective occupation as they believed that the Municipality would not evict them if they occupied places in small groups.

Conclusions

This experience in Catania is peculiar because even though the city hosts many migrant communities (sometimes homeless people),[4] houses are rarely squatted by migrants as protagonists with the support of Social Center activists. In the Bernini case, although the activists attempted to self-manage and maintain the space, they faced eviction of the palace. Taking into account the limited experience of many activists at that time and the newness of the situation for the city, some positive results can be obtained. A synchronized effort of Roma and Bulgarian squatters and native activists to share, co-exist, and self-manage the space, including sharing common debasement from the media and neighboring community – is unusual in Catania. On the one hand, activists had to accept that they could not save the palace from eviction, in spite of their arduous efforts. On the other hand, Roma squatters got motivated to take part in self-managing activities, such as attending meetings, cleaning projects, reconstruction of the palace, and maintenance of the neighborhood that they had never considered before. Working on and sharing common spaces also permitted to overcome existent internal prejudices among the Romas and Bulgarians and also tensions between the migrant squatters and native activists. Moreover, the information material produced by Aleph was the first real effort made to give visibility to Roma and Bulgarians people's living conditions and their efforts to live with the native communities. These memories are voices of the excluded communities, their earnest efforts to live like other socially accepted people, and their perspectives on their challenges are crucial to deconstruct many prejudices and sweeping generalizations that surface on these communities' unhealthy behavior or unhygienic living conditions or criminal livelihood mechanisms.

Notes

1 For further information: www.aleph.noblogs.org (in Italian).
2 This camp was partially located in a never-before-used Municipality's sport field. The Municipality made it available in May 2011, as a temporary solution for Roma who lived in another squatted building (evicted in that same period). The camp existed for more than one and half years, and hosted about 170 people, but the Municipality never had a plan to better organize the camp and to improve the way of living of Roma. In 2011, after the eviction, the Municipality did not propose the Romas with any other housing alternatives.
3 *Corso dei Martiri* is one of the streets involved in a strong process of gentrification of the San Berillo neighborhood since the 1960s. Most of neighborhood's palaces were demolished, new ones were built and the old inhabitants moved to the periphery of the city. A huge area, divided in three big ditches or holes dug under the streets, is still not re-qualified and has been squatted for years by Roma, but also by Bulgarians and some people from North African countries. Squatters lived in self-made shacks. Between 2013 and 2014, the Municipality evicted people occupying these ditches.
4 According to the XXIV immigration Report (2014) written by Caritas and Migrantes Association, Catania is the second city in Sicily (after Palermo) in terms of number of migrants: about 18.4% over the total of 162,408 people.

9 The untold struggles of migrant women squatters and the occupations of Kottbusser Straße 8 and Forster Straße 16/17, Berlin-Kreuzberg

Azozomox and Duygu Gürsel

Introduction

In West Berlin, Kreuzberg was the central district of the squatting movement in the early 1980s; almost half of the squatted houses were located there. At the same time, it was also a migrant neighborhood. The majority of West Berlin's migrant population came from Turkey and lived in Kreuzberg. Although the miserable housing situation of migrants has been narrated and analyzed through diverse perspectives, the role of migrants taking action on, initiating, participating in, and transforming the housing/urban struggles, specifically in the squatting movement, remains mostly untold. With this chapter, we aim to discuss two squatting experiences of migrant women from Turkey in the early 1980s in Berlin-Kreuzberg. Our aim is not to add a footnote to the history of the squatting movement, but rather to ask new questions and rethink the history and the future of urban struggles in light of the following questions: How did the struggle of migrants get marginalized in this narrative of urban struggles and the squatting movement? How does the squatting of migrant women reveal the limits and the possibilities of the squatting movement? How does the untold story of migrant squatting change our understanding of migration and the squatting movement? In order to elaborate on these questions, we are going to first look at the strained relationship of the radical left with migration; the link between migration and housing politics; the proposal to rethink migration as a social movement and part of the squatting movement in West Germany. We eventually present and discuss two squatting experiences of migrant women.

The tense relationship of the radical left with migration

As the desire for a social transformation emerged in 1968, carried out by the student movement and proletarian and anti-authoritarian youth who realized that they needed a partner for revolution, many of them left the universities and spread into the factories to build the front line with workers; there they 'discovered' the migrant workers, constituting a large number of mass industrial workers, as *avant guard* of the class struggle. In this context we point out the example of one of the first squats in West Berlin, the *Georg von Rauch-Haus* (occupied in 1971), where primarily pupils, young workers, trainees and runaways lived. In this house, few high school students and 10 corporate company workers lived, the occupants were serving their apprenticeships or working as toolmakers, machinists, welders, bricklayers, and alike (azozomox 2014).

But soon the desire to work and live differently came in conflict with the monotonous factory work and resulted in the dissolution of the factory action groups. The workers' strikes[1] were organized by migrant factory workers, and the joy resulting from these strikes was shared by few action groups on the left, but did not diffuse to the rest of the society. The lack of any analysis of racism within the class struggle and the racist division of labor directed the need for the initial constitution of the solidarity of German workers. The initial attempts to discuss workers' movements were not executed. Later on, when discussions were carried out on the subject of leftist politics or the role of the unions, mass industrial workers were already phasing out due to mass layoffs and the informatization of the economy. The factory action groups began engaging in neighborhoods. However, the attempted solidarity of the political relation between students and workers turned into a caretaker relationship with people in need. For example students turned into social workers while *avant garde* migrant workers turned into ghetto inhabitants to help and assist. An interview from a leftist daily newspaper in 1982 captured this moment. Four comrades – one of whom was an Italian migrant – who were a part of the group *Revolutionärer Kampf (revolutionary struggle)* and worked as factory workers and squatted a house and collaborated together with other migrants in Frankfurt/Main in the beginning of the 1970s, came together again to discuss if migration to Germany should be stopped or regulated (*Ausländerstopp*). As such, the crisis of the left and its inability to analyze the politics of migration and take a position in discussions on "foreigners" was further projected in the election of the Conservative Government. Simultaneously, the birth of an alternative movement with the Tunix Congress in 1978 in West Berlin was an attempt to revitalize the anti-authoritarian left and fight together against the diverse structures of power and repressions. Initiatives for an alternative economy were taken. Squatting also emerged in this new atmosphere, both against the privatization of housing and probing into alternative lifestyles and for the self-organization of life. The idea of not waiting for a revolution, rather revolutionizing everyday through the self-and collective management of living spaces, was exciting. The problem was to understand how extensive these ideas of collective management could be. Was it limited to left-radical self-help groups who were repressed and were now trying to get back into the labor and housing market? Or was it a political movement connecting with other movements of the wider society, such as the struggles of migrants?

Rethinking migration as a movement

To think of migration from the standpoint of "autonomy", means to emphasize the social and subjective dimensions of migration movements. It is an attempt to argue against traditional immigration theories that depict immigrants as victims of migration trapped between state and capital. Karakayali underlines that 'the subjectivity is not free of structural constraints, but there is always the potential of an "excess" that can emerge within a field of tension, that transforms the whole field' (Gürsel 2013: 220). Bojadžijev (2008), in her study on the struggles of migration in 1960s and 1970s, challenges the dominant discourse on the migration history for representing migrants as passive and defenseless victims by breaking these narratives through migrant experiences in resisting the repression at borders, in the factories and in their neighborhoods, to highlight their strengths in forming their collective and individual subjectivity.

Entanglement of migration and housing politics

After the *gastarbeiter* (guest-worker)[2] regime ended with the halt in migrant recruitment in 1973, migration continued through family reunification. Throughout this period, migrants relocated from isolated shabby guest worker hostels to apartments in the city; during this phase, housing and urban politics emerged as one of the principle instruments to limit and control migration. The Berlin Senate of 1975, later implemented in other federal states, enforced *Zuzugssperre,* moving restrictions for migrants based on nationality, specifically referring to migrants from non-European countries to certain districts, such as Tiergarten, Wedding and Kreuzberg. This can be taken as an example of the management of migration through urban politics and through the creation of internal borders across cities. Another instrument of control is the housing supervision law (Wohnungsaufsichtsgesetz), which originally protected the tenants but is now used against the family reunification of migrants since one of the clauses demands the residence permit of new migrants as an obligation to verify occupancy in a legally conforming apartment (that is an apartments with a minimum housing space of nine square meters for each adult and six square meters for each child under six years). Furthermore, racism in the housing market manifested itself through explicit housing announcements such as "only for Germans" or "not for foreigners" or an illegal additional rent. In this situation, many migrants did not have any other choice but to live in shabby abandoned buildings, which were left ready for demolition and vulnerable to the speculation of corporate builders.

Squatting in West Germany and West Berlin

In West Germany, at the beginning of the 1970s, a new squatting movement emerged in the wake of the worldwide 1968 movements for radical change of society and life. Various squatting movements that spread from West Germany differed from one another, from place to place and from time to time in their intensity and their cycles. In some places, just one house was occupied, while in other places, many spaces were squatted. For example, since the beginning of the 1970s and mid-1980s, Potsdam has experienced more than 40 squats, while Hamburg had more than 50 squats, and in East and West Berlin together roughly 645 buildings were squatted – all together more than 1,000 buildings and hundreds of wagon-like caravans, trailers, trucks and the like were squatted. The composition of the squatters varied greatly, expressing a broad diversity within the frame of anti-authoritarian, emancipatory ideas and politics and reflecting the influence and interrelation with other social, cultural, and political movements. Among the squatters, we find people with different class backgrounds and political tendencies (anarchists, anti-authoritarian activists, anti-imperialists, autonomous activists, anti-fascists, environmentalists) as well as creative artists, workers and more, but also, autonomist women, radical queer and trans-gender people. In addition, people of color, migrants, inter- and trans-nationalists and refugees have participated, though they have been the minority in the squatting movement (amantine 2012: 32)

The first big squatting movement, from 1970–1974 in West Germany in Frankfurt/Main, was against housing speculation, rent increases, demolition of buildings and gentrification and gave birth to 20 squats, which included a squat by immigrants. The occupation of *Friesengasse 5* in September 1973 was probably the first migrant squat in West Germany; it was unfortunately evicted the same day by the police. The squatting movement slowly receded by the end of 1974 with the eviction of the last squat.

The squatting movement was accompanied by large organized rent strikes from Italian, Kurdish/Turkish, Greek, Spanish and Yugoslavian migrant workers who were suffering in Frankfurt/Main under squalid living conditions and were no longer willing to pay horrendous rents.

In their first publicly announced strike of *Ulmenstraße 20* in 1971, they declared that they would not pay more than 10% of their income for rent. At the peak of the rent strike, 1,500 migrants participated in it. It was also supported and organized by groups like *Lotta Continua* or *Unione Inquilini.* The rent strike extended its initial limited criticism on housing conditions towards a broader criticism on general living conditions. At the same time, in 1972, a major strike by migrant workers was organized against Opel and VDM in Rüsselsheim, near Frankfurt. But due to state repression, with 90% of the trials lost due to non-payment of rent, the movement gradually abated.

The 1980s experienced the second big squatting wave in West Germany with around 400 squats in total and roughly 200 squats in West Berlin alone. This time again, two buildings were squatted by migrants. In November 1980, several Turkish families, who previously lived in *Forster Straße 18* under unworthy and claustrophobic conditions (for example, ten people sharing a room), squatted Forster Straße 16 and 17 with the support of local activists. On 18 February 1981, probably the first occupation of migrant women happened with the squatting of *Kottbusser Straße 8,* which was led by eight Turkish and Kurdish women, one German woman, and four children with the support of the *Meeting and Information Point for Women from Turkey* (TIO – *Treff- und Informationsort für türkische Frauen*).

Occupation of the house Kottbusser Straße 8

Since many families from Turkey and Kurdistan were living in overcrowded and small flats, many of them were supporting the squatters and had sympathy with the squatting movement in West Berlin. *Seda,* one of the squatters from the organization Meeting and Information Point for Women from Turkey (TIO) in Berlin-Kreuzberg, directed the group to squat the empty house in Kottbusser Straße 8 all together. After a short discussion, they decided to do it spontaneously without thinking it through – so eight women, four children, and one German comrade from the organization, who decided to join them spontaneously, met the following week in the organization's office and finally entered the house. Since there were renovations going on in the house, the presence of construction workers caused major trouble during the occupation of the building. The women were insulted and harassed in a fascist, racist and sexist manner, Schemme and Rosenberg narrated the following:

> Those were women and moreover Turkish women that provoked them. The German construction workers, who were renovating the house, wanted to prevent the occupation by their own means. A friend of mine was strangled and choked by them, and one of these workers aimed a scraper at her. Another woman had her hair pulled so hard that she was bleeding. That was really dramatic. You do not want to work or pay rent, but you want to squat German houses, swore the construction workers. (Schemme and Rosenberg 1981: 6–9 [Author's translation])

One of the workers saw Seda and approached her furiously, grabbing her by the arm of her fur coat and dragging her down the stairs saying, "You should have been gassed!" She fell down the stairs terrified – nothing happened thanks to the coat – and then ran immediately to the flat where the other women were waiting and locked the door from inside (Celebi-Gottschlich 2014)

The construction workers gathered in front of the locked door and shouted at them to come out. They were unprepared for something like that and scared, and they did not know what to do exactly. Shortly after police arrived, the construction workers left, and the women could surprisingly stay in the squat. Supporters also heard about the squat and came to chant for them, to express their solidarity. As the children started to get hungry and thirsty, supporters who heard the children cry threw milk and other supplies to the balcony of the flat where the group was staying. They also tossed a purple transparent with the words "Just Courage" written on it.

After 3 hours, as the women were becoming anxious, a representative from the public housing company *GSW*, who had bought the house three years earlier in 1978, came to negotiate with the women. After telling him about their housing misery for 20 minutes, they received the house key and were quite shocked and simultaneously relieved because instead of taking them to the police station the house owner granted them access to the house.

They found out later that the representative from the housing company GSW talked to the *Senator for Internal affairs, Ulrich,* and the police filed a report of property damage of one door. At night, only two women from Turkey and their two German friends stayed in the house. For some of the squatters from the radical squatting group, it was not "revolutionary enough" to get the keys of the squatted house but against their squatting honor.

Despite the construction work, the house was still in a state of decay, and it did not have electricity, running water or toilets. Therefore the renovation of the squat was the first task in the list of things to do. The women received then their first donation of 5,000 Deutsche Mark (DM) from the pharmacist *Ulf,* to start the renovation of the building.[3] Not all the migrant women who squatted the house remained. In the end there were Seda, one single Kurdish woman with three children and another single Turkish mother with three children who moved in, but the other squatters in the beginning were German. They formed a house-community, held regular meetings and also participated in neighborhood assemblies and squatting councils, where all the political issues of squatting, negotiations with the state, house raids, etc. were discussed and debated.

Several reasons contributed to taking the initiative to squat an empty building. One of the reasons was having experienced racist/sexist discrimination in finding a new flat. Especially young women, who were separated from their men, suffered greater difficulties – the housing shortage among those woman was immense, especially among single mothers from Turkey and Kurdistan.

TIO was trying for some time to find apartments for those women but failed due to their ignorance of the local authorities and landlords. One woman with four children, who was living in a one-room apartment, was thrown out by the clerk of the state housing office with the comment: "Get lost". TIO collaborated also with another two women's shelters in Berlin, where more than ten Turkish women were looking for a flat:

> It would be perfect to have a house where Turkish and German single mothers could live with their children. Because more and more Turkish families get crushed and the Turkish women are left alone with their kids and are hassled massively from their former husbands. In this context it would be a real protection if they could live together with other women and not so isolated and dispersed throughout different districts (Schemme and Rosenberg 1981: 6–8 [Author's translation])

The harassment of women who decided to separate from their husbands has always been a problem. One of the migrant women living in the squat in Kottbusser Straße 8, who got divorced before she moved to the house, had to deal with her husband continuing to stalk her. Therefore, some people from the squat went to him and demanded that he not harass his former family anymore; over time he stopped stalking her. Another single migrant woman with their children had several other struggles, which were beyond the grasp of ordinary German people. Both of the women had health problems, they were working at the factory under very harsh conditions, they had to deal with the harassment of their ex-husbands, with the difficulties of lacking German language knowledge, and with the problems which their children faced at their respective schools. They also did not have any time to join the house meetings or go to demonstrations.

From time to time, there was translation provided for them. Seda believes that they felt left out among the "alternative squatters" because of their language problems and the missing feeling of togetherness. The woman activist claimed that the other squatters within the house were overwhelmed with the problems of migrant women and did not really care much about specific problems. Encountering problems in a shared space on a daily basis brought another kind of challenge to working on social projects with migrants.

The option of negotiating with the Senate of Berlin in order to legalize squatted houses was a controversial topic within the squatting scene in general. In Kottbusser Straße 8, the migrant women were in favor of legalization because of their legal, social and politically precarious situation in general, but none of the other native squatters supported this because they did not want to be perceived as the traitor of the movement, as many of their friends were in jail because of activities and demonstrations related to squatting. And the main political demand towards the Senate of Berlin was clear: without the release and freedom of the prisoners (of whom some got prison sentences of more than one and a half years without parole), there would be no negotiations at all.

After long discussions, it was decided that the German squatters would leave and the house would be transformed into a whole migrant legalized squat. Finally the squat got legalized under very good conditions and under very inexpensive rent contracts. After the Germans left the house, radical left Turkish and Kurdish groups used the empty flats as their offices. In the meanwhile, the original squatters had all left the house and new people, also Germans, moved in. The house still functions as a house community although the radical political agenda has transformed.

Seda left for different reasons. She believed that the last remaining women, who finally left the house, had to confront the pressure of the conservative migrant community and probably left alone without solidarity and support. Solidarity and non-hierarchical collaboration were not one of the strengths of the group, which was clearly illustrated through the interview of the German colleague from TIO who had joined the migrant women in the squatting, as she claimed that without her, squatting the building would not have been successful. In addition to this paternalistic way of thinking, she went to Turkey for two weeks, and after her return, she published a book explaining how Turkish women are oppressed. But TIO itself, who supported and participated in the occupation from the first day, became a target in September 1984 when a man affiliated with the Turkish fascist *Grey Wolves (Bozkurtlar)* attacked the *Meeting Point*. The man entered the space and shot three times with his pistol, killed a woman called *Neriman*, who died later in the hospital, and critically injured another woman called *Seyran*. Seyran was shot in her neck and recovered slowly in the hospital. She recalls the unknown man who shot the women and the racist investigation practiced by German police following this event:

> The day I left the intensive care unit, two police officers approached me and asked me, if my father was the gunman and that I should not protect him.

> These policemen knew that he could have a reason to shoot because I was running away from home. I was shocked. My father just had visited me in hospital. How could they imagine that my father had done this? [...] I said no, my father did not do it, but they grinned and persisted. The police said: I should think about it twice, it could be possible, that I am afraid to say the truth. [...] They did not believe that a political reason could be the cause of the assault. (Ates 2003)

Finally the offender was arrested and propaganda material of the Grey Wolves was found in his house. Although he killed one woman, he was only charged and processed with manslaughter. And despite the fact that he was clearly identified by the witnesses as the person who shot the women, he was acquitted in a scandalous trial due to the lack of evidence. After this assault, TIO received a lot of support from the women's movement and community and the autonomous/left-radical political people. As an example, a taxi-collective showed solidarity and parked their taxis in front of their meeting-point (amantine 2011: 207)

The occupation of Forster Straße 16 and 17 in November 1980

> We have tried to live together me, my husband and three children in a 36 m^2; flat. We have waited three years for the housing office. They have showed us flats in the outer circles of Berlin for 800–900 (DM). We did not earn that much then, and we had to send money to our family in Turkey. We could not pay this amount of rent. But we definitely wanted to live in a better place. … The building next to us was vacant. I have thought, why should we live in one room, when there are flats with 4 rooms empty next to us? (Zeynep, activist and migrant-squatter, from an Interview with Kreuzberg Postasi, 1980)

Zeynep, a migrant female worker from Turkey, was living with her husband in a very small flat. As she wanted to bring their kids from Turkey to Germany, she was looking for a bigger flat but failed due to racism in the housing market (Refer to Nur and Sethman Chapter 6, in this volume). Zeynep was baby-sitting a German child next to her factory job; upon the request of the child's mother who happened to be a friend, she started attending neighborhood assemblies where they suggested she squat two empty buildings in Forster Straße 16 and 17. They had organized meetings with other neighbors and talked about how to squat the building. One night, they decided to get into the house. At 1:00 in the morning, they went to the house with candles since the electricity of the abandoned house was cut off. Immediately police came and kicked them out. They did not resist against the police and waited; then they left and went in the house again and stayed until that night. Then students with sleeping bags, came by to support the squatting action, stayed back and suggested the migrant families sleep at their houses. The group assembled in the next morning and went to the Municipality to get the tenancy agreement with the neighbors and supporters as a forceful congregate. Zeynep remembers the jolt on the faces of the Municipality workers witnessing the large local crowd with the migrants. After announcing that they were going to renovate the building and stay there until it was demolished, they succeeded in getting a temporary tenancy agreement on November 26, 1980.

Common ground

Throughout the squatting process, close contact was established among neighbors, but still a stable exchange or collaboration between German and migrant neighbors was missing. Zeynep described the situation as the existence of an invisible wall – a wall that divided the street into two groups. On the one side, there were overcrowded buildings with migrant worker families, and the other side consisted of German small families of white-collar workers or native single households. The idea to bring these two groups together was almost like bringing children together through intercultural education, to establish a self-managed Kita, which Zeynep and others initiated and worked at as kindergarten teachers.

Another potential for common ground, bringing alternative and migrant milieu together, was the politics of the governing *CDU Senate* and the immediate frontal attack of *Heinrich Lummer, Senator of Internal Affairs* (1981–1986), aimed at both squatters and migrants at the same time. Lummer carried out brutal evictions of squatted houses, one of which led to the death of a young activist, *Klaus Jürgen Rattey,* in September 1981 that issued a new decree, the *Lummer-Erlass*, also known as the *Decree against Turks.* This law aimed to deport young Turkish migrants who came to Germany during the family reunification period. The general political atmosphere was very restrictive and hostile towards migrants, expressed through the Senator's dialogue and also in legislation and laws. In 1983, roughly 1,400 migrants were deported exclusively from West Berlin. On New Year's Eve in 1984, six detainees pending deportation died in a fire in the overcrowded deportation prison at *Augustaplatz*, which was holding up to 20 persons in one cell. And only a few months earlier, in August 1983, *Kemal Altun*, a Turkish asylum seeker, jumped out of the window of the sixth floor of the Higher Administrative Court, during his deportation trial for fear of facing torture and death in Turkey if deported, when the Junta took over the country in a military coup in September 1980. Those harsh deportation practices of the West Berlin Senate were also applied to women and aged migrants, like an 80-year-old Turkish woman who was deported even though her five sons were living, working and supporting her in Berlin. And in this political and historical moment, Lummer did not hesitate to declare freely his racist philosophies, like:

> When we solve the problem of foreigners, we solve the problem of unemployment. The number of foreigners has to be reduced with all the urgency and by all means – regardless of fundamental basic rights. The Germans do feel estranged from their environment in Kreuzberg, because of all the foreigners living there and it starts with the smell of them! (*Spiegel* 1984: 78)

This new racist legislation brought together squatters and Turkish organizations for spontaneous actions and massive demonstrations. *Aras Ören*, a German-Turkish writer/poet, was excited and surprised with the heterogeneity in these demonstrations:

> All Turkish people, from the right to the left orientation, went to street for the first time together. I did not expect it. And even more, also many Germans have grasped that Berlin's migrants are more than a minority group among others and joined in the demonstration. Or stayed at home, but had finally doubts about the infallibility of decreed politics. (Böhm 1981: 15)

Euphoria, happiness, and excitement were the common expressions that described this moment. Zeynep, the initiator of the first migrant squatted house, was asked if there were any conflicts in making decisions or in meetings during the process of squatting; she replied without hesitation that there were no discussions, everybody was in solidarity, and there was an "awesome joy". Her memories are associated with the festivals they organized and the joyous experiences. She was surprised and excited with the heterogeneity of the group, who were supporting them in solidarity. As they went to the district office in Kreuzberg on foot with a demonstration in order to demand their right to stay in the house, there were 200–250 people who joined them, even older people with their walking sticks. The solidarity was also reciprocal. Zeynep said that they were also visiting the other squatted houses. She remembers bringing tea for the university students who squatted the old fire station. But it was not in a form of migrants doing the catering again. She says that she was going there with her husband and they were having food and drinks with the student squatters and listening to music and dancing with joy. Her nostalgic emphasis brings it to the point: "We were young back then". As they visited the second migrant squatted house with flowers, she was very impressed and got excited and thought, "It was totally my thing, women are capable of doing everything". The squatted houses had open doors, and they also had many visitors – journalists, students, professors from universities, and also people from West Germany, who came to support their case and do interviews for the media. For instance, a group of apprentices from an employment office supported them by tapping the wire to supply electricity for free for the all flats in the house so that they could do the renovation work.

The difference of the first migrant squat experience was that they managed to create a joyful collaboration. Zeynep mentions that they were organizing street festivals and breakfasts on the street. They were preparing everything in a collective way, and everybody was bringing something to eat or drink. These kinds of activities offered more possibility for a relaxed exchange and visibility on the street, which was a different picture than the usual demonstrations.

Forster Straße and the day care center Kita Komsu

The establishment of the intercultural kindergarten on the ground floors of the migrant squats played an important role in this encounter. The goal of this kindergarten was actually to break the invisible wall between German white-collar workers and their migrant worker neighbors by bringing their children together. Zeynep's role as the initiator and mobilizer of squatting and the kindergarten, and her role as a former factory worker with a primary school graduation degree to become a kindergarten teacher in their own self-managed Kita, was also crucial in terms of shaking the perception of migrant women in the alternative scene during a time when they could not go beyond the role of the victims. Although she was earning way more in the factory than in the kindergarten, it was her decision and wish to see all of the children playing together and doing something to achieve this to make it happen.

Intercultural education is not a recipe against racism; it is rather a learning process, wrote *Amman*, one of the teachers, explaining the concept of *Kita Komsu*, which means "neighbor" in Turkish. The basic condition for this is a mixed group of children and teachers, but in order to achieve the goal the readiness of the teacher to get to know the "other culture" and to learn from it is necessary. According to their concept, teachers should also learn to get along with other cultures, since children have a distinctive sense of how teachers approach each other and orient themselves according to this behavior. They have also developed an exchange system that allows teachers to visit different children groups for a self-control and awareness system. As Zeynep visited a children's group, upon request of one of the teachers, she confirmed the doubts of the other Turkish-speaking teachers regarding a German teacher in the group. She observed that the German female teacher behaved more aggressively towards the Turkish children. Subsequently they brought racism as a discussion topic to the meeting and warned the teacher about her behavior. After she continued to behave in the same manner, they decided to collectively dismiss her.

Different from other parents who initiated the kindergarten, they had an aspiration of intercultural education, equality among children and teachers, and grassroots democracy. *Klaus,* who was the co-teacher with Zeynep, explained that it was a challenge to follow this aspiration. There were no Turkish or Kurdish teachers with training in kindergarten education; in fact, they used to be factory workers in contrast to the German teachers, who had studied and mostly came from middle-class families.

The challenge was also distributed unequally among these structurally different groups. Whereas Klaus had to write concepts and official letters by himself, Zeynep was busy going to seminars to keep up, learning the language, working as a teacher and as an unpaid/voluntary social worker for other migrants. Additionally she was raising her own children, which meant also having many struggles with the education system and fighting against the everyday and structural racism, dealing with the health issues, problems from the factory work, and keeping social contact with the families of the children. The idea of grassroots democracy and self-management of the Kita came to its limits at the end of 1980s Klaus became the manager of the kindergarten, which still exists under his management in a bigger place close to Forster Straße with 30 teachers. After 19 years, Zeynep quit her job at the kindergarten due to health-related problems.

The texts on the migrant squats of Forster Straße 16/17 do not provide an extant narrative of the stories or perspectives of migrants. The migrants rather remain as the background actors without any identity, although they were in reality the "leading" actors. Paradoxically they are mentioned in monolithic categories such as foreigners, Turks or Kurds, except one interview with Zeynep in a Turkish neighborhood magazine, Kreuzberg Postasi. The critique is not offered in order to devalue the contribution of supporters or consider any effort unnecessary, but on the contrary their contribution offers a very important example of community organizing. However, the stories and perspectives of migrants and the production of a critical knowledge of their experiences are missing and unfortunately objectified in the available documentations of this squatting period.

Migration challenging the narratives

Pruijt (2013) offers five categories of squatters; deprivation-based, as an alternative housing strategy, entrepreneurial, conservational and political squatting. Our examples of migrant squatters urge us to revise these categories. The group of migrant women squatters was motivated both by deprivation but also by the desire for another form of life, a vision of alternative housing strategy, liberation from the oppressive family or husband. It was not only the urgent need for a house, which pushed migrants to squat, but it was also the urge to be part of a revolutionary moment. This moment was contrary to existing expectations, not realized by the male leftist migrant figure, the beloved figure of the revolutionary mass worker, but by Turkish and Kurdish women with their children. According to Pruijt's (2013) category of deprivation-based squatting, activists occupy the building for those who are in need. However, in our example of Kottbusser Straße 8, migrant women who were in need of housing occupied the building, although not intended, in the end for the political activists.

The example of Forster Straße 16/17 is a mixture of deprivation-based, entrepreneurial and conservational squatting. Zeynep quit her factory job, which was remunerated with 1,700 DM monthly, and started to work as an educator in the kindergarten, getting paid only 1,000 DM, while at the same she was attending educator's seminars. She broke also with the "migrant figure", who works to save more and more money but instead held a different representation as a person who actually cares about her neighborhood, social cohesion and children's education.

Although categories of squatting are helpful to reveal the heterogeneity of squats' motives and conflicts, squats should be also understood in terms of flows and becomings. People moving in and out of crisis and conflicts can change the direction of the squat, which contests concepts like organization and hierarchy. The squatting experience also suggests that the initial goal of squatting can change accordingly. Therefore it is important to look at squatting within the framework of historical transformations.

The squatting wave of the early 1980s is discussed in three phases; emergence, expansion and downfall (Holm and Kuhn 2011). The emergence phase is associated with the establishment of the citizen initiative *SO 36*, radical autonomous squatters, the establishment of the first squatting council in April 1980 and the coining of the idea of rehab-squatting. The expansion phase started after the street riots on December 12, 1980, following a prevented occupation in Frankelufer 48 (Kreuzberg), and accelerated with the corruption scandal and the fall of the Berlin Senate – causing a political vacuum. As Lummer (CDU) took office in May 1981, as the Senator of Internal Affairs in West Berlin, his immediate targets were posing threats to "internal security" from squatters and migrants. The downfall of the movement is marked with the death of a young squatter, Klaus Jürgen Rattay, who was trying to escape from the police violence. The Senator, at a press conference, in the evicted squat Bülowstraße 89, announced his "victory" and his declared war against the "criminal" squatters. Alternatively, the squatting in Forster Straße 16/17 belongs to the emergence phase, in which less than 20 buildings were squatted. The squatting in Kottbuser Straße 8, when 79 buildings were squatted in West Berlin, marks the expansion phase.

Although the repression led by Senator Lummer played an important role in breaking the movement, it would be misleading to think his action led to the death of the squatting movement. Besides fordist repression techniques the post-fordist forms of control and governance emerged out of negotiations. Self-help building for squatters facilitated through IBA-international construction exhibition, self-help funds for migrant organizations distributed through newly appointed commissioners for foreigners constituted the pioneers of new forms of control and governance. Through these techniques, the political position of two movements has been weakened, and their creative and transformative power got partly institutionalized, privatized or oppressed and criminalized. In the period from 1979–1984, around half of the squats were evicted, and the other half were legalized; still these movements offer an important infrastructure and experience for urban struggles.

Conclusions

Today the housing struggle takes place within the much more complex field of finance capitalism and globalized neoliberalism. The district of Kreuzberg turned from a working-class neighborhood of decaying buildings to an attractive district with cafes and galleries, in which investors do see more renovation and potential speculation to accumulate capital. Today migrants struggle against the rising rents and threatened eviction from their neighborhood which was revitalised on their labor, love and relationships. An analysis of the expulsion of migrants from the city centre while rethinking the categories of class, race, and gender together is possible if the history of urban struggles are rewritten from the perspectives, stories and struggles of migrants.

Notes

1 One of the largest episodes of wild strikes was carried out by Turkish workers at the automobile plant *Ford* in Köln-Niehl, in 1973, after the dismissal without notice of

around 300 Turkish workers, because of the unauthorized prolongation of their holidays (Karakayali 2000).

2 *Gastarbeiter* stands for immigrant worker, it refers to those migrants who came to the former West Germany between 1955 and 1973 as part of a guest worker program.

3 The squatters were supported also by a nationwide initiative of entrepreneurs and self-employed. They released a press release demanding the Senator responsible for urban development to stop the evictions and showing solidarity with the goals of the squatters. Additionally, they were also offering sponsorship as in the ad from 12 September Tagesspiegel with the title "Unternehmer und Selbständige unterstützen Instandbesetzer" (Entrepreneurs and Self-employed support Squatters).

References

amantine (2011) *Gender und Häuserkampf*. Münster: Unrast.

amantine (2012) *Die Häuser denen, die drin wohnen*. Münster: Unrast.

Ates, S. (2003) *Große Reise ins Feuer: die Geschichte einer deutschen Türkin*. Berlin: Rowohlt.

azozomox (2014) 'Squatting and diversity: Gender and patriarchy in Berlin, Madrid and Barcelona'. In SqEK (ed.) *The Squatters' Movement in Europe*. London: Pluto, pp. 189–210.

Bojadžijev, M. (2008) *Die windige Internationale: Rassismus und Kämpfe der Migration*. Münster: Westfälisches Dampfboot.

Böhm, M. (1981) 'Berlin ohne Türken?' Interview with Aras Ören. *Zitty* 26: 15.

Celebi-Gottschlich, S. (2014) 'Hausbesetzung 1981'. In Allmende, E. V. (ed.) *Mehr als 50 Jahre Migration. Dokumentation von 2 Touren zu Orten migrantischen Widerstands in Berlin*. Berlin: Geographie FU Berlin, pp. 18–25.

Gürsel, D. (2013) 'Discursive acts of citizenship: Kanak Attak in Germany'. In Gürsel, D., Cetin, Z. and Allmende E.V. (eds) *Wer MACHT Demo_kratie? Kritische Beiträge zu Migration und Machtverhältnissen*. Münster: Edition Assemblage.

Holm, A. and Kuhn, A. (2011) 'Squatting and urban renewal: The interaction of squatter movements and strategies of urban restructuring in Berlin', *International Journal of Urban and Regional Research* 35(3): 644–658.

Karakayali, S. (2000) 'Across Bockenheimer Landstraße', *Diskus* 8: 41–47.

Pruijt, H. (2013) 'The logic of urban squatting', *International Journal of Urban and Regional Research* 37(1): 19–45.

Schemme, D. and Rosenberg, B. (1981) 'Frauen besetzen Häuser. Interviews', *Courage* 6(4): 6–9.

Spiegel (1984) "Wahrer König", (9 January 1984) (2): 78–79.

Part III

Resistance to exclusion, criminalization and precarity

10 Space invaders

The 'migrant-squatter' as the ultimate intruder

Stephania Grohman

"Invasion of the migrant squatters", read the *Daily Mail* in June 2012 (Dovar, 2012). The newspaper was not referring to the surge of absentee homeowners buying up properties in London's prime neighborhoods, such as Knightsbridge, Belgravia or Notting Hill; the retro-horror-film headline pointed to an unintended side effect of foreign property investment – the increase in squatters being evicted from some of the city's most expensive mansions. The European financial crisis, according to the newspaper article, had led to a surge in economic migrants from the Eurozone, some of whom ended up squatting in buildings which were left empty and "vulnerable" by their "foreign billionaire" owners. To be sure, the *Mail* did its best to not assign blame unilaterally – the problem, this much was clear, was foreigners, regardless of whether they owned the property or squatted in it. The article, however, exemplified a discourse that has dominated the public discussion of squatting in the UK in recent years and reached a peak in the run-up to the partial criminalization of squatting in 2012: the discourse of the "migrant-squatter".

The "migrant-squatter", in this case, is not a real person but rather a discursive figure that has come to represent squatters in general in at least some parts of the public imagination. While it has only risen to prominence in the right-wing press comparatively recently, in this chapter I want to explore how this figure relates to wider discourses on poverty, migration and the status of the poor – domestic or foreign – as less than human. While neither squatters nor migrants routinely enjoy much sympathy in public discourse, the convergence of the two groups in the figure of the "migrant-squatter" creates an entirely new category, combining two distinct modes of exclusion into an unparalleled image of threat to the territorial control of citizens. Moreover, this convergence is not coincidental – the fact that, for media like the *Daily Mail*, "squatter" and "migrant" have become practically synonymous serves the interests of creating an "underclass" of social undesirables who, through their occupation of spaces where they have no business, are presumed to have brought repercussions on themselves.

Squatting and migration in the UK — fact *versus* fiction

The "squatter-migrant" made its debut in the British press during 2010/12, conspicuously close to a public conversation on whether or not squatting should be made a criminal offence. While squatting in the UK had previously been associated with different groups – from Bohemians to Hippies to Punks – one could now gain the impression that squatting was exclusively a pursuit of all kinds of shady foreigners. Nearly every week, one could read about "jobless Italians" (Cohen 2010), "knife-wielding Lithuanians" (Kisiel 2010), or "gangs of Romanian gypsy squatters" (Alleyne 2011) taking over innocent citizens' homes when those citizens had just popped out for milk. In other cases, the intruder's precise nationality was secondary – a group of "migrant-squatters" who occupied an elderly lady's house in Peterborough in 2014 were apparently sufficiently indicted by simply being described as "not from here". Tacitly implied – and well understood by the target audience – is that "migrant" does not simply refer to any kind of foreigner. As the *Daily Mail* article illustrates, while both "foreign billionaires" and squatters come from elsewhere, only those whose lack of funds clearly identifies them as undesirable are labeled "migrants". As Bridget Anderson (2014: 3) remarks, "the migrant as imagined in public debate is not the foreign born professor, financier, or architect, but the person who cleans their house", or, as in the case of the "migrant-squatter", goes on to live in it.

To be sure, the perception that there is a "really existing connection" between migration and squatting is not mistaken. Long before the 2008 crisis, Britain's "squatter's rights" – the Common Law tradition that means squatters could only be evicted from a property under specific circumstances and never under threat or violence – attracted activists and adventurers from all over Europe wanting to immerse themselves in the creative and liberal environment that was the British "squatting scene" (Common Place 2008). The music and art scenes that flourished in occupied spaces were legendary, a ready-made radical community was available instantly to the would-be revolutionary, and, at least here, Punk culture was decidedly not dead.

But there were also less political factors that made squatting an attractive option for new migrants, radical or not. For one thing – speaking from personal experience – getting a foothold in the British rental sector as a new arrival is anything but easy. When I arrived in the country in 2008 to study for a postgraduate degree, I soon realized that not only would I have to shell out considerably more for my accommodation than I was used to, coming from Central Europe, but it also turned out that money alone was not enough. As soon as I started to view properties within my modest means, estate agents informed me that not only would I need references from previous landlords – a requirement that does not exist in my country – but I would also have to provide a UK-based homeowner-guarantor in case I could not pay my rent. Since these conditions were all but impossible to fulfill for someone who had been in the country for two weeks (I am still waiting for somebody to start a business brokering homeowner-guarantors to desperate newcomers for a hefty commission), it eventually took payment of five months' rent up front, on top of the deposit, as well as a letter from my university to secure basic housing. I was lucky to be able to borrow this money, but for someone less privileged, these requirements can well be impossible to fulfill. If rented property is out of reach, even with enough funds for a deposit and the first month's rent, squatting may be left as the only viable option for less economically privileged migrants arriving in the UK (see also Di Feliciantonio Chapter 7, in this volume).

It is no misconception that migrants are disproportionately represented among those unfortunate enough to come into contact with police in the course of squatting activities. As one group of squatting activists reports, after the criminalization, "41 out of the 95 people arrested for squatting by the Metropolitan police were Romanian"[1] – evidence, perhaps, that squatters who have migrated may have relatively less experience or language skills to deal with police, and possibly evidence that police have used the new legislation predominantly to go after squatters whom they regard as easy targets. However, despite these "really existing connections" – some of which are discussed in other contributions to this volume – the discursive construction of the "migrant-squatter" has little to do with the actual migration status of actual squatters (see Dadusc, this volume, Chapter 22). Instead, it serves to describe a specific subject position (or, perhaps more to the point, "object position") in a discourse that is much older than recent media attention would imply. Its purpose, then as now, is to create a social category for those who, by virtue of their particular relation to the space they occupy, come to be exempted from the presumed "community of value" (Anderson 2013) implied in citizenship. As Anderson discusses, while much academic discourse distinguishes non-citizens (i.e., migrants) from "failed citizens" (i.e., the domestic poor), it makes sense to treat these two categories as one since they both are construed as the "other" of the "good citizen", i.e., the legitimate member of this community. The figure of the "migrant-squatter" to some extent foreshadows the convergence of these two groups, as well as the citizen's fears of them and especially of their possible alliance against the interests of the "community of value". Central to this fear is the idea of both the migrant and the squatter – and most of all, the hybrid creature that results from their combination – as invaders.

Invasion and evasion

The fear of invasion – of the "home" by non-paying strangers or of the nation-state by non-paying foreigners – runs deep in the British psyche. Perhaps not surprisingly for a culture built on several waves of conquest of a small island, as much as on that island's large-scale invasions of other places, the figure of the intruder and plunderer has a strong hold on the public imagination, especially in times of crisis. The "conquering warrior" is a staple of British mythology, be it in his incarnation as a Norman invader or, in more recent iterations, as a boardroom warlord whose profit represents the legitimate spoils of war (Grohmann 2015). But despite his recurring role as a positive role model, especially in conservative ideology, the warrior archetype also produces uneasiness – getting on the wrong side of him, the past few hundred years have shown, can be costly for those invaded in the traditional slash-and-burn sense of the term or can merely lead to exploitation for capitalist gain. It is therefore possible that the peculiar British preoccupation with invasion is a kind of collective projection, an uncomfortable compromise in which the collective psyche deals with its own unfinished business by attaching it to somebody else.

That is not to say, of course, that public hostility to poor migrants coming to steal "our" jobs and benefits is a particularly British specialty – the financial crisis as well as recent upheavals in the Middle East have sparked fears of being overrun by hordes of plundering outsiders all across Europe. But only in Britain has the ubiquitous fear of invasion of both the shared space of the nation state and the private space of the home combined into a figure whose most salient feature is the fact that he/she illicitly and deliberately occupies space that he/she should not, both in the public and the private realm. Migrants, one could say, are seen to squat the space of the country in much the same way that squatters illegitimately occupy the space of a building. But while people who are "only" migrants can, in principle, redeem themselves by becoming "hardworking taxpayers", and squatters can, in some cases, find understanding if they are poor "natives"; the "migrant-squatter", guilty of invasion in both senses, focuses the fears of large-scale and small-scale intrusion into a perfect image of threat.

Politically, the figure of the migrant-squatter as the ultimate invader serves several purposes. On the one hand, the assertion that only shady foreigners squat serves a political agenda where the only attempt at fighting poverty consists of trying to re-define the meaning of the word, such as the UK government's current plans to "eradicate" child poverty by applying a new definition that essentially turns it into a category of moral failure. When the idea that "native" Brits might be plunged into homelessness and destitution through precarity, austerity and a sustained attack on the welfare state must no longer be uttered, it becomes politically convenient to imply that a lack of shelter only affects those who have no business being here in the first place. On the other hand, the idea that "our homes", along with "our" jobs and benefits, are on the long list of things migrants come here to steal fuels support for stricter border controls and stronger deportation regimes. But besides such obvious short-term strategic purposes, the figure of the "migrant-squatter" also fits neatly into a more long-term agenda – that of dehumanizing the poor (see van Houtum and Aparna, this volume, Chapter 3), whether domestic or foreign, altogether. The imagery of invasion, both of the state and of the home, plays a central role in this strategy, and it is reminiscent of some of the darkest chapters of history.

Outsiderhood and "social death"

In his discussion of slavery, Orlando Patterson (1985) describes the social position of slaves in terms of the concept of "social death". On the one hand, this term refers to the idea that being enslaved involves being cut off from one's previous social ties and the social roles one has played, thus stripping away the very social identity that makes a human being into a (social) person. On the other hand, the thus de-personalized slave is then re-introduced to a new social order – that of the master – as a nonbeing", an object instead of a subject (Patterson 1985: 38). The legitimization of slavery therefore involves a process of removing the personhood bestowed upon a human being by his/her community and turning him/her into something less-than-human – a mere thing that is then seen fit to be bought, sold, used or destroyed. In this view, slavery is therefore not so much a matter of economics – of whether or not a human being can be property or a commodity – but a social process in which violent domination is legitimized by stripping the subjugated of their full moral status as members of the social order.

The representation and legitimization of this process, according to Patterson, took one of two different forms, depending on the main mode by which slaves were "recruited" in a given slave-holding society. If slaves were mostly acquired from outside the culture in question, this process was legitimized through what he calls the intrusive mode of representing social death: "the slave was ritually incorporated as the permanent enemy on the inside – the 'domestic enemy'… unsupported by a chain of ancestors reaching back to the beginning of time" (Patterson 1985: 39). For this mode, it is inconsequential whether or not the enslaved actually planned to invade; what matters is that once they arrived, they were regarded as enemies and potential invaders who must be crushed in a kind of pre-emptive strike. The slave, despite the likely violent way in which he/she was removed from his/her own origins, is therefore defined as an intruder, a hostile alien who has no business being where he/she is, who must be strictly controlled. It stands to reason that the majority of the enslaved would have gladly never set foot into the territory of their captors, but their continuous captivity could only be legitimized by a mode of argument along the lines of 'we got them before they got us'.

Where the dominant mode of acquiring slaves was from within the culture in question, this state was legitimized by what Patterson calls the extrusive mode of representing social death: "Here the dominant image of the slave was that of an insider who had fallen, one who ceased to belong and had been expelled from normal participation in the community […]. The destitute were included in this group, for while they perhaps had committed no overt crime their failure to survive on their own was taken as a sign of innate incompetence and of divine disfavor" (Patterson 1985: 41). Underlying this mode is an assumption that the individual has a moral responsibility toward the community to look after his/her own needs, and failing to do so is not simply an economic but, first and foremost, a moral failure, for which enslavement was seen as an appropriate punishment. As in the intrusive mode, such a person was extracted from the web of social relations that supplied him/her with an identity and purpose and was re-incorporated into society under different and entirely unequal auspices. One could say that, while in the intrusive mode the "socially dead" were seen to be less-than-persons because they were enemies, in the extrusive mode, they were seen as enemies because they had confirmed their status as less-than-persons through their own moral failure at securing their own survival.

As scholars in the field of poverty and social exclusion have remarked, slavery is not the only context in which these modes of dehumanization can be observed. Researchers in the field of homelessness (e.g., Ruddick 2002) for example emphasize that the homeless, many of whom also squat, face a very similar process of being relatively removed by force from the social system they are embedded in and face a similar objectification in the eyes of the "normal" population. As with slaves, the social death of the homeless involves a stripping-away of a person's previous identity, and they end up in a social position in which they no longer count as full moral agents and are subjected to the disciplinary regimes of the state, by means of force if need be. Studies in the social psychology field demonstrate that, in this case, "dehumanization" can be taken quite literally – from the perception of "normal" people, the street homeless actually cease to register as human persons and instead elicit neural responses more typical of a pile of rubbish (Fiske 2010).

While the social death of the homeless parallels that of slaves in many ways, their subsequent fate is less one of exploitation than one of disposal. In so far as they are "native" to the culture they operate in, their social death most often follows the extrusive mode of representation – they are former 'insiders' who have fallen. But unlike slaves, who were regarded as useful "things" by their masters, the homeless are "things" that have no value whatsoever. At the same time, homeless people who have migrated also incur the accusation of being intruders, as implied in the intrusive mode, and foreign rough sleepers. They therefore become prime targets for reprisals, such as the current UK government's proposals to deport foreign rough sleepers and ban them from re-entering the country (Dominiczak 2014). Thus, homeless migrants are not merely seen as things; they become what anthropologist Mary Douglas (1966) called "matter out of place" – "things" that must be disposed of outside the space inhabited by the "socially alive".

Insiders without and outsiders within

The category of "social death" can therefore help in understanding the importance of the idea of invasion in the context of the construction of the "migrant-squatter". As Patterson (1985) argues, both modes of representation are not so much accurate descriptions of the provenance of subjugated groups, as they are post-hoc legitimizations of their subjugation. A socially dead person does not have to be an invader in order to be enslaved – rather, the fact that they are enslaved is justified through the assertion that if they were not enslaved, they would surely attempt to invade. By the same token, the "fallen" insider is not identified as an enemy so much by their actual hostility to the social order, but their subjugation is again legitimized through the idea that if they were not pre-emptively subjugated, their status as outsiders to the moral order would make enemies of them. But whether "outsider within" or "insider without", "social death" means that they have become external to the social and moral order, and that therefore, they no longer have to be considered under the same moral standards applicable to actual people.

If the homeless, and especially homeless migrants, are prime candidates for a diagnosis of "social death", then the figure of the "migrant-squatter" – who does not only insist on occupying space he/she does not belong in, but does so on purpose – is that of a social casualty who remains strangely alive. The whole point of declaring some people "not actually alive" is, after all, to assuage the moral confusion that may overcome people when trying to decide who one has to treat under generally accepted moral rules and who one does not. If moral rules are applicable to all living persons, then declaring some to not be quite alive is a simple if not elegant solution that avoids moral conflict over why it should be considered allowable to treat them as "things". In order to avoid further confusion, it is, however, crucial that those declared dead behave as expected of the deceased and refrain from asserting any kind of agency or moral demand. The migrant-squatter, however, by virtue of his/her very existence, challenges this assumption.

One part of this challenge is that the migrant-squatter, as opposed to a "mere" migrant or a "mere" squatter, combines both roads to "social death" in one figure. In so far as he/she is a migrant, he/she is construed as the prototypical hostile intruder – a thief whose illegitimate occupation of the space of the nation state can only mean danger for the property, entitlements and – in the case of "knife-wielding Lithuanians" – potentially the life of ordinary citizens. The menace can only be controlled by subjecting the miscreant to the full force of the disciplinary apparatus, including internment and involuntary deportation. In so far as he/she is a squatter, on the other hand, he/she embodies the fallen insider or "enemy within", a social force who by virtue of its very existence threatens to overthrow the established order. Mike Weatherly, the British politician (former MP for the Conservative party) and a driving force behind the criminalization of squatting, aptly expressed this sentiment in his alarming assertion that squatters were planning to turn the country into a "medieval wasteland without property rights" (Weatherly 2013). Each of these frightful figures on its own is enough to spur the public imagination into panic mode – but combined, they turn into something even more ominous.

For the "migrant-squatter", this hybrid of the external and the internal enemy is defined precisely by his/her refusal to remain socially dead. Once thought to be safely contained within the disciplinary system of welfare provision, or kept out through the policing of borders, the enemy now not only confirms his double status as invader of public and private territory, but he/she also possesses the audacity to frame this course of action as an assertion of his/her status as a moral person. To squat, that is to foreground the basic human need for shelter at the expense of law and property, means, after all, to boldly assert one's existence as a human being whose needs take precedence over the needs of private profit and state control. It is as if the migrant-squatter, despite being declared socially dead not once but twice, continued to rub his existence into the face of the citizenry from beyond the grave, and there is precious little said citizenry can do about it. Like a kind of social zombie apocalypse, the invasion of the "migrant-squatters" redefines the boundary between the "socially dead" and the "socially living", bringing those declared non-persons into the streets and living rooms of a panicked populace to re-negotiate their moral status.

Conclusion

It would therefore appear that the invention of the "migrant-squatter" (actual squatters who have migrated notwithstanding) is at least an ambiguous move. On the one hand, by suspecting both the migrant and the squatter of being invaders who must be contained and/or kept out, this discourse serves to re-enforce their status as "socially dead" and thus not worthy of moral consideration. Where mass homelessness and poverty among "fellow citizens" would likely spark public moral outrage, if it is only social or literal outsiders who suffer, then the problem, ergo, is not that great. Promoting the perception that only foreigners would "have to" squat thus translates into widespread apathy toward the continuous surge in evictions and repossessions, as well as toward the out-of-control housing market – and re-enforces the idea that social and material abjection is something that happens to other people. Of course, the maneuver is becoming increasingly transparent, as parts of the solid middle-classes are now struggling to get a foot-hold in even the modest rental market, not to mention the actual property ladder. However, in the absence of a strong movement against evictions and repossessions such as other European countries have seen, the appeal of this worldview may be understandable. In the light of the existential insecurity the economic crisis has caused throughout Europe, fears of one's own economic and social demise can at least temporarily be calmed if only one can find a good enough reason why the poor or homeless are fundamentally different from oneself.

On the other hand, however, the discourse of the "migrant-squatter" also points toward the fear – held consciously or unconsciously – that those excluded from the "community of value" could one day cease to take their status lying down and decide to rise from the grave of social and moral abjection that the good citizens have dug for them, and thus bring about the apocalypse of the very value system that denied them their humanity. In this prospect lies the real threat of the figure of the "migrant-squatter", who may well have come to avenge him/herself on parts of a citizenry who has long secretly suspected that the moral sleight of hand by which it has declared others as less-than-human may someday come back to haunt it.

Note

1 See the website: https://rooftopresistance.squat.net/this-new-law/.

References

Alleyne R. (2011) 'Gang of Romanian gipsy squatters 'targeted several houses'. *The Telegraph* (17 August 2011). Available: http://www.telegraph.co.uk/news/uknews/immigration/8707490/Gang-of-Romanian-gipsy-squatters-targeted-several-houses.html.

Anderson, B. (2013) *Us and Them? The Dangerous Politics of Immigration Control.* Oxford: Oxford University Press.

Anderson, B. (2014) *Exclusion, Failure and the Politics of Citizenship*. RCIS Working Paper. Toronto: Ryerson University.

Cohen T. (2010) 'Hotelier leaves home for a week so it can be decorated... then 15 jobless Italian squatters move in'. *Daily Mail* (21 October 2010). Available: http://www.dailymail.co.uk/news/article-1322246/Man-leaves-home-week-decorated-15-squatters-in.html.

Common Place (2008) *What's This Place? Stories from Radical Social Centres in the UK and Ireland*. Leeds: University of Leeds.

Dominiczak P (2014) 'David Cameron: I'm ready to lead Britain out of Europe if migrant reforms fail'. *The Telegraph* (27 November 2014). Available: http://www.telegraph.co.uk/news/uknews/immigration/11259473/David-Cameron-Im-ready-to-lead-Britain-out-of-Europe-if-migrant-reforms-fail.html.

Douglas, M. (1966) *Purity and danger: An Analysis of Concepts of Pollution and Taboo*. London: Routledge.

Dovar, D. (2012) 'Invasion of the migrant squatters: Eviction orders for London mansions DOUBLE as economic crisis drives influx from overseas'. *Daily Mail* (9 June 2012). Available: http://www.dailymail.co.uk/news/article-2156833/Invasion-migrant-squatters-Eviction-orders-London-mansions-DOUBLE-economic-crisis-drives-influx-overseas.html.

Fiske, S.T. (2010) 'Are We Born Racist?' In Marsh J., Mendoza-Denton, R., Smith, J. (eds) *Are We Born Racist? New Insights from Neuroscience and Positive Psychology*. Boston: Beacon Press, pp. 7–16.

Grohmann, S. (2015) *The Ethics of Space: Homelessness, Squatting and the Spatial Self*. Doctoral Thesis submitted to Goldsmiths, University of London, unpublished.

Kisiel R. (2010) 'Knife-wielding Lithuanian squatters who move in when residents go out'. *Daily Mail* (24 September 2010). Available: http://www.dailymail.co.uk/news/article-1314526/Knife-wielding-Lithuanian-squatters-residents-out.html.

Patterson, O. (1985) *Slavery and Social Death*. Harvard: Harvard University Press.

Ruddick, S. (2002) 'Metamorphosis revisited: Restricting discourses of citizenship'. In Hermer, J. and Mosher J. (Eds.) *Disorderly people: Law and the politics of exclusion in Ontario*. Halifax: Fernwood Publishing: 55–64.

Weatherly, M (2013) *Squatting Statement*. 4 March 2013. Available: http://www.mikeweatherley.com/2013/03/04/squatting-statement/.

11 Racialization of informal settlements, depoliticization of squatting and everyday resistances in French slums

Thomas Aguilera

In his public speech in Grenoble on July 30, 2010, President Nicolas Sarkozy associated "Roma migrants" coming from Eastern Europe, with labels such as, nomads, delinquency and slums[1] [Author's translation]. If this discourse is only the visible part of the stigmatization and repressive policies toward migrants living in illegal housing since the end of the 1990s, it helps us to understand more clearly the process of the racialization[2] of poverty and informality. In France, thousands of migrants from Eastern Europe have been associated with informal slums considered their "natural" living place, building the idea that Roma people are dirty and delinquent through the "dehumanization" of migrants (see van Houtum and Aparna, Chapter 3 and Grohman, Chapter 10, this volume) and the "naturalization of poverty" (Fassin 2014). This chapter analyzes this process and argues that it has helped depoliticize the experience of squatting land and thus inhibited the disruptive political potential of illegality to claim housing, social help and more rights.

The "Roma question" has constituted an important obstacle in France for social movements to push officials to implement alternative housing and social policies because it constitutes an obstacle against the politicization both of the policies towards slums and the practice of squatting. When they claim for more than survival, slum dwellers are considered out of place by the officials, the media, NGOs and public opinion. Slum dwellers have to stay politically passive if they want to appear as the "good poor" and benefit from social welfare (Coutant 2001; Bouillon 2010; Anderson 2013). Stigmatization dynamics are so strong towards these migrants that even political squatters have trouble helping them to squat buildings. Local associations or NGOs mainly supported these populations and mobilized conventional resources to stop evictions and claim for new rights, but without using illegality as a disruptive[3] political resource, i.e., without disturbing the formal rules of the conventional political forums. By playing the traditional lobbyist game, these organizations do not violate the law and try, on the contrary, to fit in order to avoid the blame for their actions and slum dwellers.

In this chapter, I analyze: 1) how informal settlements and their inhabitants have been racialized since the 1960s by state policies and the media, and 2) how this process has impeded the disruptive use of squatting by these groups and their supporters. State policies have created a regime of exception that inhibits migrants from taking a risk in collective action. Indeed, by maintaining a permanent state of emergency as a form of governmentality (Agamben 2005) and as a mode of management of the poor in cases of humanitarian crisis (Fassin and Vasquez 2005), this regime, that claims to be temporary to legitimate its hardness, reduces opportunities for protest, collective action and policy changes. At the same time, I also show that it does not totally crush resistances and alternatives and that slum dwellers are able to silently resist constraints and exploit resources in order to survive in the core of the metropolitan area. This resistance is also subversive in the sense that it challenges the rules of housing as well as social and urban policies.

The history of French slums: the racialization of informal settlements

In France, the long history of slums has always been associated with the racialization of poverty and migrations (De Barros 2005). Spanish and Portuguese workers called by the government to be used as manpower during the 1930s, and Algerian workers during the 1960s, settled at the periphery of the French cities in slums as a consequence of the inefficient housing policies and segregation processes. In 1960s, 75,000 people were living in slums in France (62% in the Paris Region). In 1970s, after the first National Plan of rehousing, 45,000 people were still living in slums. Among them, 75% were migrants according to Internal French Ministry data (1966).

The police violently intervened inside the slums in order to repress activists.[4] The special "Brigade Z" was created, including policemen previously sanctioned for professional faults. This served to build an atmosphere of terror around Algerian migrants (Lallaoui 1993: 53). At the same time, social catholic activists and NGOs began to claim for the rehousing of the families and put pressure on the government, while squatting buildings or self-building *autoconstruction* on vacant lands. After various accidents and a strong focus by the media, the Parliament voted in a series of laws to eradicate slums and relocate families to social housing. The policy was implemented through two different instruments. Most of the non-Algerian families were directly relocated to social housing in the suburbs that constitute nowadays the so-called *French banlieues*, while Algerian families were first relocated to Transitory Camps where they were supposed to be "educated" and "re-socialized" (Tricart 1977). However, although the government declared that the last slum was evicted in Nice in 1976, many Algerian families were relocated to Temporary Camps until the 1990s, and such slums still existed until the end of the twentieth-century.

Nevertheless, slums emerged again as a massive phenomenon during the 1990s in the periphery of Paris when the end of the Eastern Bloc opened a new wave of migrations from Eastern Europe (Reyniers 1993). At first, the migrants were considered as political refugees fleeing wars and discrimination. But successive European Union and national legislations progressively removed this status and forbade Romanian and Bulgarian migrants to work in France.[5] As a consequence, the political refugees became "undocumented" migrants (Nacu 2010). Social aids were reduced, the state was no longer obliged to provide them housing, and migrants were forced to squat lands and build houses with precarious and recycled materials (Olivera 2011). In 2012, in France, the state administration counted 16,399 people living in 391 slums (41% of the slum dwellers lived in the Paris Region).[6] According to official data, in the last 10 years, 82% of the slums' inhabitants have been from Romania and 6% from Bulgaria.

These figures fit into the "Roma question" that has been built by public actors and the media who contribute to the confusion around migrants and nomads. Since 1912, there has been a law in France that assigns a specific status to and imposes police control over nomads – called *Gens du voyage*. A recent law ordered municipalities to build camps to host them, but it also allowed them to evict settlements of nomads outside these zones (Law Besson 2000). The migrants who live in French slums are not nomads and just ask for housing, but municipalities often use the Law Besson in order to justify the eviction of slums, saying that all illegal settlements outside these specific camps can be quickly evicted. This procedure is illegal when it serves the eviction of squatters.[7] Systematic forced evictions and lack of policies and economic opportunities to absorb Roma migrants have provoked an extreme socio-economic precariousness, forcing them to work in the shadow economy, confirming the old prejudice that Roma people are nomads, offenders, and burglars, and that they adopt a culture or an essence of poverty (Fassin 2014; Olivera 2015). The "Roma question" has been built by the state and media. The media plays a major role in shaping the image of the Roma migrants and poverty by diffusing prejudices that became official and by shaping the idea of race (Gilens 2003), by confirming an association across ethno-racial category, the architectural form of the slums, and extreme poverty (Legros and Vitale 2011). For instance, in 2013, 86% of the French population thought that Roma people were nomads (Mayer *et al.* 2015: 254) and thus chose to live on squatted lands and in slums.

However, Roma migrants are not nomads, and living in a slum is not a cultural choice. Moreover, all slums are not exclusively inhabited by migrants. Homeless camps have been consolidated. Travelers or seasonal workers also live in informal settlements. In the French Overseas Departments and Territories, we also find many slums that have been developed and are now entirely informal neighborhoods (Bernardot 2012). All Eastern European migrants are not Romas. And all Roma people do not live in slums.[8] Some poor migrants also squat buildings in order to get more stable life conditions. Ironically, when they do so, they are no longer considered Roma by officials and the media, but instead just migrants and squatters. This last observation confirms the artificial association of two categories: illegal slums and Roma people. In all cases, most of the slum dwellers have been living in France for 15 years but are still considered migrants by the state that evicts them twice a year, on average. In fact since 2011, according to data elaborated on by Philip Goosens and Grégoire Cousin since 2010 (Goosens and ERRC-LDH 2010–2015), each person would have been evicted 1.7 times per year. This frequency does not allow inhabitants to consolidate their houses or the infrastructure in camps (Perou 2014). The wrongly attributed nomadic status of Roma people has been created by these repetitive and systematic evictions combined with local and selective projects. The forced mobility combined with the threat of the eviction of the national territory creates illegality, instability and racialization.

As a consequence, the stigmas toward Roma migrants are much stronger than toward other communities[9] and have increased during the last five years (Mayer *et al.* 2015). In 2014, 77.5% of the French population thought that Roma migrants did not want to integrate in France and that they got their economic resources from robberies and trafficking (Mayer *et al.* 2015: 254). More precisely, 82% thought that Roma migrants constituted a separated homogeneous social group, while only 66% presumed this in 2011 (Mayer *et al.* 2015: 252).

The two faces of policies towards slums: repression and *bricolages*

Since the 2000s, the governance around French slums has been fragmented (Legros and Olivera 2014). There are two main levels of policies toward French slums (Aguilera 2016). First, the state mainly ensures the persistence of a strong repressive framework within which the prefectures command police evictions, the Ministries of Internal Affairs and of Immigration maintain a coercive climate over migrants, refugees and slum dwellers. President Sarkozy made this repression public when he launched an explicit hunt for the Roma people, described as the inhabitants of "illicit camps". Moreover, each slum eviction represents an opportunity to evict migrants from the national territory. The number of evictions has even grown with the arrival of the Socialist Party (PS) to the national government since 2012: 932 persons were evicted from their domicile in 2010; 4,334 in 2011; 5,100 in 2012; 10,778 in 2013; and 13,483 in 2014 (Goosens and ERRC-LDH 2010–2015; Aguilera 2015: 199).

The second level concerns the local administrations and mainly municipalities that are directly in contact with the demands of slum dwellers, voters and NGOs. The municipalities have layered two positions that complement each other. On the one hand, most of the French municipalities have systematically evicted slums during the last 20 years, while denouncing the state inaction in terms of migration and housing policies and asking national governments to intervene to resolve what has become one of the most visible local issues in the media. On the other hand, since the middle of the 2000s, a few French municipalities have tried to implement alternative integrative projects (Legros 2011). All these projects are selective, put strong constraints on the beneficiaries, and legitimate the eviction of the families that are not accepted in the dispositive. In the Paris Region, the first project of an "insertion village" was opened in 2007 after a fire accident and NGO protests. After this first experiment, a dozen municipalities in the Paris Region, but also in the regions of Nantes, Lille, and Bordeaux, have implemented more or less the same type of projects with the same fragmented governance.

All these projects follow the same model that has spread in a few years. After accidents (usually fire and death of children), municipalities "open" temporary zones; social workers select families on socio-economic criteria; the beneficiaries are integrated into temporal housing and have to accept restrictive living conditions. The project fixes them in order to find jobs and permanent social housing. Public policies are always selective, and resource allocation depends on the discretionary work of social workers and street-level bureaucrats (Lipsky 1980), particularly when specific measures are implemented toward migrants and Roma (Vitale and Claps 2010). In our case, the rate of integration/exclusion is decided following the financial capacity of the municipality and the partners, and the political willingness of local representatives. At the beginning of the procedure, municipal and NGO social workers distribute questionnaires in order to know the situation of each family – education, health, skills of the father, etc. On this basis, only the families who have chances of finding a job and a house within five years are selected (Aguilera 2015a).[10] But most of the slum dwellers are not integrated into the project and are evicted without any solution. The municipalities that implement such projects easily legitimate the eviction of all other slums and squats from their territory.

These temporary villages have provoked violent debates by stimulating the opposition of NGOs and intellectuals against what has been quickly equated to "concentration camps" and denouncing the perpetuation of the state of emergency, i.e., the suspension of human rights for the poor on behalf of their health. These alternative projects have been presented as innovative, when in reality they only recycle old measures coming from the 1970s and the 1990s. The "insertion villages" clearly recycles the *cité de transit* (transit cities) implemented during the 1970s by the state administrations but also experiments with temporary relocation camps in the 1990s. Their implementation has crystallized new collective action around slums that have failed to change policies and discourses at the national level.

Reformist and fragmented collective action around slums

Social movements around slums and Roma groups are very fragmented in Europe, and Roma people usually fail in organizing strong public and collective mobilization in order to make policies change (Vitale and Boschetti 2011) because of the lack of institutional opportunities that make political representations possible. Tommaso Vitale (2009: 78–79) identifies three main factors that impede the development of collective action with internal resources and their impacts on policies: migrants from Eastern Europe do not have institutional recognition because of European legal barriers; the stigma around Roma people is too strong to make voters favorable to their claims; and the fragmentation of local governance makes the identification of bureaucratic or political interlocutors impossible.

In France, the most visible collective action comes from reformist NGOs that use conventional lobbying modes of action and legal instruments as weapons (Jacquot and Vitale 2014). The actors are policy oriented and try to change policies to make slum dwellers' living conditions better by showing the indignity of their situation and the systematic violation of human rights, European and national laws, by claiming for new housing and social policies, and by denouncing the political discourse.

National NGOs have worked to dismantle prejudice and still try to change attitudes toward migrants. Some of them claim for the recognition of the Roma minority at the European level (Vermeersch 2006). For example, the association *La voix des Roms* (The Voice of Roma) contributes to animated debates by depicting Roma as a people and by claiming specific cultural traits, while others prefer, on the contrary, to struggle against the culturalization of bad housing conditions in order to avoid stigmatization. In that way, the NGO *Romeurope* works as a platform. Romeurope coordinates local collectives with national NGOs and gives them support. It also organizes conferences and workshops, and works with the media. It publishes annual reports that clarify the juridical situations, the evolution of evictions, or the public policy experiments, denounce stigmatization, and make propositions.

Many medical care NGOs have also brought the minimal vital resources. For instance, *Médecins du Monde* has been very active in slums since 2000 in the Paris Region. A local mission regularly intervenes in the north of Paris (Seine-Saint-Denis) for health watch and emergency assistance. There are plenty of organizations of this type, but there was little regional coordination among them until recently. In 2015, Médecins du Monde organized a national campaign called *25 ans de bidonvilles* (25 years of slums) in order to force politicians to recognize the existence of slums, to ask them to stop systematic evictions and to invent new dispositives for integrative social and housing policies. Conferences and cultural events were organized during the fall 2015 and social scientists, NGOs, social workers, volunteers and activists collaborated to publish collective works.[11]

Sometimes, dozens of citizen committees have organized spontaneously at the very local level. They work with very few resources to help families survive by bringing food and materials to consolidate houses, build toilets, help families send children to school, etc. Students or professors volunteer to give French classes inside slums. Lawyers also volunteer to bring juridical resources during procedures or to get social public aids. In some cases, activists and local committees have succeeded in pushing local authorities to implement alternative projects (Bruneteaux and Benarrosh-Orsoni 2012). Indeed, each time an alternative project has been implemented, it has been done under the pressure of NGOs or local groups.

Finally, architects intervene in slums in order to implement self-building projects. For instance, in *Ris-Orangis*, the association *PEROU* (Pole for the Exploration of Urban Resources) helped inhabitants to build a wooden barrack/common space at the core of a slum in order to organize parties for Christmas 2012. The municipality evicted the slum a few months after. Sometimes architects work within municipal projects and receive public subsidies. For instance, in Saint-Denis (Northern suburb) and Orly (Southern suburb), two young architects tried to implement alternative self-built houses as alternatives to the criticized "insertion villages".

These actors sometimes help slum inhabitants to find vacant lands or buildings to squat and defend slums from policemen and officials. They also use the media to diffuse counter-propaganda ideas and make the police violence visible. All these actors usually remain reformists and use conventional modes of action. Their visibility in the public arena is quite weak, and they partially failed in changing the conception of the public framing of the slum and Roma questions. Sometimes, however, activists from squats or Social Centers disrupt these modes of action by helping migrants live in squatted buildings.

Social Centers, migrants, slums: "*everyone in his place*"

While relationships between radical and alter-globalization activists and Roma migrants are not always collaborative (Vitale 2009: 80), a few initiatives demonstrate how the tension between general political struggles and the search for individual solutions can lead to durable and collective projects, as Antonelli and Perrotta exemplify on the case of Bologna at the beginning of the 2000s (Antonelli and Perrotta, Chapter 12, this volume).

During the past ten years, some French activists from squatted Social Centers have opened the doors of their spaces or squatted new buildings to provide shelter to families evicted from slums or previous squats and to organize collective activities. While this type of cooperation is rare in Paris, the capital, some groups of squatters, for instance in Montreuil (East of Paris) or Toulouse (South of France), mix radical political and cultural activities with housing and social spaces. Montreuil, in the Eastern suburbs of Paris, hosts most of the autonomous squatters in the Paris Region. Some of them have tried to open squats for the Roma families in the last ten years and have usually supported them after evictions from slums or squats. In Toulouse, since 2011, the group *CREA* (Campaign for Requisition, Solidarity and Self-Organization) has opened Social Centers but also housing in apartments or entire buildings (more than 75 squats). Similarly to Spain (Martínez 2013), these squatters also provide advice to open squats. The social networks used by political squatters often denounce the eviction of migrants' squats, calling on activists to help against the police.

The link between activist squatters and migrants is difficult, but it has also strongly developed in France. The association *Droit Au Logement* (DAL) (Housing Right Association) has opened squats for West African families since the beginning of the 1990s and has been quite successful in securing relocation to social housing (Péchu 2006). Trade unions or NGOs have also organized squats in Paris. On Baudelique Street, in 2010–2011, more than 3,000 undocumented migrants (mainly of African origin) squatted an enormous building. At the end, around 400 people got documents. More recently, Chechen migrants were tolerated by the municipality of Pantin (North-Eastern suburb of Paris) with the intervention of architects and NGOs to renovate their squatted building. NGOs usually intervene in squats occupied by migrants whenever it is opened to determine if the place is safe.

What was the difference between families living in slums from those living in political squats in 2010? Firstly, it seems that the expertise and the professionalization of activists and their political networks have overcome the juridical risks that nowadays threaten Eastern European migrants. For instance, the squatters of DAL are experts in helping migrants. Secondly, it seems that the racialization of poverty produced through the "Roma question" and the slum issue, in a context of economic crisis, is so strong that it makes the association difficult between squatters from Social Centers and these migrants.

Beyond the fact that some activists believe political activism is not compatible with non-activist housing places (at least in daily life), the main obstacles come from the fragile status for migrants who can be evicted from the national territory after being evicted from the squat. Many of these evictions are illegal and do not respect the national and European Right principles (Cousin 2011), but the French State continues these procedures in order to put pressure on migrants and their supporters.

Squatting in visible places implicates juridical risks that are higher for the Roma families than for non-migrant activists or for hidden slums. Authorities exploit these risks and create uncertainty in order to weaken and divide collective actions. Above all, authorities and the media discredit Eastern European migrants who squat buildings by circulating discourses on them such as, "They are not in their place, they have to come back to slums because Roma people live on squatted lands, it is their nature to live in camps".

The rare support of Social Centers for slum dwellers, in Paris at least, can also be explained by the *policy feedback*: policies shape their targets and the forms of resistance (Pierson 1993). Municipalities and state administrations want to avoid the radicalization of social movements around slums and prefer dealing with reformist and conventional actors about the slum issue. The alliance between Social Centers and Roma will not be accepted, and NGOs have built a monopoly on the issue, particularly because hygiene and health problems have to be treated by specialized organizations.

> Six months ago, the squatters (from a Social Center) deviated from their normal role. They helped Roma to squat a municipal building in ruins that was dangerous to live in [...]. They opened during the night but were stopped in flagrance by the police. We supported their eviction. (Deputy-Mayor, in charge of migration of a Western municipality of Paris, Interview by Author, 2011).

Silent resistance to urban policies

As James Scott (1985) has shown in his seminal work on the "Weapons of the Weak" (Scott 1985), the study of visible collective action does not have to hide the most silent forms of resistance. If slum dwellers do not seem to mobilize resources in visible collective protest action or to build stable and strong networks, it is because they find weapons and build tactics to resist inside the city. Squatting brings subversion and political meaning to daily life (Bouillon 2009).

This is just a different mode of resistance, using different resources and exploiting another side of squatting as a mode of action, as a concrete and immediate right to the city (Aguilera and Bouillon 2013).

Squatting implicates a violation of the private property right and thus is itself a form of resistance to the law and public order imposed by national and local legislations. As all informal "urban hunters" (Merklen 2009), slum inhabitants develop strategies to find vacant lands, to build houses,[12] to hide their place in order not to attract the attention of the neighbors, and they seek economic resources by doing informal work, begging or recycling wasted products. As the municipalities often forbid them to connect to water and electricity infrastructure, they have to find tactics to get these vital resources.

Slum dwellers also resist external aggressions. They resist attacks from the neighbors[13] and police eviction. During police evictions, the resistance is not as physical as it can be from activists in Social Centers. Local activists and NGO members are usually on the field but often resist in non-violent forms. In the daily life, slum dwellers adapt their behavior to the hostile policemen who harass them at the entrance of the camps where they live.

Finally, they also deploy tactics in order to deviate the few resources allocated by municipalities or NGOs (Legros and Olivera 2014). Indeed, even if they are not selected to participate in the integrative programs mentioned above, families can benefit from the solidarity of the beneficiaries and get space close to the villages, health services and other infrastructures. Resources circulate between legalized camps, slums and squats.

Conclusion: the depoliticization of the "Roma question" and the repoliticization of squatting

When they squat lands, Eastern European migrants (Roma or not Roma people) are considered "Roma migrants". When they squat buildings, they are considered "migrants" or "squatters". When they do not squat (as the large majority), they are not visible in the public debate. The fluidity of the labels underlines two mechanisms that allow us to explain the treatment of slums in France and more broadly in Europe: a process of racialization and depoliticization of informal slums.

The process of naturalization of poverty for slum dwellers reaches its pinnacle and is unique in France, more than for poor migrants coming from other zones or homeless people.[14] The extreme precariousness provoked by the cycles of police eviction, combined with intense media attention, have contributed to maintaining the permanent image of people who choose to be nomads, to live in dirty and illegal places. As a consequence, during the 2000s, as in the 1950s, the architectural form "slum" has been the support for a racialization of informal housing and poverty.

Squatted lands are themselves disruptions of public order. But this disruption is diluted into a process of racialization of poverty that impedes the subversive meaning of squatting to be visible and thus constitutes an obstacle in the public sphere to the contribution of radical activists to actions and debates around slums. This contribution is denied by representatives, officials, and NGOs who concentrate their efforts and resources in the struggle against discrimination and racism in the public debate and thus who prefer to negotiate within legal frames rather than using direct action in the struggle against the political order.

However, since 2010, NGOs have also helped repoliticize the debate as a reaction to the activation of a massive eviction agenda launched by the successive national governments that explicitly targets slums without naming them. But this repoliticization has not happened through the political use and the legitimation of squatting as a mode of protest and collective action, but by the mobilization of conventional tools in order to make slums disappear. NGOs have taken the monopoly in supporting slum dwellers, and squatting is not an end at all and is not considered as an option for political struggle.

For the activist squatters, the crucial issue would be: how do we make the disruption of public order a political resource more than a "problem" for urban societies? As usual for radical activists, the issue is how to create a mass challenge without putting the migrants in danger because of the illegality of their action (Piven and Cloward 1979; McAdam 1982). These questions are the subject of debate in Social Centers as they have always played a crucial role or spearheaded radical changes through social movements since the 1960s, but the recent crisis in Europe could bring activists to reshape the question.

Notes

1 *Slum* can be defined as an illegal occupation of land without the owner's authorization, usually accompanied by self-built housing without legal access to basic services and infrastructures. Nowadays, French officials never use the term "bidonville" – used during the 1960s – but the term "illicit camp" ("campement illicite").

2 *Racialization* can be defined as the process of construction of radical otherness on the basis of biological and culturalist figures relative to one social group supposed as being homogeneous and specific, and presenting radically different ways of life comparing to the dominant social values. This process is systemic and takes place within a constellation of actors (public, private, media, public opinion) and is rooted in historical contexts.

3 Disruptive tactics can be defined as those that "intentionally break laws and risk the arrest of participants" (Cress and Snow 2000: 1078), that "question the legitimacy of power" (Piven 2006: 20). It is not only about violence but also about going "beyond the limits of compatibility of the system in question" (Melucci 1981). Squatting is a challenge to the property right, to social and housing policies.

4 At the end of the Algerian Colonial War, the French Government was afraid by the presence of FLN (National Front of Liberation) activists inside slums to activate rebellions. There were also left-wing activists and students (Hervo 2012).

5 In 2003, the status of "unsafe country" was denied to Romania. Between 2007 and 2014, the European Union (Treaty of Luxembourg 2005) allowed eight countries (France, Austria, the Netherlands, Luxembourg, Germany, Belgium, Italy, Ireland, Spain) to adapt a flexible legislation concerning the migrations from new members (Romania and Bulgaria) to close their job markets. These transitory measures that ended in 2014 forbade these migrants to work in France without a residence permit, similarly to what happens to citizens from outside the EU (except for a list of 150 precarious jobs).

6 DIHAL (2013) "État des lieux des campements illicites", Premier ministre-Ministère de l'Égalité des territoires et du Logement.

7 For example in 2014, the University of Lille-1 went to the Judge to ask for the eviction of a slum settled in its zone, using this law. The slum was evicted, but in the end the Council of State cancelled the decision because of its illegality (Conseil d'État, 17/01/2014, n°369671).

8 Considered in its broader sense, we can consider that 400,000 Roma or Gypsies ("Tsiganes") live in France (Liégeois 2009: 29). Thus, most of them do not live in slums or squats and are invisible in the media.

9 The CNDH Annual Report on Racism and Prejudices (Mayer et al. 2015) compares opinion on the Jew, Muslim, Black, North African, Asian and Roma "communities". Among them, the so-called Roma community concentrates the worst indicators.

10 For more details on the selection procedure in the regions of Paris and Madrid see Aguilera (2015a: 540–568).

11 For example, see the special issue of the journal *Projet* (published in September 2015).

12 The literature on this topic is well developed in the global South. Since the 1960s, architects and social scientists have been fascinated by the self-building skills and expertise developed by inhabitants in informal settlements (Lara 2010).

13 In some cases, the neighbors have violently attacked slums denouncing the dirt of the place and hypothetical robberies. For example, in Marseille in 2011, a camp was attacked and set on fire. In 2011, in Paris, a squat of a vacant factory was also fired.

14 Currently, 20,000 people live in slums, while there are around 150,000 homeless people in France. The gap between the two figures is negatively correlated with the media visibility.

References

Agamben, G. (2001) *State of Exception*. Chicago: University of Chicago Press.

Aguilera, T. (2016) 'Gouverner les illegalismes urbains. Les politiques publiques face aux squats et aux bidonvilles dans les régions de Paris et de Madrid', *PhD dissertation*. Paris: Sciences Po.

Aguilera, T. (2016) 'Governare le baraccopoli a Parigi e Madrid: governance, conoscenza e costruzione metropolitana'. In Vitale, T. (ed.) *Inchiesta sui campi rom*, Firenze: La Casa Usher.

Aguilera, T. and Bouillon F. (2013) 'Le squat, un droit à la ville en actes', *Mouvements* (2)74: 132–142.

Anderson, B. (2013) *Us and Them? The Dangerous Politics of Immigration Control*. Oxford: Oxford University Press.

Bernardot, M. (2012) 'Invasions, subversions, contaminations. De quelques figures et lieux contemporains d'un Autre exceptionnel', *Cultures et conflits* 84: 45–62.

Bouillon, F. (2009) *Les mondes du squat*. Paris: PUF.

Bouillon, F. (2010) 'Le squatter, le policier, le juge et le préfet: procédures *en actes* et classements *ad hoc*', *Déviance et Société* (34)2: 175–188.

Bruneteaux, P. and Benarrosh-Orsoni N. (2012) *Intégrer les Rroms?* Paris: L'Harmattan.

Cousin, G. (2011) 'Variation préfectorale de la recette OQTF'. *Urban Rom*. Available: http://urbarom.hypotheses.org/62.

Coutant, I. (2001) 'Statu quo autour d'un squat', *Actes de la recherche en sciences sociales* (136–137): 27–37.

Cress, D. and Snow D. (2000) 'The Outcomes of Homeless Mobilization: The Influence of Organization, Disruption, Political Mediation, and Framing', *American Journal of Sociology* (105)4: 1063–1104.

De Barros, F. (2005) 'Des 'Français Musulmans d'Algérie' aux 'immigrés'. L'importation de classifications coloniales dans les politiques du logement en France', *Actes de la recherche en sciences sociales* (159): 26–45.

De Genova, N. (2002) 'Migrants, 'illegality' and deportability in everyday life', *Annual Review of Anthropology* (31): 419–447.

Fassin, É. (2014) 'La question rom'. In Fassin É., Fouteau C., Guichard S. and Windels A. (eds) *Roms et riverains: une politique municipale de la race*. Paris: La Fabrique, pp. 7–69.

Fassin, D. and Vasquez P. (2005) 'Humanitarian Exception as the Rule: The Political Theology of the 1999 Tragedia in Venezuela', *American Ethnologist*, (32)3: 389–405.

Gilens, M. (2003) 'How the poor became black: The racialization of American poverty in the mass media'. In Sanford F. Schram, Joe Soss, and Richard C. Fording (eds) *Race and the Politics of Welfare Reform*. Ann Arbor: University of Michigan Press, pp. 101–130.

Goosens, P., ERRC-LDH, (2010–2015) *Recensements des évacuations forcées de lieux de vie occupés par des Roms étrangers en France*. Report for the *Association européenne pour la défense des droits de l'homme* (AEDH) and the *Ligue des droits de l'homme* (LDH).

Hervo, M. (2012) *Nanterre en guerre d'Algérie*. Paris: Actes Sud.

Jacquot, S. and Vitale T. (2014) 'Law as weapon of the weak? A comparative analysis of legal mobilization by Roma and women's groups at the European level', *Journal of European Public Policy* (21)4: 587–604.

Lallaoui, M. (1993) *Du bidonville au HLM*. Paris: Syros.

Lara, F.L. (2010) 'The form of the informal: Investigating Brazilian self-built housing solutions'. In Hernandez F., Kellett P. and Allen L. (eds) *Rethinking the Informal City: Critical Perspectives from Latin America*. Oxford: Berghahn, pp. 23–37.

Legros, O. (2011) 'Les villages d'insertion. Un tournant dans les politiques en direction des migrants roms en région parisienne ?', *Revue Asylon* (8): juillet 2010. Available: http://www.reseau-terra.eu/article947.html.

Legros, O. and Olivera M. (2014) 'La gouvernance métropolitaine à l'épreuve de la mobilité contrainte des 'Roms migrants' en région parisienne', EspacesTemps.net: 1777–5477. Available: http://www.espacestemps.net/articles/lmobilite-contrainte-des-roms-migrants-en-region-parisienne/.

Legros, O. and Vitale T. (eds) (2011) "Les migrants roms dans les villes françaises et italiennes: mobilités, régulations et marginalités", *Géocarrefour* (86)1. Available: http://geocarrefour.revues.org/8220.

Liégeois, J.-P. (2009) *Roms et Tsiganes*. Paris: La Découverte.

Lipsky, M. (1980) *Street-level bureaucracy: dilemmas of the individual in public services*. New-York: Russell Sage Foundation.

Luiz Lara, F. (2010) 'The Form of the Informal: Investigating Brazilian Self-Built Housing Solutions', In Hernandez F., Kellett P. et Allen L. (eds) *Rethinking the Informal City: Critical Perspectives from Latin America*, New York, Oxford: Berghahn Books: 23–37.

Martínez, M. (2013) 'How do squatters deal with the State? Legalization and Anomalous Institutionalization in Madrid', *International Journal of Urban and Regional Research* (38)2: 646–674.

Mayer, N., Tiberj V., Vitale, T. (2015) 'Le regard des chercheurs sur les phénomènes de racisme, in Commission Nationale Consultative des Droits de l'Homme, La lutte contre le racisme, l'antisémitisme et la xénophobie (Année 2014)'. Paris: La Documentation Française: 153–208.

McAdam, D. (1982) *Political Process and the Development of Black Insurgency (1930–1970)*. Chicago: Chicago University Press.

Melucci, A. (1981) 'Ten Hypothesis in the Analysis of New Movements'. In Pinto, D. (eds) *Contemporary Italian Sociology*. Cambridge: Cambridge University Press, 173–94.

Merklen, D. (2009) *Quartiers populaires, quartiers politiques*. Paris: La Dispute.

Nacu, A. (2010) 'Les Roms migrants en région parisienne: les dispositifs d'une marginalisation', *Revue Européenne des Migrations Internationales* (26)1: 141–160.

Olivera, M. (2011) *Roms en (bidon)villes; quelle place pour les migrants précaires aujourd'hui?*. Paris: Éditions Rue d'Ulm.

Olivera, M (2015) 'Insupportables pollueurs ou recycleurs de genie? Quelques réflexions sur les "Roms" ou les paradoxes de l'urbanité néolibérale', *Revue Ethnologie française* (3): 503–596.

Reyniers, A. (1993) 'Tsiganes d'Europe', *Études Tsiganes* (1)1: 6–10.

Péchu, C. (2006) *Droit au logement. Genèse et sociologie d'une mobilisation*. Paris: Dalloz.

Perou (2014) *Considérant qu'il est plausible que de tels événements puissent à nouveau survenir. Sur l'art municipal de détruire un bidonville*. Paris: Post-Editions.

Pierson, P. (1993) 'When effects become causes. Policy feedbacks and Political Change', *World Politics* (45)4: 595–628.

Piven, F. F. and Cloward R. (1979) *Poor People's Movements: Why They Succeed, How They Fail*. New York: Vintage Books.

Piven, F. (2006) *Challenging Authority: How Ordinary People Change America*. Oxford: Rowman & Littlefield Publishers.

Scott, J. (1985) *Weapons of the weak. Everyday Forms of Peasant Resistance*. New Heaven: Yale University Press.

Tricart, J-P. (1977) 'Genèse d'un dispositif d'assistance: les cités de transit', *Revue Française de Sociologie* (18): 601–624.

Vitale, T. (2009) 'Politique des évictions. Une approche pragmatique', In Cantelli F., Roca i Escoda, M., Stavo-Debauge J., Pattaroni L. (eds) *Sensibilités pragmatiques. Enquêter sur l'action publique*. Bruxelles: P.I.E. Peter Lang: 71–92.

Vitale, T. and Claps E. (2010) 'Not always the same old story: spatial segregation and feelings of dislike against Roma and Sinti in large cities and medium-sized towns in Italy'. In Stewart, M. and Rövid, M. (eds) *Multi-Disciplinary Approaches to Romany Studies*. Budapest: CEU Press, 228–253.

Vermeersch, P. (2006) *The Romani movement: minority politics and ethnic mobilization in contemporary Central Europe*. New York and Oxford: Berghahn.

Vitale, T. and Boschetti, L. (2011) 'Les Roms ne sont pas encore prêts à se représenter eux-mêmes!' Asymétries et tensions entre groupes Roms et associations *gadjé* à Milan'. In Berger, M., Cefaï, D. and Gayet-Viaud, C. (eds) *Du civil au politique. Ethnographies du vivre-ensemble*. Bruxelles: Peter Lang: 403–429.

12 Emancipation, integration, or marginality

The Romanian Roma in Bologna and the *Scalo Internazionale Migranti*

Fulvia Antonelli and Mimmo Perrotta

Introduction[1]

It is the end of the summer of 2002, on the banks of the River Reno in Borgo Panigale on the outskirts of Bologna. Since the spring, dozens of huts have sprung up, built and occupied by migrants from Romania, mainly Roma. The migrants who live in them, almost all of whom have no "*permesso di soggiorno*" (stay permit), are seeking black market day work on Bologna's construction sites. On 19 September 2002, these huts are demolished by bulldozers sent by the municipality. This is only the first in a long series of demolitions. About thirty Romanian citizens, who have been driven out, undergo the repatriation process, passing through the city's *Centro di Permanenza Temporanea*, or CPT (Temporary Stay Centre, that is Migrant Detention Center).[2] The rest, who number a few dozen, are first put up in an occupied Social Center, the XM24. Then, on 16 October 2002 (Figure 12.1), with the support of militants from various political groups associated with the Bologna Social Forum (BSF), they occupy a former railway workers' hostel that has been in disuse for several years located in Via Casarini, a few yards from the railway station and the city centre. And so the *Scalo Internazionale Migranti* (International Centre for Migrants, henceforth SIM), or as the Romanian migrants more simply call it, "Casarini", was born.[3]

Over the following months, many other migrants joined the first thirty families of occupiers. At the time of the final eviction of the building in March 2005, they numbered about 300. They were Romanians, Roma for the most part. Albeit in alternating phases, the occupation represented a complex political experience: it was a place of shelter and organization for Romanian migrants in transit or intending to settle in Italy, and it also represented an opportunity for migrants and activists to build common pathways towards making claims for the right to a home, free mobility of persons, and a decent job.

The aim of this chapter is to recount and analyze this event. More than ten years ago, we were part of SIM activists, and it is through our dual role as researchers and political militants that we will attempt to retrace this story from a critical stance. After recounting the story of the occupation, we expand on three aspects in particular. First, we analyze the complex relationships between the occupants of the building and the city's construction sector, with which

La mattina del 16 ottobre 2002 abbiamo occupato l'ex ferrhotel di via Casarini 23, di proprietà di Trenitalia spa e da anni abbandonato, dando vita allo

Scalo Internazionale Migranti

Lo Scalo non è un centro d'accoglienza ma un laboratorio politico in cui praticare forme concrete di resistenza e disobbedie alla legge Bossi-Fini. Lo Scalo vuol essere un primo appoggio, uno spazio che consenta ai migrant di riappropriarsi della **dignità** e dell'**esistenza** che questa legge cancella.

Dal 27 ottobre dà vita a questo luogo un gruppo di **ottanta rumeni**, che il 19 settembre la polizia aveva sgomberato con violenza dalle loro baracche sul Lungoreno, nell'indifferenza colpevole e complice degli enti locali.

All'interno dello Scalo sono stati eseguiti i lavori di pulizia generale e di ristrutturazione necessari per poter ospitare il maggior numero possibile di migranti.

Immediatamente dopo l'occupazione, Trenitalia ha tagliato le forniture di acqua, energia elettrica e gas, cercando di impedire la crescita di questo progetto. Dopo alcuni giorni di pressione nei confronti della proprietà è stato effettuato il riallaccio dell'acqua. **Lo stesso non è accaduto per l'energia elettrica ed il gas**, nonostante l'assemblea degli occupanti abbia comunicato a Trenitalia, a mezzo lettera, l'intenzione di farsi carico delle spese.

La luce all'interno delle stanze e dei corridoi è a tutt'oggi fornita da un gruppo elettrogeno che comporta costi elevati di gestione e non può assolutamente essere utilizzato per alimentare apparecchi per il riscaldamento. Riscaldamento che, con l'avanzare dell'inverno diventa sempre più necessario, considerata la presenza di bambini piccoli e piccolissimi e di donne incinte. La mancanza di adeguato riscaldamento sta già causando problemi di salute ad alcuni degli immigrati che vivono nello Scalo, i quali sono dovuti ricorrere alle cure del pronto soccorso.

Siamo, tra l'altro, a conoscenza del fatto che anche il comune di Bologna e la regione Emilia Romagna hanno richiesto a Trenitalia il riallaccio di energia elettrica e gas per lo Scalo Internazionale Migranti.

Rinnoviamo quindi con forza e pubblicamente la richiesta per le forniture di energia elettrica e gas, nella certezza che la solidarietà e la lotta di tutte le donne e gli uomini liberi farà recedere la proprietà da questo atteggiamento inumano.

È in ogni caso nostra ferma intenzione agire per riconquistare la dignità che ci vuole essere negata.

Scalo Internazionale Migranti - bologna

Figure 12.1 Winter 2002: Flyer produced by the *Scalo Internazionale Migranti*

the occupants had a relationship both in their capacity as building workers – frequently without regular contracts – and their capacity as a cause of "decay", which made them a hindrance to any increase in property value in the area where SIM was located. Secondly, we show that the history of SIM was characterized by moments of collective mobilization and political effervescence, and times when the migrants withdrew into their family networks in search of individual solutions to the problems that arose. Finally, we discuss the ambiguous outcomes of the affair: although the Italian activists and some Romanians aspired to an *emancipation* of migrant workers from the exploitation mechanisms in which they are embedded in Italy, what actually happened, despite certain political victories on the housing front, was a contradictory process of *integration* of the occupants into precisely these mechanisms, and for some of them it meant a return to a situation of *social marginality*.

The history of the "Scalo Internazionale Migranti"

Occupation

The movement of Romanians to Western Europe began immediately after the fall of Nicolae Ceausescu's regime and grew slowly over the course of the 1990s. But it increased significantly after 2002, when Romanians were no longer required to have a visa to enter the Schengen countries. They quickly became the largest foreign nationality in Italy.[4]

Nevertheless, Italian policies on international movement and citizenship, and in particular the Turco-Napolitano Law (Law 40/1998), relegated these migrants to a situation of "illegality": they spent the first part of their residence in Italy without stay permits, in a condition of *deportability* (De Genova 2002). The occupation of SIM took place at the same time as the enactment of a new law on migration, the Bossi-Fini Law (Law 189/2002), which on the one hand strengthened sanctions against undocumented migrants, and on the other launched a "mass regularization" programme, as a result of which approximately 140,000 Romanians succeeded in obtaining a stay permit.

During the months in which it was possible to file a request for regularization, the occupation of the Casarini enabled the migrants to organize, acquire information, and perhaps proceed – with the help of the Italian activists – with reporting employers who did not wish to seek regularization of their *clandestini* workers. A home and a job, therefore, were two of the central issues in the experience of SIM.

The real estate market in Bologna is characterized by high rents and purchase prices.[5] The migrants therefore had the choice between renting run-down housing, living in the municipalities in the province furthest away from Bologna, finding places in reception shelters (which were frequently "monoethnic"), or occupying unused buildings. SIM was not the only occupation carried out by Romanians in Bologna in the years considered; but it was the only one that claimed to be a political act on the public scene and had a self-managed assembly composed of Romanians and Italians.[6]

At this point, we must take note of a dualism in the *Via Casarini* occupation between, on the one hand, the political motivation driving the Italian activists and some of the Romanian occupiers and, on the other, the "simple" need for most of the migrants involved in SIM experience to have a roof over their heads. In the best times of the occupation, these two driving forces found common ground on some concrete claims. At other times, the two processes proceeded in parallel, almost without meeting, sometimes viewing each other with suspicion: the Italian militants feared that the political relationship might become merely welfarist, while many of the occupiers noted with annoyance that activists – both Italians and Romanians – complained about low attendance at meetings. This dualism was mirrored by the names given to the building. For the Italian activists and some of the migrants, it was the SIM, with its self-management assembly: it was a political project that could potentially host migrants from anywhere. For most of the Romanians – not only the occupants, but also friends, relatives, and casual visitors – it was simply Casarini, a home more than a political project.

In March 2005, the Bologna City Council brought SIM experience to an end with an eviction: the families that had regularized their status (approximately 200 people) moved to Villa Salus, a former private clinic on the outskirts of the city (Cristea 2008), paying a rent of 130 euros a month. In the years that followed, many of these family units received aid from the municipality so that they could rent private apartments, and in some cases have access to social housing. In the meantime, the undocumented migrants returned to the shantytowns on the Reno river, created temporary groups of shacks, or occupied smaller buildings in various areas of the city. These people were occasionally evicted, a process which was justified by illegal activities by certain Romanians (illegal labor contractors, mediators and sellers of "services" relating to regularization procedures, false documents or labour contracts, rack-renters of huts, fixers, and sometimes self-appointed community representatives and spokespersons) to the detriment of their weaker compatriots or the more recent arrivals. These areas of abuse and unlawful behaviour were partly the result of exclusively repressive policies on cross-border mobility, legal only in a rhetorical sense; a situation of structural violence (Scheper-Hughes 1993; Bourgois 1995; Farmer 1996) promoted the exploitation of migrants by other migrants, and even SIM was not out of it.

During 2005, a group of occupiers continued to be active under the name of the *Assemblea dello Scalo Internazionale Migranti*. Their purpose was to draw the city's attention to issues concerning black market labour, housing, and stay permits for those without access to Villa Salus. In the autumn of 2005, the shantytowns on the banks of the Reno became a national political and media issue due to internal disputes in the city council led by Sergio Cofferati, mayor from 2004 to 2009, regarding the evictions carried out without the involvement of the social services.

The make-up of the migrant occupiers and the Roma issue

Nearly all the occupiers of Casarini were Roma from Romania, but numerous differences lay behind this apparent uniformity. In the first place, there were issues relative to gender and generation: in Via Casarini, in addition to the many young men who had migrated alone, there were also numerous families. Women, children and adolescents were the most vulnerable groups in the occupation. Among the most interesting of SIM initiatives were the women's meetings, which were held between Romanian and Italian women, and the schooling programmes for children and adolescents.

Also, the Roma of Via Casarini consisted of two separate groups from the point of view of regional and "ethnic" origin. On the one hand, there was a larger group better able to represent itself within the city's political arena; they came from the south west of Romania, from the city of Craiova and the surrounding rural areas. They were, or presented themselves as, "Romanianized Roma" or "urbanized Roma": although they acknowledged themselves as Roma, they claimed that they had been fully 'integrated' into Romanian society. In Romania, they were for the most part labourers in the building sector, the mining industry, and agriculture. As those who have studied the history of the Roma in Central and Eastern Europe have noted (Piasere 2004), this is a "subordinate" integration – but still integration – of Roma into local societies, first as slaves until the nineteenth century, and then as manual workers and citizens of the socialist states in the second half of the twentieth century.

On the other hand, there was a smaller group less able to represent itself as a community. This was made up of Roma from Romanian Moldova (mainly from the regions of Galati and Iasi) whom the first group called "original gypsies"; they were considered less integrated (and less capable of integration) into Romanian society because they had not given up certain Roma practices deemed to be "original", such as begging and nomadism.

In many respects, the Roma from the first group could not be treated any differently from the non-Roma Romanians who had arrived in Italy, because they had the same practices and objectives. Differently, those from the second group found "legitimate" jobs more rarely and took to begging more frequently; in addition, the illiteracy rate and incidence of school drop-out among minors was higher in the second group than in the first group. However, the boundaries between the two groups of occupiers were not completely clear. Most of the "urbanized Roma" used the Roma language as their first language, just as the "original gipsies". There were also illiterate individuals among the "urbanized Roma", as well as some who begged, although this was only a temporary subsistence strategy.

In any case, the occupiers were above all migrants, and they were subject to a mobility control regime that made their lives extremely precarious; they ran the paradoxical risk of returning to a nomadic life: having been obliged to become sedentary by the communist regime, the emigration context now made it necessary for them to travel again to find job and escape the risk of repatriation.

The occupiers constantly sought to present themselves to the city as construction industry workers who were exploited and underpaid, rather than as Roma, because they were aware of the racial stigmatization of their population both in Romania and in Italy. The idea that the inhabitants of Via Casarini were mostly building workers convinced the city of Bologna, at least to a certain extent, and played an important role in determining the outcome of the story, unlike what occurred with other Roma groups who arrived in Bologna. For example, although the Roma who arrived from the former Yugoslavia in the 1990s were received as war refugees, they lived through the "Roma camp" experience, and after an initial reception period, many of them suffered marginalization and denial of their rights. The path followed by the Romanian Roma ten years later was different: they were treated as economic, and not political, migrants. They did not have access to reception centres, and paradoxically, this forced them to manage their housing solutions in the city for themselves, albeit in a conflictual manner.

The eviction of Via Casarini would also once again divide the destinies not only of the two Roma groups, but also of the documented and undocumented migrants. The "urbanized Roma" with stay permits found accommodation in Villa Salus followed by rented apartments, with financial support from the municipality. The "urbanized Roma" who lacked residence permits built new shacks on the banks of the river Reno or in other parts of the city, where they were met with evictions and bulldozers. For their part, many of the "original Roma" returned to their forced nomadic existence in shantytowns and makeshift camps in Bologna and other cities.

The outcomes of this experience illustrate that various factors influenced the pathways followed by SIM Romas: the original group and its *habitus*; the legal status in the country to which they emigrated (documented or undocumented); and the employment situation (in the "legitimate" economy – formal or informal – such as the building sector, or in "illegitimate" activities such as begging). There was also the human and social capital that enabled certain Roma from the first group to start small building companies and therefore become fully integrated into the city's economy.

The "Scalo Migranti" and the antagonist movement in Bologna

In the early stages, there were diverse movements driving the Via Casarini occupation. They consisted of organizations, *collettivi*, and political groups united by a coordinating body, the BSF, which was founded in 2001 to organize the participation of all Bologna's antagonist movements in the demonstration against the G8 in Genoa.

This array of activists was generically termed by the media the "no-global movement", but in fact it was made up of a large number of political groups who decided to form a coalition on a number of issues. They were the historic, environmental, feminist, third-worlder, autonomous and student movements, the alternative trade union movement, the more politicized areas of the traditional trade union movement (the metalworkers' union), radical Catholic groups, the most socially active non-governmental organizations, and certain parties on the radical left.

After the G8 meeting in Genoa, despite the fact that the entire Italian movement was deeply affected by the violent nature of the police action, and although internal disputes within the movement's management led to disagreements between the more pacifist and more aggressive groups, social forums continued to exist locally, although they became increasingly fragmented. In Bologna, some groups raised strong opposition to the Turco-Napolitano Law and the CPTs, and later contested the draft of the Bossi-Fini Law.

The eviction of the Romanian shantytowns suddenly brought the activists into contact with a small group of migrants, and this gave them first-hand experience of their working and living conditions. The *XM24*, a Social Center that had been occupied for less than a year,[7] was sensitive to local issues and open to a plurality of political identities and groups; decided to welcome the migrants on a temporary basis to get to know them better and organize the subsequent phases of their struggle. In the meanwhile, some groups within BSF decided to occupy a building that could become a "*Scalo*", a sort of hub for the political organization of newly arrived migrants.

The occupation's self-management assembly of SIM, which was initially (in 2002) crowded with various political groups, increasingly resembled an extremely combative residents' committee above all involved with urgent debate on issues of survival: getting water and electricity supply reconnected; the right to medical treatment; access to schools for minor children; stay permits for pregnant women and new mothers; collective maintenance and management of the building; and problems of relationships with the neighbourhood. These activities, which were criticized by certain members of the movement as being welfarist and not particularly political, were believed by the activists most closely involved with SIM to be profoundly radical. In all the debates held by SIM, the central issue was not the denunciation of the undocumented status of many of the inhabitants as the product of an unjust law. Rather, it was the right to be "here and now" individuals with visibility in the city, exploited workers who had universal human rights concrete and not theoretical – rights that include the right to give birth in a hospital and to receive the necessary medical treatment, the right to an education, the right to a job performed under safe conditions, and the right to a roof over their heads – which could not be denied simply by defining them as "illegals".

Another aspect that fostered the estrangement of certain political groups accustomed to practicing politics in a "disembodied" manner, with excessively symbolic and abstract language, labels, and objectives, was the fact that, in the self-management assemblies, a group of Romanian migrants and Italian activists asked questions about practising a true, direct democracy and participation, in dispute with the groups to which they had initially belonged. They asked for a form of participation by everyone in collective decisions, without rushing ahead, and without Italians being appointed as the (disloyal) spokespersons for the needs of the Romanian migrants or as cultural and political mediators.

In schools, hospitals and women's clinics, in social services, in their workplaces on building sites, in the streets, in public places during and after demonstrations, and before the privatized water companies, Romanian migrants engaged on a daily basis with the political practice of being *de facto* citizens for all intents and purposes.

SIM assembly participated less frequently in meetings in the city organized by other groups who discussed the issue of migrants as new revolutionary subjects, applying the profiles of the 1960s workerism (*operaismo*), and proposing forms and languages of conflict very distant from migrants' everyday experiences. A number of activists, intellectuals and groups incautiously and uncritically translated some radical theoretical readings of transnational mobility developed in academic contexts (Moulier Boutang 1998; Hardt and Negri 2000; Mezzadra 2006) on the level of political action. Their political practices were closer to a presumed "should be" of migrants as actors capable of subverting global neoliberalism than to a careful analysis of local political contexts and histories of groups, such as that of the Romanians on the banks of the Reno and their complex and mobile identities, aims, challenges and daily practices.

The *Scalo Migranti* and the construction sector

Building workers with nowhere to live

In the early 2000s, there appeared to be an affinity between the Italian building sector and (male) Romanian migrants that is significant for understanding SIM question (Perrotta 2011). From the mid-1990s, and for over ten years thereafter, the construction sector underwent a constant and intense growth that drove the entire national economy and led, *inter alia*, to an enormous use of land on the Italian peninsula. The availability of migrant labour, less costly and more "flexible" than that offered by the native workforce, was one of the factors explaining this growth. When Romanians were still non-EU nationals, Italy was an attractive destination because it was relatively easy to find jobs in the informal sector, with the hope of regularization through one of the periodic "amnesties"; in those years, 40% of the male Romanian migrants employed in Italy, more than 100,000 workers, were in the building sector.

The male migrants who passed through SIM therefore found more or less casual labour in the Bologna-area construction sector, and, more rarely, in logistics and agriculture. Domestic work had a similar value for Romanian women as the building sector had for the men; some of the female residents of SIM were employed in this sector. But for the experience of SIM, this was less significant than construction because most of the occupiers were men, and women more often arrived for family reunification than for work.

The attraction exerted by the informal economy in Bologna as far away as villages in the south west of Romania – through the migrants' social networks – also explains the enormous growth in the number of occupants in Casarini. Between December 2004 and February 2005, when the eviction of the building became a reality, the occupants carried out an internal inquiry. This *inchiesta*, to which we will return later, showed that around 80 out of 300 occupants lacked a stay permit but were employed in the black market, mainly in the construction sector.

Evicted by gentrification

The eviction of SIM and a partial residential solution for its occupants, first in Villa Salus and then in social housing, were due in part to an interest in increasing the value of buildings in the area around SIM, at a time when a number of zones of the city were subject to profound urban planning changes (Collettivo Piano B 2007).

At the time of the occupation, the groups of the movement organized within the BSF were looking for a public-owned building near the city centre to illustrate the contradiction of having numerous government- or council-owned buildings left empty at a time of high demand for residential, social or cultural uses: this was a question of policy, not just of public order.

The Ferrhotel in Via Casarini is owned by the *Ferrovie dello Stato* (the State Railways). In autumn 2003, it had been temporarily loaned to the Bologna City Council, when the municipality initially planned to recognize and "regularize" the occupation. This is a residential area of the *Porto* district, adjacent to the historic city centre and close to important financial and cultural assets (such as the Mambo Museum and the Cineteca). In 2002, the area was the subject of rebuilding and planning projects for residential purposes intended to gentrify it, in part due to commercial growth – above all involving "vintage chic" bars and restaurants in the central part of the district.

The presence of an occupation, especially one started by Romanian Roma migrants in a zone where the locals were not accustomed to living with foreign residents because of the high rents, immediately provoked alarm among the inhabitants. This was also caused by a campaign on the part of certain daily newspapers – including *Il Resto del Carlino*, which is the city's most widely-read paper – which described SIM and its inhabitants in alarmist tones as uncivilized invaders backed by no-global groups intent on destroying the peace and safety of the neighbourhood. This was followed by newspapers articles on a supposed increase in thefts from apartments after the Romanian Roma arrival. The issue that caused the biggest scandal was the alleged aesthetic deterioration that SIM brought to the district; the inhabitants' complaints focused on the clothes dryers on the pavement outside the Ferrhotel, the overflowing rubbish containers near the building, the small children who hung around and played in the street, and the begging by some of the "traditional" Roma at traffic lights. This perceived *degrado* was above all the result of the unexpected visibility in public spaces of individuals who were stigmatized for their origins and traditionally lived far from the city or in its interstices. Xenophobic and extreme right-wing parties such as the *Lega Nord* and *Alternativa Sociale* demonstrated against SIM and excited a residents' committee opposed to the occupation (Figure 12.2).

What was not said in the public arena was that the municipality urgently needed to protect the interests of the building companies investing in the area against the risk of decreasing property values due to the presence of SIM. Nonetheless, the municipality was not unanimous and was split between the "legalitarian" obsession of Mayor Cofferati (nicknamed "the Sheriff" because of

ASSEMBLEA PUBBLICA
giovedì 30 settembre
presso il CENTRO SOCIALE SAFFI Via Ludovico Berti 2/7
ore 20:30

Il Ferrhotel, occupato da immigrati e delinquenti Basta! è un pericolo pubblico per i cittadini residenti in questo quartiere e per tutta la città di Bologna.

SGOMBERIAMOLO!
Difendi la tua Terra!
Difendi la tua Famiglia!
Difendi la tua Casa!

VIENI ANCHE TU !
risolviamo insieme i problemi del Ferrhotel
con una
ALTERNATIVA SOCIALE

info:338/

Figure 12.2 2003: Flyer produced by neofascists against the *Scalo Internazionale Migranti*

a series of repressive council ordinances against every type of social behaviour beyond the norm in terms of public order), and actors such as social services, which were uncomfortable with forms of action unconcerned with low-threshold reception and the prevention of social marginality.

It was within this context that SIM became a concrete proving ground and a symbolic battlefield for the type of city to be defended – was Bologna a city that had traditionally been welcoming and tolerant, the city of qualified factory workers and small artisan shops, of students and alternative cultural creativity, or was it a glossy-vintage shop-window city focused on a radical urban transformation that would turn it into an international tourist destination? The people who opened the debate on Bologna's identity at an historic moment of transition (Boarelli *et al.* 2010) were the most recent arrivals, who lacked the "right to the city" (Lefebvre 1968) and lived on its social margins; thus, they showed the border between inclusion and exclusion of some individuals and groups from the city's political, social and economic centre. Therefore, caught between diverse tensions, in part due to a lack of structures for the initial reception of migrants, the local government gradually sought to transform SIM into an experience that conformed with its own logic, above all when a significant number of its occupants were regularized by an amnesty and therefore acquired the right to exist. The administration avoided tackling the contradictions in migratory policies, and only promised an accommodation to "regular" families, seeking to divide the occupiers, calling the Italian activists "voluntary workers", and inviting them to take part in a series of round table discussions on the management of social schemes to promote women's and children's health and the registration of children at schools.

The Italian activists rejected the logic of top-down "co-management" with the institutions and did not participate in the subsequent actions, which led to the migrants' transfer to Villa Salus, preferring to start again from the shantytowns on the banks of the Reno erected by those who had been excluded from the council's reception project because they lacked stay permits.

In its search for somewhere to put the "regular" families, the local government found itself at the centre of a series of planning and political contradictions with which migrants usually had to deal on their own. The first solution proposed was to ask the *Ferrovie dello Stato* to delay the date of expiry of the loan of the Ferrhotel in exchange for a series of renovations, but the state company rejected this proposal. The administration then proposed programmes for the accommodation of the families of documented migrants at a reduced rent. This housing had to be found with the help of small municipalities across the province of Bologna; the mayors of these small towns unanimously declared their unwillingness to welcome the families. The municipality therefore turned to Villa Salus, a former private clinic on the outskirts of the city at some distance from other homes. The building was purchased for five million euros, the intention being to make it initially a transit centre for Romanian families and then turn it into a specialized medical centre. In 2007, Villa Salus was closed. Currently, it is abandoned, and the local residents frequently file complaints about the presence of new groups of squatters.

Individual solutions and collective struggles

From a political standpoint, the history of SIM unfolded in alternating phases. There were two periods during which the occupation was at its most politicized.

The first was during the initial months of the experience, when the undocumented Romanian Roma had the opportunity to obtain a stay permit through the amnesty offered by the Bossi-Fini Law.

The amnesty enabled an illegalized migrant to secure a stay permit if an employer declared that he or she had worked for that employer for at least three months since 10 June 2002, and if the employer filed an application for the employee's stay permit and made a one-off payment of approximately 800 euros for unpaid back taxes. This mechanism had various consequences for undocumented migrants and therefore for most of the people who passed through the Reno shantytowns, the *XM24* Social Centre, and *Via Casarini.* There were some whose employers decided to file the application but made the worker pay the cost. Others had employers who did not want to seek regularization; in these cases, the occupiers lodged complaints with the labour inspectorate and, following checks, either the employer was "persuaded" to regularize the employee or the employee obtained a stay permit for six months to look for a new job. Some others attempted to obtain a stay permit through a complicit businessperson to whom they had to pay very large amounts of money, some thousands of euros. Finally for those migrants who arrived in Italy after 10 June 2002, amnesty was not an option, and so they remained illegalized and therefore *deportable.* For all of these people, the weekly meetings and daily activities at SIM became important for obtaining information and discussing action strategies.

In the following months, SIM became a hub in Italy for the migrant networks of Roma from the south and east of Romania. The rooms in which a single family had previously lived became increasingly crowded; the hall used for meetings became filled with mattresses; the two large terraces were covered with wooden shacks; and even the underground areas of the building were occupied in the final months of 2004. This was used by the city council as a justification to request an eviction because the building was likely to collapse.

While in the first months the occupiers were united by shared problems and sought to deal with them together, later they all resumed the perception of their principal group of belonging as not the collective of squatters but their family and social networks. According to one of the most convincing studies on Romanian migration into Europe (Potot 2007), those networks are the social groups with which Romanian migrants identify themselves most closely.

The new arrivals, who rapidly came to outnumber the initial occupiers, did not consider Via Casarini as a building that had been conquered, and they were not aware of its history. By sheer force of numbers, they imposed dynamics that risked distorting the political nature of the occupation, causing conflicts and disaffiliation, both among the Italian activists and between the most politicized Romanian occupiers and those more concerned with their families.

A new phase of joint commitment and political effervescence arose in the winter of 2004–2005. The building was overcrowded, and there were growing public demands for eviction of the occupiers. Proposals on how to accommodate the occupants were put forward, and it immediately became clear that they only concerned families that had at least one member with a stay permit. However, the number of occupiers without stay permits who had arrived after the 2002 amnesty had increased enormously.

To fight against the eviction, a public campaign was launched. An inquiry was carried out by Italian activists and migrant occupiers, and in January 2005 a report on work and pay conditions on building sites was delivered to the municipality and presented to the general public.

One of the proposals prompted by the report was that local authorities devise legal mechanisms whereby undocumented migrants who reported their employers could apply for stay permits. To this end, the use of Article 18 of the Migration Law (*Testo Unico sull'immigrazione*, 1998) was requested. This law applied above all to victims of human trafficking and prostitution, and it permitted, for social protection reasons, the issue of stay permits to undocumented foreigners who were subject to violence or severe exploitation and willing to collaborate with the courts.

During this period, around 80 'undocumented' workers from SIM and the banks of the Reno declared their willingness to report employers and illegal recruiters to the courts in exchange for the commencement of regularization procedures. With this campaign, the Roma from SIM were once again fighting for recognition as workers forced to find employment in the underground economy because they were undocumented.

A young Roma who arrived from Craiova at the end of 2003 and who did not have a stay permit recounted his experience of black market labour during a public meeting:

> I worked here in Italy as a painter for a year and half for an employer in the province of Ferrara. He was a good person [...] for a year, and then he asked me to bring some other guys [to work]. I brought some friends and my father along. We worked, and he told us, "When we finish at this site, I'll pay you everything I have to give you". I trusted him, because he had paid me everything for a year. [...] We finished at the site, and he gave us another [job], but no money. And then, when I said, "Listen, boss, you need to pay me something [...] We don't even have breakfast in the mornings, like all Italians do", [...] He replied, "OK, I'll pay you in a month and give you all your money". Then he wrote me some cheques, and I took them to the bank where they told me, "Only one of them is good – the others are fake. If you want, we'll call the police". [...] I was scared to call the police [...] I'm *clandestino*, you can do anything to a *clandestino*. Then I said, "Look, boss, only one of these is good, but the others are fake". "No, that's not true – let's meet in a week". Some time passed, and I told him as a joke, "Boss, if I file a complaint, what could you do?" And he replied, "With a hundred euros I can get rid of you". "OK. Thanks, boss". And I was scared [...] Not only us, the Romanians, but also all the others, we didn't come here to work as slaves and then be threatened. We came to make a life like everyone else, like an Italian. (Vaniel, Male migrant, 25 years old, May 2005)

After some positive initial comments, the campaign was not supported either by the local authorities or by the Bologna trade union organizations, and the project came to a halt.[8]

Emancipation, integration, and marginality: an attempt at an evaluation ten years on

Ten years later, as activists and researchers, we can reflect on the sense and the political outcomes of SIM experience. If we look at the issues relating to work, housing and mobility, we see that, in many ways, the political goal pursued by the occupation was impossible emancipation from an exclusionary society that restricts freedom of movement within the European space for certain men and women. The experience gives us the opportunity to consider the contradictions in both the reception policies implemented by city councils and the battles fought by social movements.

With regard to the first of these aspects, SIM experience showed that the controversial nature of restrictive domestic laws on migration, such as the Bossi-Fini Law with its artificial borders between legality and illegality, had a persistent effect on local territories, and therefore on the actions of city councils, which found themselves having to manage contradictory situations. On the one hand, they had to guarantee the right of categories of persons considered vulnerable to reception and social protection; on the other, they could not concern themselves with those who were "illegal". In addition, they did not want to lose the support of citizens and economic power groups opposed to reception. The emergency rhetoric enabled local councils to manage this contradiction and make it "useful": once they became "legal", the Romanian migrants were classified as "exceptions" with extraordinary *ad hoc* resources being used (in this case, for the purchase of Villa Salus and the rent allowances) and the adoption of policies for this group of migrants alone, not for all migrants in the same situation or, more generally, for people in the city in need of housing.

On their part, the Italian and Romanian activists took advantage of the effect the issue might have in the city if it was presented as an emergency: the objective was to open up a fault line around which to discuss all the aspects of the migrants' condition that were "ordinary" or invisible because they were not treated as matters of public order. Here, the *inchiesta* into black market labour and the denunciation of high rents responded to a need to raise certain crucial social issues in the broad sense, and not just limited to the Romanian Roma.

In fact, the occupiers of Casarini achieved a number of significant political results: the reception centre at Villa Salus and the subsequent placement in "normal" private apartments, thanks to the rent subsidies that the Bologna municipality provided for several years. Nonetheless, these results did not affect the city's real estate market in any way; nor did they change local policies on housing and reception; nor, finally, did they lead to a commitment by the municipality to combat black market labour in the building sector. To a certain extent, the Roma from SIM *integrated* into the city: some of them are now factory workers, agricultural labourers, or even the owners of small building firms. Despite the fact that an association of former occupiers continues to suggest initiatives on Roma culture and conditions, however, the (utopian) process of *emancipation* from exploitation in the workplace and problems in finding housing at accessible rents has been interrupted.

If we were to stop at this point in the story, in 2007, when the families were accompanied from Villa Salus to private housing arrangements and then to social housing, the balance of the experience, with all its contradictions, might be positive. However, certain stories, as recounted by occupiers whom we met years later, give a different picture.

> I met Roxana where she is living now, in the Pilastro area. The Pilastro is a part of the San Donato district of Bologna. Following a series of planning decisions dating back to its creation in the 1960s, the high density of social housing for large families and a certain degree of isolation from the rest of the city made it a place that was first used for the reception of refugees from the wars in the former Yugoslavia and then in Kosovo. Later, it was used as social housing for families of different migratory origins who shared a high risk of social and economic vulnerability. Roxana had two small children, and I remembered she was pregnant with the older one when she was at SIM. Much more time seemed to have passed for her than for me: we were two women of the same age, but we seemed to be from different generations. Roxana told me that her family had been relocated to the Pilastro and paid a subsidized rent. Initially, everything went well because her husband was working, and the municipality paid 50% of the rent. Then, the rent support was scheduled to decrease, and they had to pay for a home at market prices, *albeit* in a market with lower prices like the houses in the Pilastro district. As the rent increased, Roxana's husband lost his job because of the economic crisis. In the meantime, she helped the family by doing casual cleaning work. Her husband was unable to find a job and gradually became extremely depressed. As a result, he no longer left the house, put on weight to the point of extreme obesity, had serious health problems and was almost unable to move. Roxana was desperate because with two children she was unable to ensure the survival of her family, and she was afraid of being evicted. Despite this, she did not want to seek social services for support because she was afraid they would determine she was an inadequate mother, and she worried that her small children would be taken away after what, for her, had been the unbearable experience of the strict checks carried out by social services at Villa Salus. Today, Roxana is a woman who is completely alone and experiencing severe difficulties.
>
> Marian was one of the people with whom the Italian activists had the most conflict during the occupation of SIM. One could never understand what Marian was thinking, and his interventions during the self-management meetings at SIM were always aimed at defending his own interests and those of his group. On various occasions, other occupants told us that Marian was involved in illegal activities outside SIM, and in fact he was one of the few who bought a used car very quickly. Despite the disagreements, Marian always attended

joint discussions and managed to become regularized despite the fact that his work was divided between black market labour in the building industry and some kind of trade that was never clearly identified. I met him in the Pilastro one morning, after almost eight years. I was walking below the large council house buildings in the district and heard my name called from a window. It was Marian, who invited me up to his apartment, as he could not come down. His apartment was very large. He had been lucky; he was assigned a newly built council dwelling with no maintenance issues and low energy costs. He told me that the rent was fifty euros plus various expenses and that he was not working because he had had a "small problem" with the law, but he was innocent. Marian had brought all his children to Italy, all of whom except one were now adults. He wanted to talk to me about his youngest son; he asked me for help, because the social services, on the orders of the Juvenile Court, wanted to send the boy to a children's home because Marian had a criminal indictment and because he himself had been deemed an inadequate father due to his problems with the justice system. His wife, who was a tiny, shy woman, had been run over by a car some time previously, had scars on her face, and was physically very run down. She was also extremely concerned for her son.

Ionut lives in the Pilastro district. He, too, was not working at this time. When I met him at Marian's house, we recalled the times at SIM with some nostalgia because today Ionut is the father of two small babies, and he is alone. His wife, who was also an occupier, died some years ago of a heart attack at a very young age; she died the day after she had been discharged from the Emergency Department where she went because of symptoms of a heart attack, she was not treated for heart attack but something else.

Niro was an adolescent when we were at SIM. He was a minor only because officially he had no relatives to act as his guardian. He belonged to the "original Roma" group and had ties to some adults who appeared to be relatives. We never succeeded in sending him to school; he was a beggar and tried to put money aside for his mother, brothers and sisters in Romania. He never talked of his father. Before the eviction, because he had no chance of being legalized, Niro asked us for a small loan to buy a ticket to return to his mother in Romania. We thought we would never see him again, but a few months later he was back in Italy. SIM had been evacuated, and Niro wandered between squats, still a minor and lacking any means of protection. Over the years, I met Niro all over the city, always focused on his begging. Seeing him four years later was a shock: he had changed greatly, from a blonde boy who was always smiling but extremely shy into a person of indefinable age. He had a deep scar on his forehead and face, and he staggered and did not seem to be completely *compos mentis*. He told me about a very serious car accident in Romania that had almost killed him, and asked me for money for medicines. I later saw him in town with a very young pregnant wife; today he has four children and lives in Romania with her. His life is a constant coming and going between Romania and Italy, on the margins of both societies. His physical condition mirrors his very poor social class, and his children must live with struggles.

Conclusion

Meeting some of the occupants of SIM after many years means being immersed in the *misère du monde* so vividly described by Bourdieu (1993). Today, their marginality is no longer associated with the lack of a stay permit: they have become EU citizens, but have discovered that the economic crisis, employment uncertainties, and discrimination have emptied their rights of citizenship. For many of them, the fragile *subaltern integration* that was proposed to the occupiers of SIM by the local authorities has progressively disintegrated over the years. The municipality, rather than work on the structural conditions of the exclusion of migrants, has tried to avoid the alliance between the specific claims of the Romanian migrants and broader social sectors through the use of limited and costly welfare policies that have provoked conflicts with Italian citizens.

The movement itself, which first entered a phase of decline and then fragmented into numerous factions, did not have a strategic vision that, with the issue of housing for a group of individuals as its starting point, presented the problem of quality of life within a more comprehensive framework. Since then, in other areas of the city, the housing question has exploded and has involved both families of migrant origin who have lived permanently in Italy for years and also Italian precarious and unemployed workers alike. This time, however, it will not be possible to 'resolve' the new occupations using solutions based on an emergency situation, but only by radically rethinking housing policies for everyone.

Notes

1 Although the chapter overall is the result of collaborative writing, the Introduction and Sections 1.1, 1.3, 2.2 and 4 were written by Fulvia Antonelli, white Sections 1.2, 2.1, 3 and the Conclusion were written by Mimmo Perrota.

2 *Centri di Permanenza Temporanea* were created in 1998 by the Turco-Napolitano Law to hold undocumented migrants for a maximum of 60 days in cases where it was not possible to expel them from Italy immediately by accompanying them to the border because it was necessary to carry out investigations into their identity or because they needed travel documents. There was strong opposition to these centres from the time they were created from lawyers, jurists, and judges, who contested the constitutionality of an administrative detention process unconnected with criminal offences, and from movements and associations that denounced the abuses, inhumane treatment, and damage to the human dignity of the people held in these centres. *Centri di Identificazione ed Espulsione* (Centers for Identification and Expulsion) replaced CPT in 2008.

3 The 2008 documentary *La Colonna senza Fine*, directed by Elisa Mereghetti and produced by Ethnos, tells the story of SIM experience.

4 On Romanian migration to Italy, see Cingolani (2009). Other cases of occupied buildings, shantytowns, and camps mostly inhabited by Roma occurred in many Italian cities. In Milan in this same period, a large building was occupied in Via Gioia, near the central railway station. Revelli (1999) has recounted an incident that took place some years previously near Turin; Vitale (2009) has analysed the story of the shantytown in Via Barzaghi in Milan.

5 On city and regional housing policies and reception facilities, see e.g., Bernardotti and Mottura (1999); Grillo and Pratt (2002); and Decimo (2003).

6 SIM is situated within the wider history of occupations by migrants in Bologna: the buildings in *Via Stalingrado* occupied by North Africans in 1990; the occupation of Basilica of *San Petronio* in 1998 by squatters evicted from a building in *Via Rimesse*; the *Lazzaretto* Social Centre occupation by Serbian Roma; and the 2014 occupation of the former Telecom offices in *Via Fioravanti* by 300 migrants.

7 The *XM24* and the other Social Centres of the 2000s in Bologna were the heirs of the young proletariat groups of the 1977 movement and of the Social Centres where a political, social and artistic counterculture had been developed in the 1980s and 1990s (such as the *Isola nel Kantiere*, which operated between 1988 and 1994; see D'Onofrio and Monteventi 2011; for a national view, see Dines 2000; Mudu 2004). In 2002, the main Social Centres in Bologna were part of the BSF.

8 Despite this, we can now reflect on it as a pioneering proposal: it was used by unions, associations and political parties in subsequent years as a proposal in the struggle against illegal recruitment and the heavily exploited labour of migrants.

References

Bernardotti, M. A., Mottura, G. (1999) *Il gioco delle tre case. Immigrazione e politiche abitative a Bologna dal 1990 al 1999.* Torino: L'Harmattan Italia.

Boarelli, M, Lambertini, L. and Perrotta, D. (eds) (2010) *Bologna al bivio. Una città come le altre?* Roma: Ed. dell'Asino.

Bourdieu, P. (ed.) (1993) *La misère du monde.* Paris: Seuil.

Bourgois, P. (1995) *In Search of Respect: Selling Crack in El Barrio.* New York: Cambridge University Press.

Cingolani, P. (2009) *Romeni d'Italia. Migrazioni, vita quotidiana e legami transnazionali.* Bologna: Il Mulino.

Collettivo Piano B (2007) 'La fabbrica e il dragone. Casaralta: inchiesta sociale su una fabbrica e il suo territorio', *Metronomie. Ricerche e studi sul sistema urbano bolognese* 34/35: 43–103.

Cristea, C. (2008) 'Minori rumeni tra percorsi migratori e rappresentazioni. Né soli, né accompagnati a Bologna', unpublished PhD dissertation, Cooperazione internazionale e politiche per lo sviluppo sostenibile: Università di Bologna.

Decimo, F. (2003) 'Gli elementi di un conflitto urbano. Questione abitativa e immigrazione marocchina a Bologna'. In Sciortino, G. and Colombo, A. (eds) *Stranieri in Italia. Un'immigrazione normale.* Bologna: Il Mulino, pp. 71–101.

De Genova, N. (2002) 'Migrant "illegality" and deportability in everyday life', *Annual Review of Anthropology* 31: 419–47.

Dines, N. (2000) 'What are "Social Centres"? A Study of Self-managed Occupations during the 1990s', *Transgressions: A Journal of Urban Exploration* 5: 23–39.

D'Onofrio, S. and Monteventi, V. (2011) *Berretta rossa. Storie di Bologna attraverso i centri sociali.* Pendragon: Bologna.

Farmer, P. (1996) 'On suffering and structural violence: A view from below', *Daedalus* 125(I): 245–260.

Grillo, R. and Pratt, J. (eds) (2002) *The Politics of Recognising Difference: Multiculturalism Italian-Style.* London: Ashgate.

Hardt, M., Negri, A. (2000) *Empire.* London: Harvard University Press.

Lefebvre, H. (1968) *Le droit à la ville.* Paris: Anthropos.

Mezzadra, S. (2006) *Diritto di fuga. Migrazioni, cittadinanza, globalizzazione.* Verona: Ombre Corte.

Moulier Boutang, Y. (1998) *De l'esclavage au salariat. Économie historique du salariat bridé*. Paris: PUF.

Mudu, P. (2004) 'Resisting and challenging neoliberalism: The development of Italian social centers', *Antipode* 36(5): 917–941.

Perrotta, D. (2011) *Vite in cantiere. Migrazione e lavoro dei rumeni in Italia*. Bologna: Il Mulino.

Piasere, L. (2004) *I rom d'Europa. Una storia moderna*. Roma-Bari: Laterza.

Potot, S. (2007) *Vivre à l'Est, travailler à l'Ouest: les routes roumaines de l'Europe*. Paris: L'Harmattan.

Scheper-Hughes, N. (1993) *Death without Weeping: The Violence of Everyday Life in Brazil*. Berkeley: University of California Press.

Revelli, M. (1999) *Fuori luogo: cronache da un campo rom*. Torino: Bollati Boringhieri.

Vitale, T. (2009) 'Politique des évictions. Une approche pragmatique'. In Cantelli, F., Roca i Escoda M., Stavo-Debauge J. and Pattaroni L. (eds), *Sensibilités pragmatiques. Enquêter sur l'action publique* Bruxelles. P.I.E. Peter Lang, pp. 71–92.

13 “We are here to stay”

Reflections on the struggle of the refugee group “Lampedusa in Hamburg” and the Solidarity Campaign, 2013–2015

Simone Beate Borgstede

In spring 2013, something new and inspiring emerged in Hamburg’s social movement landscape through a self-organized group of about 300 refugees with Italian aliens’ passports popularly known as “Lampedusa in Hamburg”.[1] These people were originally migrant workers from Libya and other sub-Saharan countries. In 2011, they became war refugees when the North Atlantic Treaty Organization (NATO) intervened in Libya, civil war escalated, and thousands of African workers were forced to flee via the Mediterranean Sea to the Italian island of Lampedusa since rebels and official military forces blocked their passage to their home countries. With Resolution 1973, the United Nations (UN) Security Council referred to the need to protect migrant workers in Libya (United Nations 2011). However, after the migrants received their right to stay and work in Italy for humanitarian reasons, outside the camps they were confronted with acute homelessness, poverty, and a competitive job market with thousands of refugees who had already arrived in Italy before them. Italian authorities even gave some of them money to leave the country and try their luck in other European countries (Gräfe 2014). Thus, they came to Hamburg, one of the richest European cities in the North of Germany where they found that their Italian work permits were not recognized because under the Dublin regulation refugees must stay in the country where they first disembarked.

In this chapter I want to highlight the history of Lampedusa in Hamburg and simultaneously detail the impact of the mobilization and the transformations this brought to the social, political and cultural atmosphere of living together in Hamburg. The focus will be on the campaign’s development and its ability to raise the question of who belongs and what it means for “a community” when not all people living in it are recognized as having equal rights and access to its resources in a time in which exclusionary programmes like PEGIDA[2] mobilize thousands of people onto the streets. Apart from this, protests against the building of shelters for refugees and attacks on refugees themselves are widespread in Germany (Hebel 2015). Furthermore, I reflect on how experiences of squatting, “right to the city” initiatives and similar solidarity networks, developed over the last decade in certain neighbourhoods like St. Pauli, can play a role in the process of organizing the daily survival of these refugees and can stabilize the conditions for political struggle.

Highlights of the Campaign

While staying in the official winter emergency programme for homeless people initiated by the Senate (i.e., government) in 2012 and 2013, the refugees came into contact with "the Caravan", a group of refugees and other activists, and started to organize themselves to fight for their right to stay (the Caravan 2015). In public, they first emerged at the Kirchentag, the main meeting place of lay people in the Evangelical church (Gerstner 2013). They addressed all political parties at the Bürgerschaft (parliament), and they are known for their attempt to enter the town hall and speak to the mayor of Hamburg about their situation in May 2013. They also tried to organize a camp, which was prohibited by the police. They could only install an information tent near the main railway station (Ludwig 2013), and they confronted the mayor at an election event with their demands (David 2013). They asked for housing, work permission and access to health and education (Laufer and Füllner 2013). However, the Senate answered via the Minister for Social Questions that they had no right to stay, only to a travel allowance back to Italy.

The negotiations between church and government broke down when the church found out that a planned option for housing was restricted to only those who allowed their fingerprints to be taken. Bishop Fehrs accused the government of taking the first step towards deportation over humanitarian assistance on this act (Zand-Vakili 2013). The St. Pauli church gave shelter to 80 people of the Lampedusa group (Gerlach 2013). One of the priests, Sieghard Wilm, became a constant commentator on the migrant situation in the media (Finger 2013; Mikuteit 2013). An increasing number of supporters of all faiths (or no faith) turned up to organize the living situation of the refugees. Neighbours came to support the refugees with food, clothes and other needed things at all hours. People from the previously squatted houses in the vicinity of Hafenstraße set up a huge tent and offered their famous wall for slogans and pictures. The football club FC St. Pauli donated jerseys and tickets for their games to the Lampedusa migrant community. A former bouncer of St. Pauli's nightclub scene provided a guard at night. German lessons were organized. Two outdoor neighbourhood parties with barbeque, music and dance were organized in Park Fiction, the nearby neighbourhood park (Park Fiction 2013a, b). Moreover, the media reported about these events – everyday and everywhere.

This was the beginning of the so-called "African Summer", when unknown waves of solidarity with these refugees swept through the city and beyond – by the Caravan and refugee solidarity groups; students, churches and mosques; union members; football fan clubs; through neighbourhood initiatives of the "right to the city" movement; and by people of all parts of society. "Lampedusa – they are here to stay" became the common slogan in addition to "refugees are welcome here". The refugees received innumerable invitations to tell their stories from schools and universities (Stadtteilschule Stellingen 2013). The Green and the Left Parties asked for a moratorium in parliament (Dey 2013). Italian lawyers were invited to explain why refugees had no economic future in Italy (Düperthal 2013).

A majority of the refugees joined *ver.di*, the union for workers of the public sector; its spokesperson asked publicly for a work permit (Hasenborg 2013). The well-known Thalia Theatre played Elfriede Jelinek's *Die Schutzbefohlenen* together with the refugees, first in the church and then in the theatre. The person responsible for Human Rights in the government of the Federal Republic visited the refugees and spoke of the possibilities of a humanitarian solution (NDR.de 2013a). Not only in Hamburg, but also in the leading newspapers and journals in Germany, reports on the struggle appeared, and the group was even recognized internationally (Brück 2013; Voigts 2013; Wir sind Lampedusa 2013; BBC 2013; Chidi 2013).

By June, Lampedusa in Hamburg had already written an open letter to the Senate and asked for a group solution based on Paragraph 23 of the Residence Law, which would give every member of the group permission to stay for humanitarian reasons. But the Hamburg government refused to accept them as a partner in negotiations of a political solution. When a rickety boat capsized and drowned 270 refugees near Lampedusa in the beginning of October, it became even clearer how much the intentions of the government and "the people" towards the Lampedusa refugees differed. The police terrorized them through controls based on racial profiling especially in the vicinity of the church (Dolzer 2013). Many refugees were arrested, their fingerprints and photos were taken forcibly, and some were driven to accept "*Duldung*" (Behörde für Inneres 2014). *Duldung* is a temporary suspension of deportation for a period needed to apply for a legal residence status. However, *Duldung* does not provide the refugees any assurance of permanent stay (Lampedusa in Hamburg 2013; Appen and Stenzel 2013).

This racist campaign by the authorities led to a solidarity movement with Lampedusa in Hamburg on the streets. People spontaneously observed police actions, or blocked the roads to stop the police moving through the traffic chaos (NDR.de 2013b). Others found more militant forms of boycott. The activists of the campaign *Flora bleibt unverträglich*[3] put an ultimatum to the Senate to stop the racist controls (Flora bleibt 2013). Responsible politicians like the mayor were tracked wherever they went, and the mayor was even pursued at his residence. Graffiti was sprayed and banners hung out of windows. Spontaneous demonstrations took place every day, the biggest with about 8,000 people after one of the home matches of the local football club FC St. Pauli (Agenturen/nd 2013). The Senate's racist policies also led to an international appeal by academics, artists and other well-known citizens (signed by Etienne Balibar, Judith Butler, Catherine and Stuart Hall and Elfriede Jelinek, among others) to stop racial profiling and recognize the refugees' right to stay and work (Lampedusa in Hamburg 2013).

On 2 November 2013, the solidarity mobilization peaked with a huge demonstration of around 15,000 people. Diverse banners showed that the refugees already belonged to the city. Speakers included those of Lampedusa in Hamburg and other refugee groups from all over Germany, such as a representative of the VVN, the organization of the victims of National Socialist (NS)-fascism; Rolf Becker (a well-known actor); the chair of the teachers' union; and the second chair of ver.di. The atmosphere was more like a festival than a demonstration, with music, chanting, and dancing (Appen 2013a).

The refugees renewed their demand for direct talks in October. In response, the Minister of the Interior suggested they were criminals who obscured their identity, and therefore it was a violation of immigration law to support their stay in Germany (Herwatz 2013). The Minister of the Interior offered a Duldung for the refugees in his negotiations with the Bishop, which she welcomed. Duldung does not provide any permanent resident solution; instead, by accepting Duldung, the refugees risk losing their Italian passports and, in the case of failure, risk deportation to their countries of origin because these legal procedures would not consider their statuses as refugees of war in Libya. Usually, refugees from countries like Ghana, which are considered "safe countries", are denied their refugee statuses. Some refugees decided to try Duldung, but most members of the group refused to give up their Italian papers because this provided them with no secure solution (Appen and Kaiser 2013). Meanwhile, 73 refugees accepted Duldung; 11 cases have gone through the procedures of the *Ausländerbehörde* (authorities for immigration), and so far none of the cases have had any success. The refugees instead need to appeal these decisions to the courts. Afterwards, they can appeal to a *Härtefallkommission* (hardship commission); here, the first cases were not going to be debated before October 2015.

Lampedusa in Hamburg again gained public recognition when 111 lawyers of Hamburg supported their demand for a political solution based on Paragraph 23 of the Residence Law (Anwältinnen 2013). Another highlight was the school strike with a demonstration of several thousand pupils for Lampedusa in Hamburg and for different refugee politics in December 2013. To protest and demonstrate in the streets of Hamburg together with all the young people who fought with them for their right to stay clearly signalled to the refugees that this city was a possible place to build a new life.

Winter had come once more, and the Senate continued to ignore the situation of the refugee group. Besides the prefabricated huts provided by some church congregations, many housing projects and centres created a private winter emergency programme to host the refugees. Nevertheless, there was not enough space for all the refugees, so they were either without a shelter or cramped in small spaces under very harsh conditions. But the struggle continued with weekly demonstrations and manifestations (Hellmuth 2013). Members of Lampedusa in Hamburg and Bernadette La Hengst performed the *Universal Schattensenat* where the refugees presented their ideas for an alternative government (Thalia Theater 2013). Moreover, the group intensified their networking with other refugee groups in Germany and in other European countries. Shortly before Christmas 2013, police stopped a huge demonstration in solidarity with "Rote Flora", the inhabitants of the Esso-houses[4] and Lampedusa in Hamburg only 200 meters from the starting point (taz, 2013). Some participants were severely attacked, others started to fight back, and the whole demonstration was stalled. Afterwards, there was an aggressive media campaign against the autonomous left in Hamburg. In addition, the police installed danger zones where they arbitrarily controlled whomever they wanted. In spite of this, every evening hundreds of people met in these zones and ridiculed the police. For example, people carried things like the toilet brush they hid under their jackets as a "weapon" to counter the police if attacked; funnily, the toilet brush became the new sign of protest for this movement (Wickert 2013; Wierich 2014).

In February 2014, two conferences with African and German academics and activists attacked neo-colonial structures in Africa as causes of migration and flight, the NATO-intervention in Libya as against international law and the camp system in Germany. In March, thousands of people participated in the Lampedusa in Hamburg parade (Appen 2014a). A report of the academic services of the *Bundestag* (Federal Parliament) supported the demand of a group solution to stay via Paragraph 23 of the Residence Law against the claims of Hamburg's Senate that this was not a possibility (NDR. de 2014).

Spring also brought new forms of actions. Lampedusa in Hamburg participated in common refugee initiatives like the transnational freedom march (Mauer 2014). On 1 May 2014, a former school was squatted and claimed as the refugee welcome centre in the Karolinenviertel, one neighbourhood of St. Pauli, after a parade initiated by "right to the city movement" activists. When the police threatened to evict the welcome centre the same evening, the squatters which included Lampedusa in Hamburg activists, decided to leave the house with a demonstration through the quarter to avoid criminalization. The place in front of the house, though guarded by police, was used as an assembly point for the following weeks to keep the need for a refugee welcome centre in the public eye (Schäfer 2014; Jung 2014). In the summer, one Lampedusa demonstration and the demonstration of the Squatting Days, an event organized to discuss experiences and perspectives of squatting, ended here (Appen *et al.* 2014).

Lampedusa in Hamburg organized a silent, peaceful gathering in front of the town hall to signify the lack of progress on their situation. Police brutally ended this peaceful action (Brück 2014). The state strategies to control and stall people's demonstrations, protest marches, squatting buildings or occupying squares, however, also clearly signified the difficulty to develop political pressure with symbolic actions when people are without citizenship rights, because even the passive resistance can be criminalized, all of which raised serious problems. Again, eminent people in Hamburg and in other parts of the country published a manifesto, which insisted on a political solution (Hier eine Zukunft 2014).

However, the solidarity movement had lost some of its enthusiasm since the summer of 2013. The refugees were frustrated that they had not been recognized by the Senate as war refugees and that their demands were still not settled after more than a year of struggle. They felt betrayed by the withdrawal of the Nordkirche (Northern church) when most of them decided against taking Duldung, and they had to face the loss of the hyper media interest. More radical supporters criticized Lampedusa in Hamburg for not escalating its campaign and risking criminalization; they said they did not want to provide social care. Moreover, some people and groups in the solidarity movement shifted their energy to organizing solidarity for the Yezidis and other Kurds threatened by ISIS. Lampedusa in Hamburg was no longer the main focus of interest in town.

Nevertheless, the mobilization progressed in diverse forms. The professions project opened up a new discursive space through which refugee and other activists showed that refugees had a life before they were forced to flee. Marily Stroux's photos and interviews highlighted the work experiences of the members of Lampedusa in Hamburg in their respective countries of origin and in Libya. At a press conference in June 2014, Peter Bremme, the spokesperson for refugees without full citizen rights in ver.di, stressed how unreasonable the Hamburg Government acted towards the refugees who had already made Hamburg the centre of their lives. And at a time when workers were needed in different economic sectors, it was imprudent to send away the people who are highly motivated to work. At the press conference, several refugees spoke of their work experiences, their challenges related to seeking jobs in Italy, and the job offers received in Hamburg. Their stories, challenges, experiences and everyday struggles received widespread media coverage. The underlying message was that these ignorant politics went against all common sense (Professions Project 2014). Arguably, why should the same refugees who could work legally in Hamburg if employed by an Italian company be restricted to work legally for any other local job?

Members of Lampedusa in Hamburg in their professional clothes handed out flyers convincing the local people to address the Senate to provide the refugees with work permits if they needed a chef or tailor or auto mechanic (Appen 2014b). In the beginning of September, a camp was set up in Park Fiction by *Schwabinggrad Ballett* (Schwabinggrad Ballett 2014). Besides Lampedusa in Hamburg, refugees from Turin and Berlin participated in three days of discussion and artistic interventions, such as music, films, dancing and communal meals. The notorious *Schanzenfest*, which had been cancelled the previous year because it used to end in clashes between young people and the police, was now a refugee welcome festival (Refugees Welcome 2014). Members of Lampedusa in Hamburg temporarily squatted the foyer of the Kurt-Schumacher-Haus, the residence of the Social Democratic Party, to reiterate that there was still no recognition of their situation and rights to work (Kurt-Schumacher-Haus 2014). In October 2014, Lampedusa in Hamburg celebrated "Emancipation Days" with self-organized theatre and a discussion among academics and refugee activists about the causes and effects of war and flight in Africa (Emancipation Days 2014). The artist group "Baltic Raw" built the "eco-favela" in the area of the art factory Kampnagel and devoted this project to the Lampedusa group as a space for working and living (Baltic Raw 2014; Schipkowski 2014). However, the Hamburg Government suggested to the public that there was no need to act – most of those Lampedusa refugees still in the city had taken Duldung, and others would leave before the winter (Balasko 2014).

In this situation, in the winter of 2014–2015, a new alliance was set up in Hamburg which included Lampedusa in Hamburg and other refugee activists and groups working with refugees – such as the refugee council, the Medi-Büro (medical office) and Café Exil – but also political organizations such as the

Interventionist Left, after the ver.di youth, the students of the teachers' union GEW, activists from the "right to the city" movement (like "St. Pauli selber machen") and student groups (like the council of the University of Applied Sciences) (Never mind the papers 2015). The alliance started to organize a huge demonstration under the slogan "Right to the City – Never Mind the Papers!"

Together with people from housing projects living with Lampedusa refugees, one host family and one of the refugees discussed the situation of the group (many members were still on the streets) in the main local newspaper, the *Hamburger Abendblatt*. They demanded a political solution for the Lampedusa group. Private housing and support could not offer a perspective without the right to work – the refugees had the right to decide where and how they wanted to live (Mikuteit 2015).

On 31 January 2015, 8,000 people turned up to support a political solution for Lampedusa in Hamburg and a real welcoming culture towards refugees which would give everybody in the city equal rights (David 2015). New faces and new slogans emerged like "papers for all or for nobody!" Activists were amazed by the diversity and the positive energy of the demonstration. It showed clearly that the movement was not dead but had started to rejuvenate and re-commence.

Before the elections to the town government on 15 February 2015, Lampedusa in Hamburg invited representatives from the Green Party to a public discussion about their election promises and were assured that they would go for a reliable political solution. Though the Social Democratic Party lost the absolute majority, the Green Party did not succeed with this in the coalition negotiations. The "Right to the City – Never Mind the Papers!" alliance organized several small actions under the slogan "Greens, we are watching you!" (Never Mind the Papers 2015; Grüne Hamburg 2015). Union activists and members of Lampedusa in Hamburg demonstrated for the right to work for all people living in the city independent of their legal status at the "Jobmesse Hamburg", where many companies of the city were looking for employees (Recht auf Stadt – never mind the papers 2015). The struggle of the group and the fate of its members has not only become the interest of common refugees and other refugee groups transnationally, but has also become the issue of academic research (Meret and Corte 2014; Meret and Jørgensen 2014; Borgstede 2015). Moreover, it gained worldwide recognition through an artistic presentation of Tobias Zielony in the German Pavilion at the Biennale 2015, Venice (Ausbruch 2015).

Some reflections and analysis

In this section, I address the question the kinds of volatile force of this struggle held. Mobilization stood on two novel factors which are diverse yet connected:

1. the emergence of a huge, self-organized group of refugees, its diverse direct approaches to the public via actions, discussions in schools and universities, and open letters; and
2. the involvement in the solidarity and public discourse on refugee politics of a broad range of people throughout society.

First of all, through their self-organization, these refugees were recognized as self-determined agents by the public. The focus was no longer on how the refugees were treated or how they were victimized and criminalized, but on their struggle for the right to live and work like any other native or non-native person with citizenship status (see Dadusc Chapter 22, in this volume). Of course, their being victims of war, especially a war in which Germany was involved via NATO membership, was underscored. But it was not the main issue that organized solidarity across many groups. The struggle was broadly a refugee initiative, with local and international support, that publicly articulated the refugees' demands. Their main slogan "We are here to stay" was accepted as both a rightful claim and a description of their current situation. It showed that the European politics of making Europe a fortress were not only outrageous but impractical. Even thousands of refugees drowning in the Mediterranean Sea would not stop the migrants from coming to Europe. The Dublin Regulations that required people to stay in the countries where they had initially arrived proved to be obsolete. Finally the discourse around refugee politics in Hamburg connected with real migrant subjectivities through their voices.

Undoubtedly, this was not the first time there had been people's struggles in solidarity with their fellow comrades who were threatened by deportation. But this was the first refugee group who fundamentally questioned the entire European Union politics around immigration regulations, who had the self-confidence to claim that they could economically contribute to the society of which they wanted to become a part, and who were heard by the public, if not by the governing politicians. The refugees were able to challenge the hegemonic discourse and politics of "the West and the Rest", as Stuart Hall called the Eurocentric system of representing "the other" as inferior (Hall 1992, see Anderson's Foreword, in this volume). Their initiatives and public interventions challenged stereotypical images of refugees, especially Africans, as "passive" or "docile". The autonomous struggles functioned as an "eye-opener" to many in regard to the racist and colonial structures and ideas which still persist in their guises in European societies. People started to organize not only for things needed but also for themselves. They developed new perspectives around organization, alliance building and demonstration tactics. The everyday lives of those involved in the refugee protests were fundamentally changed. The public discourse based on a widespread consensus to provide humanitarian help was politicized by the perspective that humanitarian help, if one took this seriously, could be only secured through political change.

The breakdown of the alliance with the church temporarily blocked a further broadening of this solidarity. When the Bishop welcomed the Senate's offer of Duldung without consulting the refugees, it was necessary for them to clarify that nobody should speak in their name (Kaiser 2013b). Political disputes followed; the struggle was threatened to fall back on a limited grassroots mobilization, lost visibility and thus terrain in its fight for making a new welcome culture in Hamburg's public sphere hegemonic.

But the problem of working with very different agents of political, social and cultural life also provided the insight that instead of homogenization of the struggle a more pragmatic course of "live and let live" was needed. The long fight without gaining physical success and the power obtained for the organization of everyday necessities – housing, food, money for tickets and the journeys to Italy to renew the refugees' Italian passports – seemed to have reduced the power for political activities. Some of the refugees found jobs, though under illegalized conditions and with very low pay (Hennig 2015). However, this everyday struggle also brought new people into contact with the refugees and thus intensified the direct influence of their struggle.

St. Pauli's history of protest culture and the 'Right to the City' Movement

Not all of these activities were based on the neighbourhood of St. Pauli, and private housing was provided for the refugee group in different parts of Hamburg such as Volksdorf and Wilhelmsburg, Hamm and Ottensen. So what made St. Pauli important?

St. Pauli had always been a part of the town where different rules were followed. Living in an area before the town walls, its inhabitants were not recognized as citizens with equal rights (St. Pauli Archive 1990). St. Pauli's population is heterogeneous. Here, what was illegal by law was considered legitimate, and legitimacy was conceived differently over the rest of Hamburg. The squatters of St. Pauli Hafenstraße had already used this in the 1980s (Borgstede 2010). They were able to circumvent a dictate of the Social-Democratic Senate that no house would be left occupied for more than 24 hours, which kept the squatters' movement in Hamburg on a low level. One of their strengths was their heterogeneity. They could never have survived the five years of more or less militant struggle, which at its peak included five days of barricades around the houses, without a left alliance and the solidarity and support of huge parts of their neighbours and many others. Only after squatters, local people and local institutions – like the school, the church and the community centre (GWA) in St. Pauli – became collectively involved in highlighting the existing problems in the public eye related to poverty, housing and the need for a specific public infrastructure, was a housing association set up, bringing 15 years of struggle to a successful end (Borgstede 2013). In a period of contemporary struggles, other squats and housing projects prospered. In the late 1980s, militants and inhabitants of the vicinity of Schanzenviertel, another part of St. Pauli, prevented the construction of a huge music hall by squatting the old theatre building and transforming it into a community centre. Squatters named it the *Rote Flora*. The Rote Flora is still a squatted building where different groups can volunteer to organize events. It stands as a non-commercial, anti-capitalist, anti-racist, and anti-sexist site of performance.

Meanwhile, St. Pauli gained a diverse background of inhabitants. Many people living in the neighbourhood were of Turkish origin. Common experiences of being attacked by neo-fascist football hooligans led to solidarity in defending the area from such attacks. Fans of the local football team FC St. Pauli organized against racist and fascist propaganda in the football arena (Nagel and Pahl 2009: 261). Their fanzines (magazines) supported protests against processes of gentrification. Thus, the township discourse became increasingly politicized. This also included solidarity with Roma people, who had found shelter in the then-occupied houses in Hafenstraße and later at the church Friedenskirche, in the north of St. Pauli (taz 2011).

Even local clubs and pubs articulated their solidarity with the former squatters of Hafenstraße when eviction again threatened (Hellwach 1991). In 1991, a public event in the arena of the FC St. Pauli brought hundreds of activists together to organize a huge music festival under the slogan "Viva St. Pauli – from the FC to Hafenstraße". Celebrated German bands like Die Toten Hosen performed as well as the choir of the sixth form of the neighbourhood school. The popular rock musician Rio Reiser played the piano while "Muttchen Wulf", one of the oldest Hafenstraßen neighbours, sang a self-made song (Borgstede 2013).

These political, social and cultural events also provided personal relationships between people and social groups of diverse backgrounds in St. Pauli and beyond. Self-esteem developed, and the protesters understood that not every decision from the government had to be accepted. Though these struggles were not always on the same level of coverage or efficacy over the decades, the relationships founded in them shaped the experience, the expectations and opportunities for new people who moved into the neighbourhood. When it went public that the last big window towards the Elbe should be closed through a massive housing-construction, this was immediately refused by the neighbouring population. In the years following the Hafenstraßen success, it was well-known that this struggle became the locus for common resistance. Artists joined the activities of the *Hafenrand Verein*, an initiative of inhabitants and "allies" of Hafenstraße, together with neighbours and all social institutions in St. Pauli-Süd (Hafenrand Verein 1992). A *Wunschproduktion* (production of desires) was initiated, and young and old were asked to participate in the formulating of ideas for a park. Through this struggle, a self-planned park emerged at this central site. Gezi Park Fiction, as it is nowadays called, still serves as a public space of play and leisure and a meeting place for common political, social and cultural activities of the people of the area (Wieczorek 2006). Here, one of the principles of the "right to the city" movement works effectively – that is, to collectively take possession of communal spaces for collective purposes, to create and manage new "commons" (Hardt and Negri 2010).

Not all these struggles were as successful. Some old houses were demolished instead of going to housing groups that campaigned for them through frequent occupations and other actions; however, a *Kinderhaus* (a home for children who were not treated well in their families) was opened (Prömmel 2013). St. Pauli lost a central infrastructure, the very much loved and needed *Hafenkrankenhaus* (hospital); the town government did not dare sell the area but acknowledged the struggle against the closure of the hospital, which started with the occupation of its first closed ward to support the development of a *Gesundheitszentrum* (health centre).

The successful taking into possession of the *Gängeviertel* in 2009 is seen by many as the initial action for the "right to the city" movement (Füllner and Templin 2011). The activists opened the historical buildings of the Gängeviertel for a weekend of art exhibitions and music, highlighting the fact that there was not enough space for cultural production in the city, and they stayed when this was welcomed by the public. They did not announce their activities as occupation and, through this tactic, avoided direct eviction by the police. They developed a solidarity culture towards refugees who frequently found space in its rooms. Historically, the backbone for the development of the "right of the city" movement is the successful struggle for Park Fiction, and the foundation was provided by anti-gentrification initiatives in St. Pauli. What did not happen here is a planning process with the consent of the neighbourhood. This is one issue the activists are still fighting for in and around the *Initiative Esso Häuser*. Since these houses could not be saved, the question is what the future of the area will be. The planning process includes an organized participation of the people of St. Pauli, structured on experiences from Park Fiction and supported by the district government (Planbude 2015). The initiative belongs to a wider self-organization, "St. Pauli Selber Machen". At the regularly organized township assemblies, activists from Lampedusa in Hamburg report on their situation. Demonstrations of the group are being supported. The newly launched newspaper of St. Pauli Selber Machen reports on their campaign and some of its activists work in the alliance "right to the city – Never Mind the Papers!" (Zauels 2014).

This rough sketch of the protest culture and history of St. Pauli points to the network that delivered opportunities linking the local people and resources in solidarity with Lampedusa in Hamburg. This does not mean that there were not activities in other parts of Hamburg important for the development of the solidarity movement. On the contrary, the new communication centre koze in Münzviertel, a quarter near the main station, which was squatted, evicted and then legalized via a short-term contract last summer but still is under threat of eviction provided space to people from Lampedusa in Hamburg to open an office for advice-seeking refugees (Schipkowski 2015). This part is an attempt to highlight the links and parallels to other societal struggles from the bottom in that geo-political room the author knows most intimately. Further it points to pivotal factors behind the actual refugees protests and directs to the developments that made the refugees feel part of the larger German community.

Results and perspectives

Beyond the obvious differences there are several aspects which Lampedusa in Hamburg and the right to the city movement have in common (Wilde 2014). Despite depending on political decisions from town and district governments, their campaigns are highly self-supporting. They do not ask for what they need, they know what they need and try to gain it themselves. They both are concerned, though in different ways, with the right to have rights, as Hannah Arendt put it so pertinently (Arendt 1998; Bot 2014). They are about making a living to self-determined conditions. Their campaigns start with the assumption that their existence shows the need for change. They operate on the basis that their issues and practices are legitimate, and there is a broad societal support for these because of "common sense".

Due to the experiences of collaborating festivals and protest marches and co-managing once-squatted places, activists in St. Pauli and beyond cannot separate refugee and migrant issues from other issues of social injustice and exclusion. What has always theoretically been an issue in the discourse of the right to the city movement is that "the city" meant everybody, and rights were here not dependent on money, social status or citizenship status (Lefebvre 1972; Recht auf Stadt – Never Mind the Papers 2015). Now there was a practical goal and urgency; "private" and "public" necessities fell together in the Lampedusa group's slogan, i.e., 'We are here to stay!'

In these two years of struggle, babies were born. One of the group, Francis Kwame, has died (Die bewegende Trauerfeier 2014). Some of the refugees have started speaking German, others have initiated new projects like the Refugee Radio or the No Border Academy (Refugee Radio Network 2015; No Border Academy 2015). Lampedusa in Hamburg shows its solidarity with the new-founded Roma association Romano Jekipe Ano Hamburg in its struggle against deportations (Never Mind the Papers 2015). Friendships have developed. Work is shared. Refugees of the Lampedusa in Hamburg group cook regularly in *Volxküche* (communal kitchen) of the Hafenstraßen houses. Their football club, the FC Lampedusa, trained by three women connected to the FC St. Pauli at a training place provided by the bouncers of the club scene, has joined the leisure league (Fromm 2014).

The support for the 1,200 refugees, who were recently accommodated in an exhibition hall in Karolinenviertel, is overwhelming. At the first neighbourhood assembly, sixteen groups were set up to organize necessities from clothing and medical support to German lessons and common leisure activities; at the second, more than 1,200 people turned up. Some of the questions that came to the fore in the Lampedusa solidarity gain new actuality: How can we join the refugees' struggles and support them directly without letting the state slip out of responsibility? In many ways, the struggle of Lampedusa in Hamburg has already changed not only St. Pauli but also the city of Hamburg.

Instead of accepting this development and giving consent to a solution that will allow the Lampedusa refugees to work and to access all the resources of the city, however, the new SPD/GRÜNE coalition is still only offering an individual Duldung. Therefore, whether some of them accept this in order to achieve better living conditions or oppose it, the fight for their and all refugees' right to stay and work wherever they want in the EU will go on.

Notes

1 The author is part of the solidarity movement and has witnessed, in-person, what she analyzes in this chapter.

2 PEGIDA is a new racist organization, which at the moment mobilizes thousands of people weekly to the streets of Dresden, calling itself 'Patriotische Europäer gegen die Islamisierung des Abendlandes', or Patriotic Europeans against the Islamization of the Occident.

3 The name is a play on words, meaning that the Red Flora, a community centre that was squatted in the struggle against a music hall at this place in the end of the 1980s, exists without a contract and is controversial.

4 The Esso-houses were tower blocks in St. Pauli which became a symbol of the struggle against gentrification through the long fight of their tenants and the neighbourhood.

References

Agenturen/nd (2013) 'Hamburg: Tausende zeigen Solidarität mit Flüchtlingen', *Neues Deutschland* (26 October 2013). Available: http://www.neues-deutschland.de/artikel/837244.hamburg-tausende-zeigen-solidaritaet-mit-fluechtlingen.html.

Anwältinnen und Anwälte fordern Bleiberecht für die Gruppe (2013) Lampedusa in Hamburg: Pressekonferenz am 15. November 2013 in Hamburg'. Available: http://www.schattenblick.de/infopool/buerger/report/brrb0023.html.

Appen, K. v. (2013a) 'Soli-Demo für Flüchtlinge in Hamburg: 15.000 fordern Bleiberecht', *taz* (2 November 2013). Available: http://www.taz.de/!126714/.

Appen, K. v. (2013b) 'Politikunterricht auf der Straße', *taz* (12 December 2013). Available: http://www.taz.de/Schuelerstreik-fuer-Lampedusa/!5052784/.

Appen, K. v. (2014a) 'Humanitäre Lösung ist möglich', *taz* (3 March 2014). Available: http://www.taz.de/1/archiv/digitaz/artikel/?ressort=ha&dig=2014%2F03%2F03%2Fa0152&cHash=6c4ef33b7ff9557d774bea0a4e1b64e1.

Appen, K. v. (2014b) 'Knackpunkt Arbeitserlaubnis', *taz* (7 July 2014). Available: http://www.taz.de/1/archiv/digitaz/artikel/?ressort=na&dig=2014%2F07%2F07%2Fa0017&cHash=e675ca8e193f9fb3e6809abb31364f91.

Appen, K. v. and Kaiser, L. (2013) Hamburger Flüchtlingsstreit: Die Fronten bleiben verhärtet. *taz* (29 October 2013). Available: http://www.taz.de/Hamburger-Fluechtlingsstreit/!126466/.

Appen, K. v., Schipkowski, K. and Speit, A. (2014) 'Squatting Days in Hamburg. Hausbesetzer machen Wirbel', *taz* (31 August 2014). Available: http://www.taz.de/!5034272/.

Appen, K. v. and Stenzel, A. (2013) 'Lampedusa Flüchtlinge in Hamburg: Kein Ende des Konflikts in Sicht', *taz* (27 October 2013). Available: http://www.taz.de/Lampedusa-Fluechtlinge-in-Hamburg/!126321/.

Arendt, H. (1998) *Elemente und Ursprünge totaler Herrschaft. Antisemitismus, Imperialismus, totale Herrschaft*. München: Piper.

Ausbruch aus der Kunstwelt – Migration der Bilder (2015) *Titel, Thesen, Temperamente, ARD* (11 May 2015). Available: http://www.daserste.de/information/wissen-kultur/ttt/sendung/br/deutscher-pavillon-100.html.

Balasko, S. (2014) 'St.-Pauli-Pastor greift Flüchtlingsorganisation an', *Hamburger Abendblatt* (29 December 2014). Available: http://www.abendblatt.de/hamburg/article135819650/St-Pauli-Pastor-greift-Fluechtlingsorganisation-an.html#.

Baltic Raw (2014) *Ecofavela Lampedusa Nord*. http://www.kampnagel.de/de/programmreihe/ecofavela-lampedusa-nord/?programmreihe=11.

BBC (2013) 'Hamburg blames Italy over 300 homeless African refugees', *BBC* (28 May 2013). Available: www.bbc.co.uk/news/world-europe-22694022.

Behörde für Inneres und Sport (2014) Hintergründe: Sachstand zu den Flüchtlingen aus Libyen'. Official Hamburg Website. Available: http://www.hamburg.de/senatsthemen/fluechtlinge/.

Borgstede, S. B. (2015) 'Geschichte ist immer offen: Denken und kämpfen mit Stuart Hall' in: Brunow, D. (ed.) *Stuart Hall. Aktivismus, Pop und Kultur*, Mainz: ventil verlag, 66–73.

Borgstede, S. B. (2013) 'Der Kampf um das Gemeinsame: St. Pauli Hafenstraße' in Baer, W., Dellwo, K-H. (eds): *Wir wollen alles – Hausbesetzungen in Hamburg*. Hamburg: Laika, 105–149.

Borgstede, S. B. (2010) 'Der Kampf um die Herzen und Köpfe der Menschen. St. Pauli Hafenstraße von 1981–1987' in *Das Argument* 289: 849–858.

Bot, M. (2014) 'Etienne Balibar's Reading of Arendt's "Politics of Human Rights"', Hannah Arendt Centre - Bard College (24 February 2014). Available: http://www.hannaharendtcenter.org/?p=12607.

Brück, D. (2013) 'Lampedusa in Hamburg. Nagelprobe für Hamburgs Flüchtlingspolitik', *Stern online* (18 October 2013). Available: http://www.stern.de/politik/deutschland/lampedusa-in-hamburg-nagelprobe-fuer-europas-fluechtlingspolitik-2065513.html.

Brück, D. (2014) 'Lampedusa in Hamburg besetzt Rathausmarkt – Polizei räumt gewaltsam', *Hamburg Mittendrin* (5 June 2014). Available: http://hh-mittendrin.de/2014/06/lampedusa-in-hamburg-besetzt-rathausmarkt/.

Chidi, K. (2013) 'Stranded African-Refugee in Hamburg', *Modern Ghana* (19 June 2013). Available: http://www.modernghana.com/news/469522/1/stranded-african-refugee-in-hamburg.html.

David, I. (2015) 'Demo in der Innenstadt: Es geht um Menschen, nicht um Papiere', *Hamburg Mittendrin* (31 January 2015). Available: http://hh-mittendrin.de/2015/01/demo-in-der-innenstadt-es-geht-um-menschen-nicht-um-papiere/.

David, I. (2013) 'Flüchtlingsproteste: Scholz verteidigt "striktes Vorgehen"', *Hamburg Mittendrin* (1 June 2013). Available: http://hh-mittendrin.de/2013/06/fluchtlingsproteste-scholz-verteidigt-striktes-vorgehen/.

Dey, A. (2013) 'Grüne und Linke: Libyen-Flüchtlinge nicht abschieben', *Hamburger Abendblatt*. Available: http://m.abendblatt.de/hamburg/article117045673/Gruene-und-Linke-Libyen-Fluechtlinge-nicht-abschieben.html#.

Dolzer, M. (2013) 'Afrikaner unerwünscht', *Junge Welt*, Available: https://www.jungewelt.de/loginFailed.php?ref=/2013/10-17/045.php.

Düperthal, G. (2013) 'Überlebenskampf in der Praxis zu hart'. Gespräch mit Loredana Leo, *Junge Welt*. Available: www.kein-mensch-ist-illegal-hh.blogspot.de/2013/08/uberlebenskampf-in-der-praxis-zu-hart.html.

'Emancipation Days' 5.10.2014 http://lampedusa-hamburg.info/de/program-emancipation-days/.

Finger, E. (2013) 'Flüchtlinge in Hamburg: Letzte Zuflucht', *Zeit online* (18 July 2013) Available: http://www.zeit.de/2013/30/fluechtlinge-libyen-st-pauli.

Flora bleibt (2013) Ultimatum an den Hamburger Senat Alle auf die Straße – Schluss mit der rassistischen Machtpolitik. florableibt.blogsport.de (14 October 2013). Available: http://florableibt.blogsport.de/2013/10/14/ultimatum-an-den-hamburger-senat-alle-auf-die-strasse-schluss-mit-der-rassistischen-machtpolitik/.

Fromm, C. (2014) 'Pässe spielen auch ohne Pass', *NDR* (4 October 2014) Available: https://www.ndr.de/sport/Wenn-die-Paesse-zaehlen-nicht-der-Pass,fclampedusa100.html.

Füllner, J. and Templin, D. (2011) 'Stadtplanung von unten. Die "Recht auf Stadt"-Bewegung in Hamburg'. In Holm, A. and Gebhardt, D. (eds) *Initiativen für ein Recht auf Stadt. Theorie und Praxis städtischer Aneignungen*. Hamburg: VSA, 79–104.

Gerlach, R. (2013) *Lampedusa auf St. Pauli*, film (83 min.), http://www.filmfesthamburg.de/de/programm/film/lampedusa-auf-st.-pauli/10472.

Gerstner, D. (2013) 'Lampedusa in Hamburg', *Brot und Rosen*. Available: http://www.brot-und-rosen.de/detail.details+M589652e2787.0.html.

Gräfe, I. (2014) 'Fluchtpunkt Hamburg: Lampedusa in Hamburg. Schicksale und Rechtslage von Flüchtlingen', *Ringvorlesung Friedensbildung Universität Hamburg* (13 November 2014). Available: https://www.znf.uni-hamburg.de/media/documents/peacebuilding/ws2014-2015/friedensbildg-ringvorlesung-2014-15-graefe-2014-11-13.pdf.

Grüne Hamburg (2015) *Wahlprogramm 2015*. Available: https://hamburg.gruene.de/wahl2015/wahlprogramm.

Hafenrand Verein (1992) für selbstbestimmtes Leben und Wohnen in St. Pauli e.V. *1992 einschnitt*. Hamburg.

Hall, S. (1992) 'The West and the Rest: Discourse and Power'. In Gieben, B. and Hall, S. (eds) *The Formations of Modernity: Understanding Modern Societies. An Introduction*. Cambridge: Polity, 184–227.

Hardt, M. and Negri, A. (2010) *Common Wealth. Das Ende des Eigentums*, Frankfurt: Campus.

Hasenborg, V.T. (2013) 'Asyl: Hamburger Libyen-Flüchtlinge werden Gewerkschafter', *Hamburger Abendblatt* (10 July 2013). Available: http://www.abendblatt.de/hamburg/article117886347/Hamburger-Libyen-Fluechtlinge-werden-Gewerkschafter.html?service=mobile.

Hebel, C. (2015) 'Angriffe auf Flüchtlingsheime: Die Botschaft lautet Hass', *Spiegel online* (5 May 2015). Available: http://www.spiegel.de/politik/deutschland/brand-in-fluechtlingsheim-troeglitz-ist-ueberall-a-1027138.html.

Hellmuth, I. (2013) 'Lampedusa-Flüchtlinge: Ein Sack Stroh reicht nicht mehr', *Hamburger Abendblatt* (14 December 2013). Available: http://m.abendblatt.de/meinung/article122923668/Lampedusa-Fluechtlinge-Ein-Sack-Stroh-reicht-nicht-mehr.html.

Hellwach (1991) 'munter und taufrisch. Supportact gastronomischer Betriebe auf dem Kiez für den Erhalt der Hafenstraßenhäuser, deren Kneipen, Cafés und Veranstaltungsräume', leaflet Hamburg.

Hennig, P. (2015) 'Was wurde aus den Lampedusa-Flüchtlingen?' *NDR* (24 March 2015). Available: https://www.ndr.de/nachrichten/hamburg/Was-wurde-aus-den-Lampedusa-Fluechtlingen,lampedusa308.html.

Herwatz, C. (2013) 'Polizei kontrolliert gezielt Schwarze. Flüchtlingsgruppe spaltet Hamburg', *n-tv* (19 October 2013). Available: http://www.n-tv.de/politik/Polizei-kontrolliert-gezielt-Schwarze-article11572611.html.

Hier eine Zukunft (2013) Manifest für Lampedusa in Hamburg. Available: http://hafenvokue.blogsport.de/2014/06/18/manifest-fuer-lampedusa/.

Kaiser, L. (2013b) 'Politische Zwietracht: Lampedusa-Gruppe greift Kirche an', *taz* (19 November 2013). Available: http://www.taz.de/!127812/.

Kinderhaus am Pinnasberg (2015) Über uns – Leitbilder und Ziele. Available: http://www.kinderhausampinnasberg.de/seiten/ueberuns.htm.

Kurt-Schumacher-Haus (2014) 'Lampedusa-Flüchtlinge belagern SPD-Parteizentrale', *Hamburger Morgenpost* (19 September 2014). Available: http://www.mopo.de/polizei/kurt-schumacher-haus-lampedusa-fluechtlinge-belagern-die-spd-parteizentrale,7730198,28460682.html.

Lampedusa in Hamburg (2013) *Offener Brief an den Senat* (20 June and 16 October 2013). Available: http://www.fluechtlingsrat-hamburg.de/.

Laufer, B., Füllner, J. (2013) 'Afrikanische Zuwanderer: Ohne Ende auf der Flucht', *Hinz & Kunzt* (20 May 2013). Available: http://www.hinzundkunzt.de/ohne-ende-auf-der-flucht/.

Lefebvre, H. (1972) *Die Revolution der Städte*, München: Anton Hain.

Ludwig, K. (2013) 'Flüchtlinge protestieren: Ein Camp auf der Verkehrsinsel', *taz* (21 May 2013) Available: http://taz.de/Fluechtlinge-protestieren/!116641/.

Mauer, M. (2014) 'Bayern, Berlin, Brüssel: Der lange Marsch der Flüchtlinge', *Blätter für deutsche und internationale Politik* (7)2014: 13–16. Available: https://www.blaetter.de/archiv/jahrgaenge/2014/juli/bayern-berlin-bruessel-der-lange-marsch-der-fluechtlinge.

Meret, S. and Jørgensen, M. B. (2014) 'From Lampedusa to Hamburg: time to open the gates', *Roarmag.org* (10 July 2014). Available: http://roarmag.org/2014/07/lampedusa-hamburg-europe-refugees/.

Meret, S. and Corte, E. D. (2014) 'Between exit and voice: refugees' stories from Lampedusa to Hamburg', *Open Democracy* (22 January 2014). Available: https://www.opendemocracy.net/can-europe-make-it/susi-meret-elisabetta-della-corte/between-exit-and-voice-refugees-stories-from-la.

Mikuteit, H. L. (2013) 'Menschen 2013: Pastor Wilm ist die Stimme der Männer aus Lampedusa', *Hamburger Abendblatt* (31 December 2013). Available: http://www.abendblatt.de/hamburg/article123426552/Pastor-Wilm-ist-die-Stimme-der-Maenner-aus-Lampedusa.html#.

Mikuteit, H. L. (2015) 'Saseler nehmen Lampedusa-Flüchtlinge auf', *Hamburger Abendblatt* (30 January 2015). Available: http://www.abendblatt.de/hamburg/article136929749/Saseler-nehmen-Lampedusa-Fluechtling-auf.html#.

Nagel, C. and Pahl, M. (2009) *FC St. Pauli. Das Buch. Der Verein und sein Viertel*. Hamburg: Hoffman und Campe.

NDR.de (2013a) 'Streit über Zukunft afrikanischer Flüchtlinge'. Available: http://www.ndr.de/nachrichten/hamburg/fluechtlinge433.html.

NDR.de (2013b) Wochenende im Zeichen von Lampedusa. 28 October 2013. Available: http://www.ndr.de/nachrichten/hamburg/fluechtlinge519.html.

NDR.de (2014) 'Flüchtlinge: Gutachten widerspricht Senat'. Available: https://www.ndr.de/nachrichten/hamburg/Fluechtlinge-Gutachten-widerspricht-Senat,lampedusa199.html.

Never Mind the Papers (2015) *Never Mind the Papers website*. Available: http://nevermindthepapers.noblogs.org/.

Never Mind the Papers (2015) *Never Mind the Papers on facebook*. Available: https://www.facebook.com/pages/Recht-auf-Stadt-never-mind-the-papers/373524372808420.

No Border Academy (2015) *No Border Academy on facebook*. Available: https://www.facebook.com/NoBorderAcademy.

Park Fiction (2013a) 'Lass' uns zusammen was essen', (13 June 2013) Avaiulable: http://park-fiction.net/lass-uns-zusammen-was-essen-freitag-14-juni-ab-17-uhr/.

Park Fiction (2013b) 'Lass' uns mal wieder zusammen was essen 9.8.2013', (9 August 2013). Available: http://park-fiction.net/lass-uns-mal-wieder-zusammen-was-essen/.

Planbude (2015) *Planbude website*. Available: http://planbude.de/.

Prömmel, E., (2013) 'Hausbesetzungen in Hamburg. Eine Einleitung'. In Baer W., Dellwo K-H. (eds) *Wir wollen alles – Hausbesetzungen in Hamburg*. Hamburg: Laika, 23–31.

Professions Project (2014) *Lampedusa in Hamburg Professions blog*. Available: http://lampedusa-in-hamburg-professions.blogspot.de/2014_06_01_archive.html.

Recht auf Stadt - never mind the papers (2015) *Recht auf Stadt - never mind the papers website*. Available: http://www.rechtaufstadt.net/recht-auf-stadt/aufruf-und-sofortprogramm-des-netzwerks-recht-auf-stadt-never-mind-papers.

Refugee Radio Network (2015) *Refugee Radio Network on facebook*. Available: https://www.facebook.com/refugeeradionetwork.

Refugee Schulstreik Hamburg (2015) *Refugee Schulstreik Hamburg on facebook*. Available: https://de-de.facebook.com/schulstreik.

Refugees Welcome (2014) Schanzen Fest 28 09 2014. Available: http://rwsf.blogsport.de/2014/08/26/aufruf/.

Rote Flora (2015) *Rote Flora website*. Available: http://m.roteflora.de/html/geschichte.htm.

Schäfer, D. (2014) 'Flüchtlinge fordern: Das ist unser Haus!', *St. Pauli Blog*. Available: http://st.pauli-news.de/fluechtlinge-fordern-das-ist-unser-haus/.

Schipkowski, K. (2014) 'Ein Passivhaus für Flüchtlinge', *taz* (6 June 2014). Available: http://www.taz.de/1/archiv/digitaz/artikel/?ressort=ha&dig=2014%2F12%2F06%2Fa0335&cHash=6ffda49c50bb70f9b62eafc5ee299893.

Schipkowski, K. (2015) 'Entspannung im Münzviertelstreit. Die Friedenspfeife geraucht', *taz* (10 June 2015). Available: http://www.taz.de/!5203495/.

Schwabinggrad Ballett (2014) *Schwabinggrad Ballett on facebook*. Available: https://de-de.facebook.com/pages/SchwabinggradBallett/270858082609.

Stadtteilschule Stellingen (2013) 'Lampedusa in Hamburg' (18 March 2013). Available: http://stadtteilschule-stellingen.hamburg.de/index.php/article/detail/6925.

St. Pauli Archive V. (ed) (1990) *'Im Schatten des großen Geldes'. Wohnen auf St. Pauli. Geschichte St. Paulis vom Anfang bis zur Gegenwart*, Hamburg.

St. Pauli Selber Machen (2015) *St. Pauli Selber Machen website* Available: http://www.st-pauli-selber-machen.de/wordpress/index/.

taz (2011) 'Zeitleiste Hamburg', *taz* (1 October 2011). Available: http://www.taz.de/1/archiv/digitaz/artikel/?ressort=hi&dig=2011%2F10%2F01%2Fa0094&cHash=bba9fe0f15cb03d46e2efce6a4caa01c.

taz (2013) 'Polizei setzt Wasserwerfer ein', *taz* (21 December 2013). Available: http://www.taz.de/!129827/.

Thalia Theater (2013) *L'Universal Schattensenat*. Available: http://www.thalia-theater.de/de/spielplan/repertoire/launiversal-schattensenat/.

The Caravan (2015) *The Caravan website*. Available: http://thecaravan.org/.

United Nations (2011) *Resolution 1973 (2011)*. Available: http://www.nato.int/nato_static/assets/pdf/pdf_2011_03/20110927_110311-UNSCR-1973.pdf.

Voigts, H. (2013) 'Flüchtlinge Hamburg Lampedusa: Verpasste Chance', *Frankfurter Rundschau* (18 October 2013). Available: http://www.fr-online.de/meinung/fluechtlinge-hamburg-lampedusa-verpasste-chance,1472602,24665684.html.

Wickert, C. (2013) 'Pressemitteilung der kritischen PolizistInnen: "Rote Flora, Politik und Polizei" zu gewaltsamen Auseinandersetzungen in Hamburg seit dem 21.12.2013', *Criminologia* (21 December 2013). Available: http://criminologia.de/2014/01/pressemitteilung-der-kritischen-polizistinnen-rote-flora-politik-und-polizei-zu-gewaltsamen-auseinandersetzungen-hamburg-seit-dem-21-12-2013/.

Wieczorek, W. (2006) 'Park Fiction', Hamburg'. *urbanmatters* (14 October 2006) Available: http://urban-matters.org/wp/wp-content/uploads/parkfiction-ww-14.10.20061.pdf.

Wierich, A. (2014) 'Die Wirkungsmacht der Klobürste', *der Freitag* (11 January 2014). Available: https://www.freitag.de/autoren/andrea-wierich/die-wirkungsmacht-der-klobuerste.

Wilde, F. (2014) 'We're all staying', *JACOBIN* (2 July 2014). Available: https://www.jacobinmag.com/2014/02/were-all-staying/.

Wir sind Lampedusa (2013) 'Von Jan Delay bis Fatih Akin - Solidarität für Flüchtlinge'. *Stern online* (23 November 2013). Available: http://www.stern.de/kultur/musik/plakataktion-wir-sind-lampedusa-von-jan-delay-bis-fatih-akin-solidaritaet-fuer-fluechtlinge-2073005.html.

Wood, G. (2013) 'Hamburger Rathaus: Demo von libyschen Flüchtlingen vor dem Bürgermeisterbüro', *Hamburger Abendblatt* (23 May 2013). Available: http://m.abendblatt.de/hamburg/article116433265/Demo-von-libyschen-Fluechtlingen-vor-dem-Buergermeisterbuero.html.

Zand-Vakili, A. (2013) 'Libyer: Flüchtlinge entzweien Kirche und Hamburger Senat', *Hamburger Abendblatt*. (3 June 2013). Available: http://m.abendblatt.de/hamburg/article116751610/Fluechtlinge-entzweien-Kirche-und-Hamburger-Senat.html.

Zauels, F. (2015) 'St. Pauli: Ein Stadtteil wehrt sich', *Hamburg Mittendrin* (9 February 2014). Available: http://hh-mittendrin.de/2014/02/st-pauli-ein-stadtteil-wehrt-sich/.

Part IV

The difficulties of defining and arranging diversity among heterogeneous subjects

14 Sacred squatting

Seeking sanctuary in religious spaces

Serin D. Houston

Contextualizing Sacred Squatting

In mid-October 2014, Arturo Armando Hernández Garcia sought sanctuary in a basement room at the First Unitarian Church of Denver, Colorado. Facing deportation orders and impending separation from his wife and two daughters, the youngest of whom is a United States (US) citizen, Hernández Garcia found temporary refuge in a sacred space, a setting where Immigration and Customs Enforcement (ICE) would most likely not arrest him because of its mandate to avoid raids in schools and places of worship. This "sacred squatting," as I call it, reflects a last resort option undertaken to stay, at least a bit longer, with family and community in the US. This action also signaled the first time an illegalized migrant in Colorado had sought sanctuary in a religious space and connected Hernández Garcia with at least nine other illegalized migrants who have secured sanctuary in Chicago, Tucson, Tempe, Phoenix, Denver, Philadelphia, and Portland-Oregon since June 2014 (Dinan 2014; Griego 2014; Lo 2015; Sanctuary 2014, 2015). Several members of this group have successfully achieved a stay of removal (usually for six months or a year) in large part due to the political pressure exerted during their time in sanctuary. After nine months of living in sanctuary, Hernández Garcia learned that he was no longer a priority for ICE so he returned to his home and family (McGhee 2015). He has not yet received legal status in the US so the possibility of separation from his family through deportation still remains a distinct possibility.

The offering of sanctuary and act of living in sanctuary reflect the goals and work of the New Sanctuary Movement (NSM), a national faith-based immigrant rights social movement. Chapters of the NSM exist in many large US cities, such as New York City, Boston, Philadelphia, Los Angeles, Chicago, and Milwaukee. A few state-based organizations, such as the Interfaith Movement for Immigrant Justice in Oregon, are also members of the NSM. Alongside vocally calling for changes to federal immigration policies and enforcement practices, the NSM supports sanctuary and advocates for an extension of legal status to all members of mixed-status families, particularly those facing impending separation due to deportation orders (Houston and Morse 2015). Bader (2014) suggests that over 120 churches and religious communities around the US have now committed to the ideals and practices of sacred squatting for illegalized migrants in mixed-status families. The current strategy of securing some safety, at least temporarily, in places of worship has older roots in the earlier 2007 iteration of the NSM, and in the 1980s and early 1990s work of the Sanctuary Movement for Central Americans (Dyrness and Irazábal 2007). In all of these situations, sanctuary activists, called by their faiths to "welcome the stranger" (Sanctuary 2014, 2015), defend the provision of sanctuary by stating that faith communities are not harboring or concealing illegalized migrants, which would be legally punishable, because all of the migrants who have lived in sanctuary are visible in popular press. Furthermore, activists note that they are adhering to higher moral codes and upholding the tenets of social justice (Griego 2014; Sanctuary 2014, 2015).

The provision of sanctuary is not a new phenomenon. In Judeo-Christian tradition, stories exist about the cities of refuge that Moses built to provide refugees fleeing quid pro quid vengeance a place of asylum (Pirie 1990). The "right to sanctuary" principle, defined as the right to seek refuge in sacred places for a short period of time, further locates contemporary sanctuary within this religious lineage. The examples that I examine indicate how these practices are currently mobilized for illegalized migrants and how sacred squatting simultaneously energizes religious spaces as pro-immigrant terrains of struggle and limits the discursive space available for articulating the complexities of immigration policies and migrant experiences. I address examples from the US context, but sacred squatting is also evident in Canada and the United Kingdom.

Spaces of sanctuary

Deciding to offer sanctuary to illegalized migrants is not something that faith communities rush into (Griego 2014; Ollstein 2014; Asmar 2015a). The decision emerges through a sustained decision making process undertaken by the community (Sanctuary 2014, 2015). The recasting of religious space as the living place for an individual or a family for an indefinite period of time requires not only a physical retrofit of part of a building, but also physical, emotional, and political support for the sanctuary seeker(s). Through offering sanctuary, places of worship transition from sites for routine gatherings to express faith to overt spaces of political contestation and engagement. Massey (2005: 99) notes that "space is the dimension which poses the question of the social, and thus of the political" because it is the "sphere of relations, negotiations, practices of engagement, power in all its forms." This description aptly characterizes sacred squatting as sanctuary families participate in and are supported by the church community, interact with lawyers and activists, and highlight the trauma that separation through deportation would cause (see also Borgstede Chapter 13, in this volume). Simultaneously, places of worship become more visibly political spaces as awareness raising events occur, activists discuss strategies, and increased public attention arises as reporters provide updates on sacred squatting (Asmar 2015b). These expressions of political action are both intensely localized in the narrative and experiences of the individual sacred squatter, and also national in the form of the sanctuary network. Illegalized migrants living in sanctuary have conference calls to share news and provide support and solidarity in the often lonely and highly scrutinized experience of sacred squatting (Asmar 2015b; Lo 2015).

While the provision of sanctuary for Hernández Garcia was a first for the Denver church, other places of worship have much longer histories with sanctuary. In August 2014, Rosa Robles Loreto, for instance, sought sanctuary in Southside Presbyterian Church in Tucson, Arizona. She was the second illegalized migrant to seek sanctuary in this church in the course of a year. The first, Daniel Neyoy Ruiz, received a stay of removal after a month of sacred squatting (Lo 2015). For Robles Loreto, her journey of sanctuary was much longer. In November 2015, after 15 months of living in sanctuary, Robles Loreto learned that her status as a low priority for ICE would provide some immunity against a future deportation. Thus, she left the Southside Presbyterian Church and returned to her family (Prendergast 2015). This church had a history of providing sanctuary for migrants during the Sanctuary Movement for Central Americans, so it had a physical space prepared for such activities. It also had familiarity with this kind of activism, so positioning the church as a site for political struggle was something quickly supported by the congregation.

Robles Loreto received deportation orders in July 2014 after she accidentally turned into a construction zone. Although she did not receive a ticket for this minor traffic infraction, the local authorities decided to check her immigration status and summarily reported her to ICE. Rather than be separated from her family and returned to a birth country that she knows only in distant memory, Robles Loreto sought sanctuary in the Southside Presbyterian Church (Lo 2015). Robles Loreto's children and husband visited often, and her boys lived with her in the church during their school vacations (Taracena 2015). Reverend Alison Harrington said that Robles Loreto could stay at the church until she could safely return to her daily life (Dinan 2014). The intersecting of space and solidarity is crucial in such statements and in the daily efforts of people supporting and living out sacred squatting.

For some people who have sought sanctuary, this act of political resistance has produced a stay of removal or a stated acknowledgement of being a low priority for deportation. For others, such as Luis Lopez Acabal, the outcome is hazier. Acabal returned home in December 2014 after more than 100 days of living in sanctuary at the University Presbyterian Church of Tempe, Arizona. Residing in two windowless rooms in a church basement, away from family and in an unfamiliar community, proved taxing. The stalling on the part of ICE to close Acabal's case also raised questions about how long one should squat in sanctuary. In other words, when does this strategy of occupying space to raise awareness, advocate for political change, and ensure some measure of safety no longer prove viable? For Acabal, even though he still cannot get a work permit and has no guarantee of his legal status, 100 days was long enough (Ollstein 2014). The realities of confinement, the separation from family and community, and the hyper visibility that sacred squatting engenders posed challenges to this form of activism.

Scripting sanctuary

Other aspects of sacred squatting demonstrate additional constraints. The scripting of those participating in sacred squatting, for instance, suggests a certain disciplining and flattening of life experiences. Furthermore, sacred squatting is not available to all illegalized migrants. A NSM tool kit from 2007, for instance, outlines the criteria for "appropriate" sanctuary seekers. Principally, "families with citizen children that have a good work record and a history of contributing to their community" are most welcomed (NSM Tool Kit 2007: 2). The tool kit continues, "It is also helpful when families can speak from the heart about their love for their children, their neighborhood, their community and this country, as well as their religious faith" (NSM Tool Kit 2007: 2). The tool kit further delineates the need for sanctuary families to have clean legal records or distant and minor infractions (NSM Tool Kit 2007: 2–3). Individuals and families holding these characteristics were described as more understandable and relatable to non-immigrants. This is crucial as the sanctuary seeker relies upon a broader faith community to maintain daily life. Sanctuary seekers also need to be prepared for media attention, so the tool kit recommends supporting those who are willing and able to undergo such public scrutiny and visibility (NSM Tool Kit 2007: 3; see also Yukich 2013, and Houston and Morse 2015).

The tool kit's articulation of how to craft and sustain sacred squatting is illustrative for several reasons. First off, rather than catalyzing religious spaces as sites of politics and social justice, as stressed in public narratives from faith communities, this internal document highlights the need to recruit people into sanctuary who are seen as comprehensible for and relatable to the broader faith community. While not explicitly stated, what this focus often translates into is a heteronormative emphasis on an illegalized migrant in an opposite sex relationship with a child or children, a "stable" life in the US, and a pending deportation situation that is not clouded by "significant" legal infractions (Houston and Morse 2015). Such a sanctuary seeker becomes visible as a proto-neighbor or community member and therefore deemed worthy of sacred squatting. Since the advocacy for sanctuary seekers tends to fall along these particular lines, not all sanctuary seekers are invited to take refuge in a place of worship. Moreover, the bounding of who can access sanctuary encourages a scripting and framing of life to fit within the dominant overarching narrative of sanctuary seekers, which flattens the dynamism of life stories. For instance, stories about Hernández Garcia emphasized that he tried to do everything correctly.

> He got a driver's license the last year it was legal for undocumented immigrants to do so before the law changed again in 2014, and renewed it regularly. He paid his taxes and never used false documents or a stolen Social Security number. He learned English and eventually started his own flooring business, negotiating to win jobs installing tile and ceramic floors in big apartment complexes (Asmar 2015a).

A 2010 felony charge of menacing with a weapon from a workplace altercation is described moment by moment in articles about Hernández Garcia with the subsequent jury verdict of not guilty brought to the forefront (Asmar 2015a). The ways in which the distance between citizen and noncitizen is bridged through depicting Hernández Garcia as an average person – a heterosexual, married, father and business owner – play a significant role in the representations of sanctuary seekers.

The vitality and complexity of people's lives recede in the face of intense scrutiny about who should be supported in sacred squatting. There is little room for the complexities of immigration policies and life experiences to emerge in these representations. Furthermore, only certain illegalized migrants fit within the expressed focus of sanctuary seekers so this form of political resistance is not available to all. Finally, it is unclear how much autonomy migrants have in the experiences of sanctuary. For the most part, it seems that the faith communities decide the terms of the sanctuary and the sacred squatter accepts these conditions in exchange for material, spiritual, and legal support. In an effort to catalyze religious spaces in a more sustained manner and produce wider solidarity, I think sacred squatting needs to evolve to generate even greater collaborative resistance to the racist immigration policies of the US. There is much more work to be done to counter the criminalization and deportation of illegalized migrants.

References

Asmar, M. (2015a) 'A Denver church joins a nationwide movement to provide sanctuary to undocumented immigrants', *Westword*. Available: westword.com/news/a-denver-church-joins-a-nationwide-movement-to-provide-sanctuary-to-undocumented-immigrants-6578160.

Asmar, M. (2015b) 'Immigrants in sanctuary: Man in Denver one of three not granted relief', *Westword*. Available: http://www.westword.com/news/immigrants-in-sanctuary-man-in-denver-one-of-three-not-granted-relief-6578719.

Bader, E. J. (2014) 'New Sanctuary Movement seeks to protect undocumented immigrants', *Truthout.org*. Available: truth-out.org/news/item/27672-new-sanctuary-movement-seeks-to-protect-undocumented-immigrants#.

Dinan, S. (2014) 'Church network offers sanctuary to illegal immigrants to avoid deportation', *The Washington Times*. Available: http://www.washingtontimes.com/news/2014/sep/24/sanctuary-2014-church-network-helping-illegal-immi/?page=1.

Dyrness, G. and Irazábal, C. (2007) 'A sanctuary for immigrants', *LA Times*. Available: http://www.latimes.com/opinion/la-op-dyrness2sep02-story.html.

Griego, T. (2014) 'He lived in the US for 15 years. Only church sanctuary saved him from deportation', *The Washington Post*. Available: http://www.washingtonpost.com/news/storyline/wp/2014/11/06/5606/.

Houston, S. and Morse, C. (2015) 'Framing immigrant sanctuaries in the US: Encounters with the ordinary and extraordinary', *Unpublished manuscript*. Available upon request.

Lo, P. (2015) 'Inside the New Sanctuary Movement that's protecting immigrants from ICE: Can a network of churches fight deportations?' *The Nation*. Available: http://www.thenation.com/article/206545/inside-new-sanctuary-movement-thats-protecting-immigrants-ice#.

Massey, D. (2005) *For Space*. London: Sage Publications.

McGhee, T. (2015) 'In metro Denver illegally, Arturo Garcia worked and lived in shadows', *The Denver Post*. Available: http://www.denverpost.com/news/ci_28766138/metro-denver-illegally-arturo-garcia-worked-and-lived.

New Sanctuary Movement Tool Kit (2007) 'New Sanctuary Movement tool kit. *Interfaith Movement for Immigrant Justice*', Available: imirj.org/new-sanctuary-movement-toolkit-for-congregations/.

Ollstein, A. (2014) 'The harrowing story of how an undocumented man found sanctuary in an Arizona church', *Thinkprogress.org*. Available: http://thinkprogress.org/immigration/2014/09/29/3572783/sanctuary-luis-arizona/.

Pirie, S. H. (1990) 'The origins of a political trial: The sanctuary movement and political justice', *Yale Journal of Law and the Humanities* 2(2): 381–416.

Prendergast, C. (2015) 'Rosa Robles Loreto leaves sanctuary of Tucson church'. *Arizona Daily Star*. Available: http://tucson.com/news/local/rosa-robles-loreto-leaves-sanctuary-of-tucson-church/article_2e0ae3c4-8891-11e5-b998-173e7444fbbf.html.

Sanctuary 2014 (2015) *Sanctuary 2014*. Available: sanctuary2014.org/.

Taracena, M. (2015) 'Undocumented mom hits 9th month in sanctuary: the "eeny, meeny, miny, mo" nature of ICE'. *Tucson Weekly*. Available: http://www.tucsonweekly.com/TheRange/archives/2015/05/26/undocumented-mom-hits-9th-month-in-sanctuary-the-eeny-meeny-miny-mo-nature-of-ice.

Yukich, G. (2013) 'Constructing the model immigrant: Movement strategy and immigrant deservingness in the New Sanctuary Movement', *Social Problems* 60 (3): 302–320.

15 Beyond solidarity

Migrants and squatters in Madrid

Miguel Martínez

Introduction[1]

Politics entails power struggles. Most of these struggles occur in public spheres and, to some extent, involve state institutions and the most recognized power holders in rule-making and rule-breaking (Piven and Cloward 2005). However, power conflicts are also inherent to private domains. They tend to remain hidden unless made visible by different political activists and social movements. For instance, family issues, gender and race relations, individual control inside "total" institutions such as prisons, psychiatric hospitals, garrisons and even schools – these are the kind of concerns where "personal becomes political" due to purposive collective action and public claims. The same applies to political squatting and migrants involved in squatting. The label "political" is adopted at their convenience by different social groups given the different power struggles in which they are embedded.

Squatters move (Owens 2013; Owens *et al.* 2013). Migrants move too. Most squatters and migrants are forced to move although they both develop their respective capacities of agency in order to figure out where to move, how, with whom, through which particular social networks, etc. They face the basic dilemma of which course of action to take given the different alternatives available and their context. The original structural and, usually, institutional violence that obliges them to move is also the same that confronts their desire to remain, when this is the case (Manjikian 2013; Chattopadhyay 2015). Squatters tend to defend the permanence of the places they have occupied. Migrants strive for the legal documents that certify their right to abode, an equal condition of citizenship, belonging and the right to stay if they wish. Since they both may face criminalization and prosecution (Martínez et al. 2014; Dadusc and Dee 2015), their decisions to move are made within a strategic framework in which every move is a nomadic tactic in the interplay of the external forces that exert violence over them.

Why, then, go public and "political"? Is it an essential part of a broader strategy aimed at claiming the right to squat and migrate? The obvious answer is that many squatters and migrants are aware of their broad political role in challenging many assumptions taken for granted in capitalist societies. The rigid or absolute right to private property and the arbitrariness of state boundaries are continuously questioned by both squatters and migrants. Moreover, what is turned into a political contention is the fact that unequal distribution of property and unequal access to national citizenship are in direct conflict with the right to decent and affordable housing, on the one hand, and the right to be granted a decent migrant or asylum status, on the other. Whenever this underlying conflict comes to the surface, squatting and migration become a key component of current politics. No matter how secret the actions of squatting and migration are, they all participate in a given situated political struggle about agenda, policies, priorities, needs, rights, representation, governance, etc. The will of remaining silent, hidden or clandestine corresponds to tactical operations or to their social belonging to explicitly political scenes. In accordance, nomadism is seldom a lifestyle or a desired state of permanent change (Bookchin 1995), but a social condition for those pushed to move within a specific harsh environment, although nomads may turn its meaning upside down and adopt the label in a positive sense to empower themselves, like the queer and hacker activists. Just as a convention, then, "social" squatting tends to represent the cases where political ideology is loose or not at the fore because the urgent economic needs are emphasized above all (this is usually the situation of most "migrant squats", see Aguilera 2013; Mudu 2014). "Political" squatting, alternatively, refers to the dense networks of activists where a structured political discourse contributes to their identity and cohesion – this can be applied not only to the explicit and restricted category of "political squatting", according to Pruijt (2013), but also to the other configurations of "deprivation-based", "alternative-housing", "entrepreneurial" and "conservational" squatting.

In order to understand the specific forms and contexts of interaction between migrants and squatters, we also need to distinguish their social conditions of life that do not necessarily overlap:

1 Some migrants may be privileged in their ways of living if they are rich or hold passports from worldwide dominant countries (usually, based on military and economic grounds);
2 Some migrants do not squat themselves or attend squats due to the additional risks that they may endure in case they do not hold legal documents to stay in a particular country;
3 Some migrants can squat houses or even run occupied Social Centers not only as a way to satisfy their housing and social needs, but also as a political tool to claim for their citizenship rights;
4 Migrants and squatters may meet in squatted spaces as equal members of the collectives who occupy, but it is more frequent that their coexistence occurs as a consequence of "solidarity" actions with migrants organized by squatters;
5 Many squatters have experienced similar situations of badly paid jobs, social exclusion, police brutality, fascist attacks and spatial displacement, like most migrants (especially those undocumented ones), although fewer squatters are so frequently harassed by the police due to their ethnic outlook or arrested due to the lack of a residence permit;

6 Most squatters are part of wider, safer, wealthier, denser and more local social networks with friends, relatives, job mates, political comrades, etc. than those to which migrants belong.

When the squatting scene is politicized and nurtured by leftist, autonomist and anarchist perspectives, solidarity with those in need and the oppressed, which include many migrants, is a founding principle. This is just the initial step to call for full social, economic and political equality all over the world, beyond the official boundaries of every state. However, the most outspoken political squatters in Europe since the 1960s (SqEK 2013; Martínez 2013a; SqEK, Cattaneo and Martínez 2014; Steen *et al.* 2014) had to face the historical circumstances of the aftermath of World War II and also the crisis of welfarist policies some decades later. This led, first of all, to the development of defensive struggles against fascism in all its dimensions – not only neo-fascist groups and political parties but also the ideological roots of fascism in mainstream politics and mass media, the authoritarian model of family and work relations, war and migration policies, police repression against anti-capitalist political dissidence, etc. Anti-fascism, then, became one of the main identity pillars for those squatters whose politics pointed beyond the walls of their living spaces. Accordingly, the opposition to racism and xenophobia was considered a logical consequence of that stance since fascist politics is based on ethnic supremacy, conservative nationalism, hatred of those seen as inferior (not only ethnic minorities and foreigners, but often homosexuals, disabled and homeless people, punks and hippies, communists, and even women) and a pervasive violence against them. Once more, the threat of fascism and racism surrounding the squats urged squatting activists to adopt open discourses to de-legitimize them and defensive actions to halt their squads and proponents. As a result of this attitude, a rich anti-fascist and anti-racist iconography, information flows and specific activities (workshops, sit-ins, border camps, etc.) have been produced by squatters and sometimes migrants involved in squats too (Wilhelmi 2002; Moore and Smart 2015).[2]

In this chapter, I will underscore the social and political affinities that both squatters and migrants shared by looking at their mutual interactions in a specific urban setting. My focus is the city of Madrid where I was an active participant in several squatted Social Centers for six years, although some examples are also taken from other Spanish cities due to their general impact in the public debates about squatting and the images and policies related to immigration. My own observations and research on squatting, since the mid 1990s, in Spain and Europe (Martínez 2013b; Seminario 2014) provide additional evidence and perspectives to make sense of this topic. More in detail, the collection of mass media news over the last two decades and eight semi-structured interviews have been used as empirical sources for the present analysis. Through this research, my purpose is to shed light on the following questions: Are migrants under-represented in squats? Or just under-reported? Do squatters assist migrants? Do migrants participate on equal terms with other non-foreigners in the self-management of squats? Why are migrants interested in squatting? Is there any clear pattern of interactions between squatters and migrants in the city of Madrid? As I will argue in the following section, the "solidarity" approach lingers all over the period of analysis, but other forms of interaction (such as "autonomy", "engagement" and "empowerment") gained political significance due to specific contextual circumstances.

When migrants squat alone

In this section I show how migrants took genuine autonomous initiatives to squat on their own, first of all, even before their numbers rose significantly in Spain in the mid-1990s. Squatting in Madrid started as a public action run by natives in the years of the transition to liberal democracy, after dictator Francisco Franco's death in 1975 (Seminario 2014), although the most discernible case of a squat similar to other autonomous spaces in Europe dates back to 1985 (Seminario 2014). In 1992, there was a notorious incident that is worth mentioning here in order to understand the existence of usually stealth squatting by migrants. A black Dominican woman, Lucrecia Pérez, was murdered while having dinner collectively in a squatted, although almost ruined building – a former fancy discotheque. She was 33 years old and had a daughter. She had neither residence nor work permit. Another Dominican man was shot in the same raid. According to the judicial sentence, the perpetrators of the crime were four men, three minors and one policeman aged 25. They all were known in Madrid for hanging out with fascist gangs and far-right football hooligans, and they also had a record of previous violent assaults. The judge included hate, xenophobia and racism as motivations for the shootings (Calvo 1993). The building was located in an upper-class area of the metropolitan area of Madrid, called Aravaca. There was neither electricity nor running water in the occupied premises, which contributed to the image of marginality and decay associated with squatted places. As interference to those stereotypes, I was part of a group of autonomist students in Sociology and Political Science who in 1993 occupied a building on the University Computense of Madrid campus and named it Lucrecia Pérez (Caravantes García *et al.* 1995: 32).

Starting in the mid 1990s, international flows of migration to Spain rose at higher rates than ever before (Martín-Pérez and Moreno-Fuentes 2012). At its peak, 2005, the figure of undocumented and non-authorized migrants was estimated at 1.2 million (Clandestino Project 2009), while the total number of people born abroad but settled in Spain reached 6.7 million by the end of 2010. It is worth noting that one million of them held Spanish citizenship and more than 600,000 were of British, German, Italian and Portuguese origin. In relative terms, foreign population in Spain shifted from less than 2% of the population in the 1990s to 12% at the end of the 2000s. Notwithstanding, in some neighborhoods (such as those in the city center of Madrid, where squatting was practiced the most) the concentration of migrants meant rates above 20%. For example, in 2011 the rate was 27% in the Central District (Schmidt 2012: 2). According to the official figures released by the government in 2011, the four main countries of origin for immigrants to Spain are Morocco, Romania, Ecuador and Colombia.

The above indicates a regular increase of migrants in Spanish cities until the late 2000s and the global financial crisis. As a consequence, expectedly, migrants participated in squats at a similar pace. We lack an accurate calculation of the proportion of migrants in squats. Stemming from my own observations since the 1990s and from other records, I would argue that the presence of migrants in political squats increased over the years (refer to Nur and Sethman Chapter 6, in this volume). However, not all the foreign nationalities were equally represented, and their percentage was in general lower than the local average. For example, according to my observations in squats in Lavapiés and Madrid city center at large, especially since 2007, migrants rarely represented more than approximately 10% of the members or attendants. In the 1980s and early 1990s, there were even less migrants attending squatted Social Centers in Madrid.

Latin American immigrants were the first to arrive in Spain and the ones who faced easier adaptation given their mastering of the Spanish language – although the racist or xenophobic attitudes of some home owners determined their ethnic exclusion in the rental market once their American accent was noticed. Political interest for revolutions and different struggles in Latin America also caused many squats to organize events and invite people from that world region. Therefore, Latin American migrants were the first to be seen in political squats. More likely, when migrants squatted buildings on their own, they tended to do it apart from the squatting political scenes, such as in the case of Aravaca. In this first stage, only occasional solidarity actions from political activists in general, and from political squatters in particular, occurred. This pattern is also evident in the following three examples:

- 2002–2004. Around 1,000 people (most of them illegalized migrants from Africa and Eastern Europe) occupied the abandoned military barracks known as *Cuarteles de San Andreu*, in a peripheral working-class neighborhood of Barcelona. Their living conditions were very harsh without water, electricity, toilets and waste management. After the police evicted the place, around 100 residents were rehoused by municipal agencies and the Red Cross. The occupation was an example of the initiative of homeless poor people to get shelter, but they were not able to run any collective self-organization of the place. Different NGOs, local associations, neighbors and some squatters from the same area (I walked in the barracks, for example, along with an activist from the *CSOA El Palomar*: González 2001) helped them occasionally and rallied to stop the clearance of the barracks. The wide mainstream media coverage of this case tended to associate the term "okupa" [squatter] with "illegal immigration", "lack of hygiene" and the official discourse expressed by the local authorities about the priority to build social housing, a hospital and public gardens on the land once the barracks were demolished (Blanchar 2004).
- 2008. "Palacete okupado" in the working-class district of Carabanchel (Madrid). One day after being reported on TV and in some newspapers, the police arrested 12 illegalized migrants out of 20 who lived in the building, although there was no official eviction. Most of them came from Senegal

and to the Canary Islands in a *pátera* (fragile boat). After their eviction from the squatted building, they were going to be deported. The multi-story house had been squatted for more than a year and a half. One of the squatters was a Spanish citizen who claimed to have signed a rental contract, which he found later to be a scam. They all enjoyed water supply but no electricity. The house was collectively managed, and the neighbors got along well with them, according to the news (Herráiz 2008). A formal association, COIN (*Coordinadora de Inmigrantes*), supported them with legal advice and professional training.

- 2010. Since 2000, a luxury development that did not meet the planning regulations had remained empty, and the street floors were occasionally used by homeless people, drug addicts and youngsters. At some point, an estimated 100 Latin American migrant families (mainly from Ecuador, Bolivia and Dominican Republic) occupied every apartment in the building because they found the original keys in the basement. The replacement of the previous occupants granted the new ones a relative support by the neighbors of the area – an upper-class municipality in the metropolitan region of Madrid, Majadahonda. The squatters even claimed for water and power supply from the town hall, but the local government tried to evict them. The squat got media attention in 2010 and still remained occupied until 2014 given the absence of any judicial lawsuit against the squatters (Medrano 2010). Mass media used the term "okupas" to name the particular migrants, although no direct connection of them with political squatters was known.

Alliances between migrants and squatters along the global justice movement

An early exception to that usual practice of migrants squatting alone started to occur when political squatters approached migrants in order to show solidarity in forms of social support, raising funds, demonstrations, etc. This happened especially after the development of a massive campaign in 2001 when hundreds of illegalized migrants strived for their legal rights to remain in Spain and a regular status (work and residence permits) by resorting to locks-in or self-confinement in different buildings (*encierros*) across 14 Spanish cities (Nodo50 2001). In the case of Barcelona around 800 migrants occupied 48 churches. Their main geographical origins were Morocco, Pakistan, India, Bangladesh, Senegal, Algeria, other African and Eastern Europe countries, and a few from Latin America. One anarchist union (CGT), some religious authorities and a group in solidarity with migrants (*Papeles para Todos y Todas*) supported the occupations, demonstrations, collection of signatures, hunger strikes and negotiations – partially successful for some of the migrants (Aparicio 2001). Some squatters joined that campaign, but most of the squats in Spanish cities were not fully involved. Nonetheless, the demonstrations and solidarity actions created a strong precedent for further cooperation. The campaign coincided with the Global Justice Movement, the activity of the "white overalls/invisibles" against the war in Iraq, as well as a slight decline in the squatters' movement after the implementation of the Penal Code of 1995, criminalizing the act of squatting (Martínez 2013b). In Madrid, at least some squatters from *El Laboratorio 2* (located in the neighborhood of Lavapiés) and from CSO Seco (located in the intersection of Retiro and Puente de Vallecas) were involved in one of the "*encierros*", although the dominant role in the negotiations with the state authorities was played by the major labor unions and a formal non-governmental organization (*SOS Racismo*).

This new wave of solidarity and closer ties may be illustrated by a case from September 2002, when one building occupied by migrants on Murcia street (Madrid, next to Lavapiés) was set on fire. Ten residents were evicted after the police arrived. Protesters who gathered in solidarity with the migrant squatters (*Coordinadora de Inmigrantes* and political squatters from *La Biblio* and CSOA *El Laboratorio 3*) assumed the fire was intentionally caused by racist-fascist groups (reported by Upa-Molotov and Indymedia Madrid, not available online). In particular, *La Biblio* was a long-lasting, grass-roots library that operated in different squatted buildings where, in addition, they offered free Spanish language lessons to migrants as a way "to fight against the laws on migration and the social exclusion of migrants" (La Biblio 2002). After the frustrating experience of several evictions, the collective decided to rent a place in Lavapiés to continue with their usual activities up to the present day.

This alter-globalization trend gave birth to the ODS (*Oficinas de Derechos Sociales*, Social Rights Offices) in the mid 2000s (Toret *et al.* 2008; Arribas 2012; La Villana 2013). At least 10 organizations became the initial members of this network of activists in different Spanish cities (Seville, Málaga, Terrasa, Zaragoza and Barcelona), and four of them were rooted in Madrid (ODS in the squatted EPA *Patio Maravillas*, ODS in the formerly squatted and later legalized *Centro Social Seco*, ODS Carabanchel and *Punto Mantero/Asociación de Sin Papeles*, which was partially transformed into *Ferrocarril Clandestino*). The origin of the ODS is not limited to activist solidarity work with migrants because they intended to investigate, devise shared strategies and act up around the general circumstances of "precarious living" in the productive, reproductive, social and political spheres. All of them affected both the migrant and the native populations. The ODS tried to set up an alternative to both the welfarist policies based on a discourse of assistance, subsidies, hierarchy of rights and state discipline; and the neoliberal policies designed to protect state borders, drive individuals into a flexible job market and criminalize any ethnic and political difference. Thus, the ODS emphasized the defense of social rights of citizenship for all, autonomous capacities and the self-organization of those disempowered by the context of precarious living in the continuum of metropolitan production and consumption. Social or urban syndicalism was considered, then, as a replacement to the old-fashioned and exclusive trade unionism in the labor market or the solidarity groups with migrants. In sum, the general purpose of the ODS was to build a solid alliance of all the people subject to the neoliberal regime of precarity, regardless of their national origin. Universal and unconditioned income and the dismantling of the detention centers for migrants were, for example, two of their specific demands. However, it has been admitted that their daily practice focused more on the concrete needs of migrants who joined the ODS – this is to say, access to essential legal documents that allow them to work and reside, provision of attorneys and solicitors to deal with their arrest and threats of deportation, help with their economic initiatives even those based on informal selling of goods in the streets and teaching of Spanish language (Arribas 2012). Nonetheless, over more than one decade, the ODS has unfolded a continuous activism that included demonstrations, workshops, legal and professional assistance, cooperative projects, music performances, etc.

Regarding the relationship of the ODS with squats, the former were, first, a solid bridge to connect the latter with other autonomous but not squatted Social Centers. These ties also challenged the dominant squatting identity where migration and precarity did not represent a central aspect of the squatters' political discourse. Instead, the members of the ODS network called to "exit the ghetto", to question the squatters' imagery of resistance without any compromise, their purely theoretical anti-fascist and anti-racist symbols without a consistent practice, and the acceptance of nomadism as a given fate. The so called "second generation" of Social Centers was thus more prone to negotiate with the local authorities to find a legal status to their squats, and even to use state subsidies to develop their activities. In Madrid, two squatted Social Centers where the work with migrants was a key area of activism, *Seco* and *Eskalera Karakola* (in the latter case, as a feminist Social Center, they usually were engaged with migrant women earning their living as domestic workers and prostitutes: Precarias 2004), succeeded in their claims of legalization. A third, *Patio Maravillas*, also attempted to do so but with little progress from the city hall after various squatted locations and several years of trying (the first occupation took place in 2007). Other new ODS were located in self-managed but rented Social Centers such as the ones in La Piluka, Properidad and Carabanchel. Anyhow, migrants and their specific living conditions became more visible and well recognized among the squatting scene of Madrid due, mainly, to the work done by the ODS.

The ODS also had a deep impact on the next generation of squatted Social Centers and houses where migration, feminism, precarious living and even institutionalization gained more political attention by all kinds of autonomist activists, squatters included. "Bangla Thursdays" in the CSOA Casablanca are an example of this ODS influence, where a group of Bangladesh migrants made an income by preparing and selling dinner, in addition to showcasing films and music from their cultural background. The multiple activities in the CSOA La Enredadera de Tetuán are another example, such as hairdressing, "Women's Saturdays", Spanish classes, teaching about computers, workshops, the "free shop", etc. where Latin American and African migrants took part frequently (La Enredadera 2011). On the opposite side of the spectrum, some ODS have been criticized for often reproducing, unintentionally or without sufficient resources to avoid it, a hierarchical and professional mode of assistance to migrants – white, European, middle-class and highly educated activists being the main social composition among the most permanent activists. They were also criticized for failing to involve all kinds of precarious natives and migrants, instead focusing only on the most urgent problems faced by some of the foreigners (Arribas 2012: 222–224).

Another expression of these developments was the *Mundialito Antirracista* (Anti-racist Little World Cup) in 2006 and the *Campaña contra el Racismo* (Campaign Against Racism) in 2007, which were launched by political squatters. The first one was a yearly event in which activists from all over the metropolitan region of Madrid shared a day of sport "competitions" (basketball, football, running, etc.) and an informal gathering for fun more than for coordinating any action or campaign. The Mundialito was organized by the squatters from CSO *La Eskuela Taller* (based in Alcorcón, a working-class municipality next to the city of Madrid). Their aim was to stop the growing racism they observed when fascist groups tried to systematically book public sport facilities in order to prevent migrants from using them. They also wanted to question sexism and competition—so the teams are mixed, with women and men, and the first prize is awarded not to the winning team but to the one that better represents the values defended by the organizers. They also expressed their intention to unite "natives and foreigners, anarchists, communists, autonomists, postmodernists and separatists" (Mundialito 2011). The Mundialito was celebrated over nine years in spite of the obstacles erected by the local authorities. Thus, it was not only squatters with different backgrounds and political styles who took part in the event, but I also witnessed the attendance of other types of radical activists, migrants' associations and political groups (mainly anti-fascist ones) without an explicit involvement in squatting. In 2014, there was a replication in another Mundialito, which was organized by CSOA *La Gatonera* in the working-class neighborhoods of Carabanchel and Aluche (Madrid) (La Gatonera 2014). In a similar vein, the Campaign Against Racism gave priority to cultural activities and collective meals over more militant and overt political actions (Rivero 2007).

During my own participant observation in Madrid since 2007, I also noted a more frequent presence of Latin American migrants in squats (not only in Social Centers, but also in squatted communes just for living such as *La Barraka* or *Cambalache*) than those from Africa and Eastern Europe. Here are just two examples of how these ties of interaction occurred, beyond sharing the needs of housing and the political interest in the self-management of squats:

a) The well-established tradition of international solidarity, which connected autonomist movements with, for instance, the Zapatista uprising (Martínez 2013b), was manifested in the organization of groups such as the RAZ (*Red de Apoyo Zapatista de Madrid*), whose members usually lived in collective squats (*La Juli*, for instance) and participated in squatted Social Centers to raise funds and spread information about politics in Mexico (at CSOA *Malaya*, CSOA *La Mácula*, CSOA *La Enredadera*, etc.). At the same time, they served to incorporate Latin American migrants in the activist life of Madrid;
b) Some Spanish activists even decided to marry illegalized migrants they knew well after years of mutual cooperation in order to halt the threats of

deportation (sometimes these weddings were celebrated in squatted Social Centers, like one that happened once in CSOA Casablanca).

In 2009 squatters and other activists formed the *Brigadas Vecinales de Observación de Derechos Humanos* – BVODH (Local Citizens Watching Human Rights, similar to the US group Cop Watch) – and made their public presentation in the CSOA *La Mácula*. Wearing reflective vests and walking around in groups of more than ten, they surveyed, documented and denounced the police implementing identity controls on the streets, at subway exits and front doors of associations, and while lining up for institutional appointments. These controls mainly addressed poor, non-regular and non-white immigrants. Although the police inspections are technically illegal if applied based solely on ethnic criteria, the police raids were systematically orchestrated, and even some internal instructions to conduct them were disclosed in the mass media. Activists collected data about the stop controls, informed migrants about their rights and, if possible, interfered with their arrest. In parallel to the claims made by other formal organizations such as Amnesty International and *SOS Racismo*, the BVODH got the media to report on the violation of migrants' human rights by the police. They also released yearly documents with the results of their watch and analysis (BVODH 2014). On one occasion, while I was attending the weekly assembly at the CSOA Casablanca, someone told us that a call from the BVODH was launched to help them chase a police raid against migrants in the Plaza de Lavapiés. All the members of the assembly went down to the square, and we joined hundreds of people shouting at the police. As a consequence, all the police cars drove away. That was a rare symbolic victory in which the union of natives and migrants screamed together "get out of here" and "we don't want police in our neighborhood". In particular, this empowering event occurred some months after the boost of the 15M movement, in May 2011.

Over the 2000s decade, in sum, solidarity actions between political squatters and migrants were developed in extension and depth. However, the most significant trigger of new forms of interaction (engagement and empowerment) came up with the experience of the ODS and their whole alter-global approach to precariousness and social rights. This opened up a ground for equal cooperation between political squatters and migrants. In spite of the practical shortcomings of the ODS, they had a strong influence on other autonomist activists and political squatters, attracted migrants to the squatted Social Centers, and empowered them with valuable skills and social networks.

Squatting together during the cycle of struggles opened up by the 15M

After the turning point of the huge mobilization that took place in May 2011, the political context changed for both migrants and squatters. On the one hand, the conservative central government excluded thousands of migrants (and also Spanish people and migrants in regular situations who were away from the country for more than 90 days or who became unemployed) from the free services of the public health system (YoSí 2015). This affected, in the first place, but not exclusively, the entire half a million undocumented migrants at that time, 2012 (BVODH 2014: 16). On the other hand, since 2013, the Campaign Demanding the Shut Down of the Detention Centers for Migrants (*Campaña Estatal para el Cierre de los CIEs*) had intensified the criticisms of the abuses, deaths, privatization, absence of public scrutiny and illegitimacy of the detention and deportation centers. Many organizations were involved in this long-lasting campaign, but after 2011 the number of participant groups, demonstrations, actions of solidarity, follow-up on individual cases and investigative reports made and spread by different grass-roots collectives increased significantly. Squats in Madrid like Patio Maravillas (and, in particular, the special committee of *Ferrocarril Clandestino*) actively took part in the campaign, and even the traditional anarchist march to the jail, which was called on the last day of every year over the past decades, was replaced by a march to the detention center of undocumented migrants (CIE) located in the area of Aluche (for evidence of the involvement of various squatted Social Centers such as *Patio Maravillas*, *La Gatonera*, *La Quimera* or *La Matriz*, see the news feed of the campaign).[3]

A third dramatic circumstance that occurred after 2011 was the soaring number of foreclosures of home owners unable to pay mortgages and the eviction of tenants unable to pay rent, in addition to the absence of any emergency solution provided by the almost disappearing public housing system. Migrants were among those most affected due to their recent incorporation into the housing market over the previous speculative construction bubble. The protests organized by the PAH (*Plataforma de Afectados por las Hipotecas* or People Affected by Mortgages) started in 2009 and continued up to the present, but they gained a wider support and media resonance after May 2011. In Madrid, an association of migrants from Ecuador was the first in joining the PAH. Members of neighborhood associations and experts such as lawyers, economists and psychologists also became regular participants in the PAH. With a similar methodology of civil disobedience at the time of the evictions and the will to negotiate feasible solutions in each case with the banks or local authorities, other "housing groups" connected to the 15M Popular Assemblies extended this wave of protests. Some of these groups also formed PAH nodes and, as I verified in one of the meetings of PAH-Centro, many of their members were former or present political squatters. In 2011 the PAH also launched the campaign *Obra Social*, which involved the occupation of abandoned buildings, especially those owned by banks and real estate developers. Instead of naming them "squats", they preferred to use the adjectives "recuperated" and "liberated" buildings. Therefore, they tried to move away from some prevailing stereotypes about the squatting movement such as its strong emphasis on a "left-libertarian" political ideology and the manifold goals of the squatted Social Centers. The new occupations focused on the housing needs of evicted families at the same time they claimed for "affordable rents" to the owners of the buildings and for "emergency alternatives" to the state authorities. Pre-15M squatters also started attending the calls by the PAH and the housing groups to stop the evictions and gave their support through their own "liberations" of buildings (De Andrés *et al.* 2015; Abellán 2015).

All of this implied a shifting point that shook the whole squatting movement as we knew it. Negotiations and mass media coverage became more accepted by the most militant activists. Squatting for housing became more visible, politicized and collectively supported. Poor families with children, homeless people and migrants became more engaged in the organization of the protest actions and the self-management of the occupied buildings. The old imagery of squatting did not vanish at all (not even in some of the new takes on it) and many old-school squats remained quite active, but most of them supported this emergent housing movement and, above all, many of them were also actively implicated in the new squatting initiatives. As was the case with most of the participants in the PAH groups, in many of the occupations branded as part of the *Obra Social* or while taking a similar political perspective (Sebastián Elcano, La Cava Encendida, La Manuela, Las Leonas, Corrala La Charca, Calle Cadete 7, Calle Argente, Calle Callejo, La Dignidad de Móstoles, etc.), there were many Latin American migrants involved although without any explicit distinction pointing to that identity (see, just as an indication, the American accent of activists in the videos of the PAH Vallekas 2014 and Cadete 7 2014). One of the squatted blocks in Sierra de Llerena, in the district of Vallecas in which many of the residents have an African origin, remained out of the public eye for two years. Afterwards, they went public and joined the Obra Social of the PAH, which was likely motivated by the attempted eviction by the owners of the building.

The 15M movement also stimulated other forms of activism in the field of migrations, and these were narrowly connected to the squatted Social Centers in use or promoted by the 15M Popular Assemblies in the different neighborhoods (Martínez and García 2012). In particular, in Lavapiés a "group on migration", *Migrapiés*, was created early on. They focused their work on the police raids, legal assistance to migrants, collective support when in need of using the public health system and, above all, the setting up of economic means of subsistence for them. Their cooperative for providing meals, for example, operated in squatted Social Centers such as *Casablanca* and *La Quimera* before renting their own self-managed Social Center (*Mbolo Moy Dole*) and extending their projects to organic agriculture, cleaning and moving services, painting, catering and alternative tourism (Diso Press 2014). Migrants, mainly with an African (sub-Saharan) origin, worked in a horizontal and consensual manner with Spanish indigenous activists, and mutual aid and support among them was the main approach instead of just providing help to the less skilled and resourceful migrants (Méndez 2012).

In short, the recent cycle of struggles also created the conditions for new forms of interactions between migrants and natives in the field of squatting. The global financial crisis and the austerity policies made clear that solidarity and isolated autonomous experiences were not able to face the dramatic circumstances of foreclosures, cuts and continuous repression. In this period the campaigns launched by the PAH and other housing groups engaged impoverished migrants and old-school political squatters in a new wave of squatting, mostly for housing purposes. Their deeper cooperation and the political context after the 15M movement in 2011 helped them to frame squatting in terms of social needs and will to negotiate, but also as an overt and visible claim in a political fashion.

Conclusions

The aims of my research focused on the relationship between migrants and native political activists in the practices of squatting. As I have shown in the prior analysis, their mutual interactions in the city of Madrid took different forms at different periods of time and in different political contexts. These interactions have been distinguished in four specific dynamics:

1. Autonomy – when migrants squat alone without the initial help of political and native squatters although some cooperation may occur later on;
2. Solidarity – either migrants or political squatters launch protest campaigns, actions or events in which the issues of migration, citizen rights, police ethnic control, etc. are the main claims at play, both groups cooperate with each other, and the squatted spaces are used to develop these ties;
3. Engagement – migrants participate in the activities and the self-management of political squats, usually squatted Social Centers run by natives, with different degrees of involvement and in different numbers in each case;
4. Empowerment – when political squatters help migrants to squat and they both may occasionally cohabit in the occupied building.

Obviously, these forms of interaction may overlap. The striking observation is that, beyond the theoretical expressions of ideological solidarity or the tendency to hide deprivation-based forms of squatting among migrants, other forms of interaction have prevailed in different historical periods. The "empowerment" forms, in particular, were developed in the aftermath of the 15M movement. "Engagement" has increasingly occurred according to the rising numbers of migrants in Spain, but also given the crucial influence of some initiatives such as the ODS that paved the way for more profound mutual aid and involvement. "Autonomy" and "solidarity" modes have been continuously present in both the migrant and squatting scenes, but their knowledge, public visibility and political support grew up in parallel to the higher social recognition and legitimation of squatting.

An additional consequence of the evidence provided above is that the political squatting networks have been relatively consistent with their left-libertarian discourse in order to add the migrants' struggles to the range of their concerns. This draws a distinctive line of separation in respect to the rare – but highly commented on by the mass media – cases of neo-fascist groups that occupy buildings while aggressively excluding and attacking migrants, ethnic minorities, non-heteronormative individuals, beggars, and alike.

Furthermore, it is worth noting that poor migrants and political squatters do not necessarily share the same socio-structural conditions of living, so every interaction between them is mediated by this unbalanced starting point. Although both categories may have experienced similar situations of badly paid jobs, social exclusion, police brutality, fascist attacks and spatial displacement, most of the native squatters are neither so frequently harassed by the police due to their ethnic outlook, nor are they arrested due to the lack of a residence permit, and they rather tend to belong to a wider, denser and wealthier social network. This partially explains the predominance of "solidarity" and "autonomous" interactions but also helps to understand the hindrances for further developments of "empowerment" and "engagement".

Regarding the limitations of the interplay between migration and squatting, we should remember, on the one hand, the issues related to every squatted project: how convenient is it to remain hidden or to go visible once a building has been taken over, what is the way of making collective and democratic decisions, how individual or collective is the organization, to what extent is there a viable self-management of the collective affairs, in what constructive-physical condition is the building and how many investments and repairs are needed, what is the legal status of the building (and the occupants) and who are the owners, what kind of threats and attacks (and by whom) are experienced and how to respond, etc. These issues urge us to reconsider the usually taken-for-granted homogeneity of the squatting situations in which migrants are active. On the other hand, the interaction between migrants and squatters is also limited by various issues that deserve a further examination, such as the vertical or hierarchical relations that tend to occur when migrants just ask for help from native political activists and no autonomous organization emerges out of their interaction (a complaint which is usually expressed by PAH members as much as it was by ODS activists before), the language barriers that oblige to exhausting exercises of translation and tend to disengage migrants from struggles when natives are dominant (thus, as an exception, Latin American immigrants are more prone to participate in squats), the gender roles and relations within some ethnic and migrant minorities which are incompatible with the egalitarian views (Azozomox 2014) of the political squatters, the troubles derived from informal economic activities (drug trafficking, for example) in which either migrants or squatters may be implicated, etc. The occurrence of these problems, as mentioned by some interviewees and verified by my own observations, highlights the structural inequalities at play and how difficult the interactions under examination have been.

The relevance of the social context is nuanced with some key political dimensions and cycles as I have suggested in prior sections. In particular, the rise of Madrid metropolis as a global city, the mobilizations called by the global justice movement and the new waves of internationalism, such as the *Zapatismo* in Mexico, gave birth to questioning the dominant borders policies for people (while not for capital) and the devastation created in poor countries by the capitalist modes of production, consumption, debt, exploitation of natural resources, etc. Furthermore, the new restrictions set by the European governments in order to limit the admission of asylum refugees' requests, the military control of borders, the use of detention Guantanamo-like camps where numerous illegal practices and violations of human rights are reported, and the deportation flights of undocumented migrants, engendered more risky forms of migrants' mobility and deadly tragedies over the last decades. These situations were combined with hypocritical institutional policies of "integration" and "multiculturality" along with the neoliberal privatizations and cuts in public services (Avila and Malo 2010), including those that specifically addressed the migrant population – and Spanish natives who ended up unemployed and migrated abroad.

Therefore, the incorporation of debates around these issues during the early 2000s in the political squatting scenes prepared the grounds for more intense and practical forms of interaction with migrants. Accordingly, starting from an initial poor interaction between them and given a certain image of marginality over the migrants' squats, the political squatting scene has evolved into a deeper concern for migration issues, its involvement in the migrants' struggles and a tighter cooperation in the practice of squatting. After 2008, the global financial crisis and the 15M movement, the occupation of houses by migrants, political squatters and other activists have boosted and challenged the criminalization which generally applied to squatting. Thus, at least in Madrid and other Spanish cities, more favorable media coverage and a shift in the tactics of negotiation turned this cooperation into a more fruitful one.

Acknowledgements

This article is one of the outcomes of the research project MOVOKEUR "The Squatters' Movement in Spain and Europe: Contexts, Cycles, Identities and Institutionalisation" #CSO2011-23079 funded by the Spanish Ministry of Science. I am grateful to Julia Lledín and Fernando Arozarena for their assistance with part of the fieldwork done on the topic of squatting and migration.

Notes

1 CSOA stands for *Centro Social Okupado y Autogestionado*, Occupied and Self-Managed Social Centre.
ODS stands for *Oficina de Derechos Sociales*, Office of Social Rights.
BVODH stands for *Brigadas Vecinales de Observación de Derechos Humanos*, Local Citizens Watch On [Migrants'] Human Rights.
PAH stands for Plataforma de Afectados por las Hipotecas, *(Anti-Eviction) Platform of People Affected by Mortgages*.
15M stands for May 15 (2011) or *Indignados* Movement.

2 The emergence of a few neo-nazi squatted social centers in Italy and Spain implies, thus, a frontal opposition to that trend. By taking over empty buildings fascist groups try to generate confusion among their neighbours and the public at large. While they embrace hierarchy, violence, racism and xenophobia, they also claim they are anti-capitalists and provide as many benefits to the communities around as other squatters do—preservation and rehabilitation of abandoned buildings, food delivery, venues for cultural performance and so forth. Therefore, the ideology and practices of fascist squatters are not only an unfortunate exception in the long tradition of emancipatory squatting, but they are also far away from the well-established anti-authoritarian, egalitarian, (direct) democratic and self-managed experience which has prevailed in the more politicised squatting movements all over Europe (Birdwell 2012; Tetuán de Todas 2014).

3 Refer to https://cerremosloscies.wordpress.com/

References

Abellán, J. (2015) 'Ciudad, crisis y desobediencia: una aproximación a las luchas por la vivienda en Madrid'. In Hidalgo, R. and Janoschka, M. (eds) *La ciudad neoliberal. Gentrificación y exclusión en Santiago de Chile, Buenos Aires, Ciudad de México y Madrid*. Santiago de Chile: Pontificia Universidad Católica de Chile, 257–274.

Aguilera, T. (2013) 'Configurations of squats in Paris and the Ile-de-France Region'. In SqEK (ed) *Squatting in Europe: Radical Spaces, Urban Struggles, New York: Autonomedia*, 209–230.

Aparicio, M.A. (2001) *Los encierros de inmigrantes sin papeles: la experiencia de Barcelona*. Bilbao: Fundación Betiko'. Available: http://fundacionbetiko.org/wp-content/uploads/2012/11/los-encierros-de-inmigrantes-sin-papeles-la-experiencia-de-barcelona.pdf.

Arribas, A. (2012) 'Sobre la precariedad y sus fugas. La experiencia de las Oficinas de Derechos Sociales (ODSs)', *Interface* 4(2): 197–229.

Ávila, D. and Malo, M. (2010) 'Manos invisibles. De la lógica neoliberal en lo social', *Trabajo Social Hoy* 59: 137–171.

Azozomox (2014) 'Squatting and Diversity: Gender and Patriarchy in Berlin, Madrid and Barcelona', in SqEK, Cattaneo, C. and Martínez, M. (eds) *The Squatters' Movement in Europe. Commons and Autonomy as Alternatives to Capitalism*, 189–210. London: Pluto.

Birdwell, J. (2012) *CasaPound: The New Face of Fascism?* Available: http://www.opensocietyfoundations.org/voices/casapound-new-face-fascism.

Blanchar, C. (2004) 'Punto final a los cuarteles de Sant Andreu, refugio de los 'okupas'. *El Pais*, 10 February 2004. Available: http://elpais.com/diario/2004/02/10/espana/1076367625_850215.html.

Bookchin, M. (1995) *Social anarchism or lifestyle anarchism: an unbridgeable chasm*. Edinburgh: AK Press. Available: https://libcom.org/library/social-anarchism--lifestyle-anarchism-murray-bookchin.

BVODH. (2014) 'Persecución y acoso policial. La persistencia de los controles de identidad por perfil étnico'. Available: http://brigadasvecinales.org/wp-content/uploads/2015/05/Tercer-informe-BVODH.pdf.

Cadete 7 (2014) CADETEvsSAREB. Available: https://www.youtube.com/watch?v=8IVEbZEAvhk.

Calvo Buezas, T. (1993) *El crimen racista de Aravaca*. Madrid: Popular.

Caravantes García, C. et al. (1995) *La okupación como analizador*. Madrid: Peligrosidad Social. Available: https://distribuidorapeligrosidadsocial.files.wordpress.com/2011/11/la-okupacic3b3n-como-analizador.pdf.

SqEK, Cattaneo, C., and Martínez, M. (eds) (2014) *The Squatters' Movement in Europe. Commons and Autonomy as Alternatives to Capitalism*. London: Pluto.

Chattopadhyay, S. (2015) 'Squatting as an Alternative to Counter Migrant Exclusion'. In Moore, A and Smart, A (eds) *Making Room: Cultural Production in Occupied Spaces*, p. 294–297. Barcelona: Other Forms and the Journal of Aesthetics and Protest.

Clandestino Project (2009) *Project Results*. Available: http://clandestino.eliamep.gr/project-results/.

Dadusc, D, and Dee, E. (2015) 'The criminalization of squatting: discourses, moral panics and resistances in the Netherlands and England and Wales'. In O'Mahony, F.L.,

O'Mahony, D. and Hickey, R. (eds) *Moral Rethoric and the Criminalization of Squatting. Vulnerable Demons?* Oxon: Routldge, 109–132.

De Andrés, E. Á., Campos, M. J. Z., and Zapata, P. (2015) 'Stop the evictions! The diffusion of networked social movements and the emergence of a hybrid space: The case of the Spanish Mortgage Victims Group', *Habitat International* 46: 252–259.

Diso Press (2014) *Migrantes de Lavapiés crean la cooperativa Mbolo Moy Dole como solución de autoempleo.* Available: https://www.diagonalperiodico.net/movimientos/23845-migrantes-lavapies-crean-la-cooperativa-mbolo-moy-dole-como-solucion-autoempleo.

González, R. (2001) *El Palomar de Sant Andreu, una okupación abierta al barrio. Entrevista a un okupa del Palomar.* Available: https://www.academia.edu/7919939/El_Palomar_de_Sant_Andreu_una_okupaci%C3%B3n_abierta_al_barrio._Entrevista_a_un_okupa_del_Palomar.

Herráiz, P. (2008) 'Inmigrantes y 'okupas' ...Y del palacete a la comisaría', Available: http://www.elmundo.es/elmundo/2008/03/13/madrid/1205398170.html.

La Biblio (2002) '¿Se acerca el día del juicio final? La Biblio amenazada de desalojo', Available: http://www.sindominio.net/labiblio/web1/doc/juicio_final.htm.

La Enredadera (2011) ¿Qué hacemos en La Enredadera?' Available: https://www.youtube.com/watch?v=PEJ2s5wX81o and http://laenredaderadetetuan.blogspot.com.es/search?updated-min=2011-01-01T00:00:00%2B01:00&updated-max=2012-01-01T00:00:00%2B01:00&max-results=7.

La Gatonera (2014) '1 Junio. Primer Mundialito Anti-Racista de Carabanchel'. Available: https://csolagatonera.wordpress.com/2014/05/22/1-junio-i-mundialito-de-carabanchel-cartel-definitivo-info/.

La Villana (2013) 'Proyecto Político y Empresarial Centro Social La Villana de Vallekas. Oficina de Derechos Sociales', Available: http://www.lavillana.org/wp-content/uploads/2013/05/villanadefinitivo2.pdf.

López, S; Martínez, X and Toret, J. (2008) *Oficinas de Derechos Sociales: Experiences of Political Enunciation and Organisation in Times of Precarity. European Institute for Progressive Cultural Policies.* Available: http://eipcp.net/transversal/0508/lopezetal/en.

Manjikian, M. (2013) *Securitization of Property. Squatting in Europe.* New York: Routledge.

Martín Pérez, A. and Moreno Fuentes, F. J. (2012) 'Migration and Citizenship Law in Spain: Path dependency and Policy Change in a Recent Country of Immigration', *International Migration Review* 46(3): 625–655.

Martínez, M and García, A. (2012) 'The Occupation of Squares and the Squatting of Buildings: Lessons From the Convergence of Two Social Movements', Available: http://www.miguelangelmartinez.net/?Occupy-the-Squares-Squat-the.

Martínez, M. (2013a) 'The squatters' movement in Europe: a durable struggle for social autonomy in urban politics', *Antipode* 45 (4): 866–887.

Martínez, M. (2013b) 'The Squatters' Movement in Spain'. in SqEK (ed.) *Squatting in Europe: Radical Spaces, Urban Struggles*, p. 113–138. Wivenhoe: Minor Compositions.

Martínez, M, Azozomox and Gil, J. (2014) 'Unavoidable Dilemmas: Squatters Dealing with the Law". In SqEK, Cattaneo, C. and Martínez, M. (eds) *The Squatters' Movement in Europe. Commons and Autonomy as Alternatives to Capitalism.* London: Pluto: 211–236.

Medrano, C. (2010) ''Okupas' de lujo en una ciudad rica'. *El Mundo*, 22 February 2010. Available: http://www.elmundo.es/elmundo/2010/02/22/madrid/1266828519.html.

Méndez, J. (2012) 'Solidaridad y ayuda mutua. El grupo de Migración y convivencia de la Asamblea Popular de Lavapiés', *Teknocultura* 9(2): 267–286.

Moore, A and Smart, A (eds) (2015) *'Making Room: Cultural Production in Occupied Spaces'*, Barcelona: Other Forms and the Journal of Aesthetics and Protest.

Mudu, P. (2014) ' 'Ogni Sfratto Sarà una Barricata': Squatting for Housing and Social Conflict in Rome'. In SqEK, Cattaneo, C. and Martínez, M (eds) *The Squatters' Movement in Europe. Commons and Autonomy as Alternatives to Capitalism*, p. 136–154. London: Pluto.

Mundialito Antirracista de Alcorcón (2011) 'Comunicado". Available: http://mundialitoantirracista.blogspot.com.es/p/comunicado.html.

Nodo50 (2001) 'Encierros y movilizaciones contra la Ley de extranjería en el 2001', Available: http://www.nodo50.org/derechosparatodos/Encierros.htm.

Owens, L. (2013) 'Have Squat, Will Travel: How Squatter Mobility Mobilizes Squatting', in SqEK (ed.) *Squatting in Europe: Radical Spaces, Urban Struggles*, Wivenhoe: Minor Compositions, 185–207.

Owens, L., Katzeff, A., Lorenzi, E., and Colin, B. (2013) 'At home in the movement', in Fominaya, C.F, and Cox, L (eds) *Understanding European Movements: New Social Movements, Global Justice Struggles, Anti-Austerity Protest*. Oxon: Routledge 172–186.

PAH Vallekas (2014) *Obra Social PAH*. Available: https://www.youtube.com/watch?v=ffnoqJCNHvU.

Piven, F.F., and Cloward R. (2005) Rule Making, Rule Breaking, and Power. In Janoski, T., Alford, R., Hicks, A.M and Schwartz, M. (eds) *The handbook of Political Sociology. States, Civil Societies and Globalization*, New York: Cambridge University, 33–53.

Precarias a la Deriva (2004) *A la deriva por los circuitos de la precariedad femenina*. Madrid: Traficantes de Sueños.

Pruijt, H. (2013) 'Squatting in Europe'. In SqEK (ed.) *Squatting in Europe: Radical Spaces, Urban Struggles*. Wivenhoe: Minor Compositions, 17–60.

Rivero, J. (2007) *Una campaña para combatir el racismo*. Available: https://www.diagonalperiodico.net/movimientos/campana-para-combatir-racismo.html.

Seminario de Historia Política y Social de las Okupaciones en Madrid-Metrópolis (2014) Okupa Madrid (1985–2011) *Memoria, reflexión, debate y autogestión colectiva del conocimiento*. Madrid: Diagonal.

Schmidt, H. (2012) *Lavapiés. Fenómeno migratorio y claves de la convivencia*. Madrid: EPIC-CAM.

SqEK (ed) (2013) *Squatting in Europe: Radical Spaces, Urban Struggles*. Wivenhoe: Minor Compositions.

van Steen, B., Katzeff, A. and van Hoogenhuijze, L. (eds) (2014) *The City Is Ours: Squatting and Autonomous Movements in Europe from the 1970s to the Present*. Oakland: PM Press.

Tetuán de T. (2014) *Los nazis regalan hambre en Tetuán*. Available: https://tetuandetodas.wordpress.com/.

Toret, J., Sguiglia, N., Fdez, S., Lama, P., and Lama, M. (eds) (2008) *Autonomía y metrópolis. Del movimiento okupa a los centros sociales de segunda generación*. Málaga: ULEX-Diputación Provincial de Málaga.

Wilhelmi, G. (2002) *Armarse sobre las ruinas: Historia del movimiento autónomo en Madrid (1985–1999)*. Madrid: Potencial Hardcore.

YoSí Sanidad Universal (2015) *Desmontando mentiras*. Available: http://yosisanidaduniversal.net/mentiras.php.

16 Narrating the challenges of women-refugee activists of Ohlauer Straße 12, International Women's Space (IWS refugee women activists), Berlin

Azozomox and IWS refugee women activists

In December 2012, a group of refugees and activists occupied a nearly vacant school in *Ohlauer Straße 12*, Berlin, and named it as "Refugee Strike House". These refugee squatters had been occupying the public square in Kreuzberg (Oranienplatz) since October 2012, after participating in the "Refugee Protest March to Berlin" from September 8 to October 6, 2012, against racist immigration and asylum laws. The square Oranienplatz was completely evicted in April 2014 when one portion of the refugees agreed to move to other facilities provided by the authorities and another group occupied the vacant school.

One floor of the occupied school in Ohlauer Straße 12 was converted into a woman-only refugee space called the *International Women's Space*[1] (IWS). The school, including the International Women's Space, was largely evicted on June 24, 2014. Some of the squatter refugees who resisted the eviction continued to stay in the building. Right now, year 2015, there are around 24 refugees at one level of the squatted building. Before eviction the building housed more than 250 refugees, while currently most of the space in this building is used by a private security service.

In June 2014, the IWS activists participated in a protest that continued for nine days when they blockaded the occupied school facing massive objection from state police. However, a group of the squatters agreed to go to a temporary shelter offered by the district. An agreement was signed with the local district after nine days of relentless protest where the group organized some meetings to plan the continuation of their activities after eviction. Shortly after, a room was found in a building of a different women's organization where this evicted group continued with their activities, such as holding meetings regularly every Monday and Wednesday. The group has compiled a brochure, published in July 2015, with only testimonies of refugee women in Germany. In December 2014, the IWS suffered one of the biggest setbacks with the death of Sista Mimi, a Kenyan refugee and an activist who played a crucial role in self-managing and self-organizing the space. Below are the interviews[2] of several women activists whose original names have been changed. They were all interviewed on February 15, 2014, in the Kollektivbibliothek in New York/Bethanien, four months before the June 2014 eviction. Interviews were carried out in German, English and Spanish.

Why did you decide to live in the squatted school at Ohlauerstraße?

MARISSA: I wanted to live in Berlin, and I am not allowed to rent a house in Berlin. Even if I was allowed, I do not have the money to rent a place. So I needed a place to stay, because I am from the Heim, from the Lager[3], outside of Berlin. And in the Lager we do not have many things that we need but in Berlin you have all those facilities, such as schools where you can interact with people. So finally my friend who was already squatting the school got me here. At first it was not easy because we shared toilets with men who did not maintain cleanliness. In addition there were no bathrooms. We did not have a kitchen or provisions to cook, so we were forced to eat out. It was expensive and cumbersome. The only option open to me was to return to the Lager, though I liked staying at the squatted school. Later, we put a toilet and a bathroom in place, cleaned the building and made it livable. Here we met people, activists and organizations, interacted with them, and living was better than the Lager. In Berlin, we could visit the hospital without any documents, and when we were sick, the doctors came to visit us. Here we are also more involved in politics than in the Heim, whereas in the Lager, with the lack of interaction, awareness of the politics, or involvement with any activities, I felt psychologically ill due to inactivity. It was very boring and stressful, and the people were not very friendly; racism was stark — I sensed an unwelcoming feeling. Berlin is a multicultural city, there are many foreigners, and the feeling of acceptance is strong. The only problem is that I feel threatened because the building is squatted, the feeling that we have to leave one day is evident. I live every day with the hope that tomorrow we will still have this shelter to sleep in.

MIRANDA: The mayor of the district Kreuzberg-Friedrichshain, Franz Schulz[4] from the Green Party, mentioned that we could stay for three days and then some decision would be reached. I came one day after the occupation, and there were not many of us inside the school. We had to think of what we were going to do for the following days. This building is so huge that we were still trying to discover it. We were trying to make a plan. It was uncertain. We started with laying the information on a table at the entrance, doing shifts for 24 hours, 7 days a week. Many people came by to help us secure the place from being taken over by the city.

GABRIELA: In the first week, we started with 30 people; two weeks later we were 63 people.

MIRANDA: We had a very old computer and first tried to create some kind of infrastructure for data storage but also some security so we could stop the police from evicting us. But it was a lot of work. At first, we had many problems to gather people for securing the door. All of a sudden many people and activists started to join in with us to protect the building. We never found a solution on how to establish a strong security at the main entrance to keep off the police. We instead kept the door open and let the supporters come in to protect the building with us from eviction. It was difficult to accommodate all of them. We had long discussions.

What were the reasons for establishing this International Women's Space?

GABRIELA: There were not many women in the refugee movement at Oranienplatz. An activist proposed to secure a place inside the school only for women. She went to a plenum at Oranienplatz to announce our plan to take a part of the school and keep it as women's space. The men said, "Well, we have to discuss this, because we are not sure if we need that". But activist woman said, "I am not asking you for permission – I am announcing that we are going to keep a space for us". That was really empowering, and it made me think. At the first assembly of the Women's Space, there were more than 30 women, mostly German activists with some experience in being part of a women's group. The Women in Exile,[5] a group that had already existed for some time, joined the Women's Space group.

A crucial incident happened that involved the people in Oranienplatz. A woman wrote in Indymedia[6] about her experience of being sexually attacked, which created pressure on the Women's Space to take a position. But we knew nothing about what had happened, nor did we know the woman or the man involved. Due to this incident, some women activists left our organization as they did not feel comfortable in such a situation, and we did not know how to move forward. This was the first time we realized that we had a problem and that we had to find ways to cope with such a situation – such as what to say when a woman reveals that she has been sexually abused. Those were really the sensitive moments that's when our group could have fallen apart.

MIRANDA: We wanted to do the work outside, mobilizing the women in the Lager, but the internal issues took a lot of our energy. We learned to divide the internal and the external activities. There are different levels of internal work related to the women because we have to self-manage the space, and simultaneously we are part of the bigger movement, which fights for the right of all, women and men. Sometimes it seems to be quite stifling when unforeseen situations and challenges arise because among the women activists there are also conflicts owing to many differences, in terms of our backgrounds, understandings and ideas to cope with a certain challenge for some women supporters and activists, the stress of self-managing school was too intensive.

Is the Women's Space now completely accepted by all the squatter occupants of the school? Were there any discussions or debates continuing on sexism?

MARISSA: Sometimes you do not have to wait for people to accept it; sometimes you have to enforce it. And that is what we have really done. Men are not going to sit back and give space to the women. Men have always tried to come inside the Women's Space; they have sneaked in from time to time. We had to be firm and ask them to leave, mention that, "this place is for women. Men are not allowed here." One time a man came in when it was snowing. He laid his stuff and blankets at the corridor of the Women's Space, right next to our office. I came along when he was sleeping and asked, "Why are you sleeping here?

This is the Women's Space". He said: "What? Have I disturbed you? Have I entered your room? Get out of here!" He was very aggressive. When I called my partners, and they wanted to have a dialogue with him, he still refused to cooperate. All other women came out. Seeing the number of women around him, he left. While leaving he abusively said, "You are chasing someone out, who has been supportive of you in securing this place from the beginning".

It is often difficult to convince the men or to make them understand, that this is the only place for women (one wing on the second floor) in a huge school, which has four floors. It is not easy; the men, not all, think we, the women, are a bit selfish or something. But that is not the case. We had many discussions and this is something we are doing every day. There came a time when we began to lock the main door to the corridor of the Women's Space, and they began to realize that this is a Women's Space and that we are affirmative. In the beginning we had, besides the lock, a wooden bar as well to secure the door because the men tried to break-in the lock. Several times we had to chase the men out.

GABRIELA: We have to understand that the school is an open space, where you have new people arriving every day, and they don't necessarily have any idea about what is going on. The women who live there try to make sure the new people are informed about the women's space. Once we hung posters on the walls with slogans against sexism. Some men did not understand it. When they saw the posters, they thought we were offering to sell sex.

MARISSA: Some people do not even know the concept of "sexism".

MIRANDA: But also we have to add that there are men who do understand what is written on the posters and still took the posters off of the wall.

How many of you are members of the Women's Space? How are you organized? What is your structure like? How do you make decisions and what is actually happening in the Women's Space?

MARISSA: We are about seven living in the Women's Space, but what happens is that women come and go when they find a better place. The school is not the best place to live in. Don't forget that people live here because they do not have another alternative. Immediately, when they get a better option, they leave. We had a woman with three children who left after finding something more stable. We had another lady, who went to her mother. At present, in the room where I am staying is shared by three women. In another room there are three more women, and two occupants in another room. Usually two or three women share each room. We are seven or eight women right now.

GABRIELA: In the Women's Space, we have one room to hold our meetings, another room for German classes and workshops, where we also have a free shop to place donations of clothes, etc. and mattresses for emergency use. One room is kept exclusively for women who need a shelter for a shorter while, say a couple of nights or a bit longer. The women here also decide how many women can share the space with them.

MIRANDA: For emergencies, it is acceptable. For instance, right now we have a Roma woman and her three kids sleeping there for two or three nights.

What kind of activities do you have here?

MIRANDA: The German classes are held two times per week, and it works very well. Apart from the women living here, there are women coming from the Lager to attend the classes. For a time we had a regular sewing workshop.

GABRIELA: There is a group of lawyers who come every other Monday. We try to organize two workshops per month, and we just finished a small brochure translated into many languages (Arabic, French, English, Russian, etc.) about the legal rights of migrants, which we are going to distribute in the *Lager Eisenhüttenstad.*[7] Some workshops are recorded, so that we can later publish the content. So far we have done workshops about sexism, legal support, self-defense, domestic violence, where we have used testimonies and books like "Let me Speak!" by Domitila Barrios de Chungara, a Bolivian labor leader; and "I, Rigoberta Menchu" by a Guatemalan Quiche Indian woman Rigoberta Menchu as references.

How do those workshops work?

GABRIELA: We usually invite someone from a group who works on the subject in which we are interested. For example, for the workshop on "sexual violence against women", we invited a woman speaker from LARA.[8] LARA offers free administrative assistance for women and girls who have suffered rape, sexual offence or assault or sexual harassment.

MIRANDA: The workshop on "undocumented work" was with a group from the syndicate ver.di together with the lawyers (see Borgstede on ver.di in Chapter13, in this volume). Usually we try to pay for the transportation and tickets for women who come from the Lager so they can attend the workshops of their interest. Sometimes we bring flyers to the Lager to invite them. We had workshops with 15 to 20 women. This was on a self-defense course we organized at the beginning.

GABRIELA: We have many activities, but what really consumes time is the everyday life in the school. We struggled for a long time to get all the permissions from here and there for the Roma children and adolescents to get accepted in a school. Later we had to find out how to pay for their transportation to the school. Altogether there are around 15 children/adolescents living there, and most of them now go to school.

So you also have some support from hospitals that offer treatment for free?

GABRIELA: Every Saturday a group of doctors come to the school. We also have some doctors we can call in case of an emergency, who have also tried to admit patients in the hospital even for the undocumented people.

MIRANDA: But every situation is different. Sometimes you go to a clinic, which treats migrants for free but you can receive very racist treatment. They speak German or little English, and then they can be very rude; they do not have patience to hear from the patient and do not let the person speak. They do not want to sort it out, and it demands a lot of attention, stressing the patient. One time I was there, and I tried to explain that someone was outside, and that I could help translate for that person. But they didn't make any effort to understand what I was saying. They made it so difficult that the medical problem was not sorted out. And then they came out and told the person that she should bring a translator. This person felt discriminated and maltreated and did not want to go back to this hospital, her pride was hurt. So this is just one example. Sometimes the hospital is very far away and difficult to reach or the person cannot go alone, needs money for the transportation and like.

Do you make all decisions collectively?

GABRIELA: Yes, exactly. Every Saturday at 2 pm we have our plenum (assembly).

MIRANDA: We have written down the structure of our plenum on the wall of our meeting office. We take the points, list the announcements and the external topics brought up from newcomers. After that is finished, we go on discussing our internal issues.

GABRIELA: We have many differences among ourselves. We know that for many women it is difficult to come to a space very much dominated by men. This is now clearer to us, one year after the occupation. Now we understand that many won't work inside the school on a daily basis but will be available for specific activities when they are invited. But we have been getting support from many women's organizations in Berlin, especially in this district of Kreuzberg. We have a great network with some groups.

Do you also have connections to other house-projects, (ex-)squats?

MARISSA: Not really.

GABRIELA: We have good connections with X-B-Liebig 34 (an Anarcha-Feminist, FLT Collective and Social Living House-project), where we threw a solidarity party, and New York in Bethanien (Social Center and House-project), which is also open to us.

MARISSA: We get support from other organizations, like churches or the Muslim community, who often bring clothes and a lot of food. Especially the Muslim community has brought a lot of stuff and has been very helpful. Once they brought so much food that we could share it in the whole house.

What do you think about the German left, the anti-racist movement?

GABRIELA: I do not like this theory of privilege explaining everything, feeling guilty of having privileges, often used by the German left because they are white, have a passport, can travel, and work. We are not fighting to get the privileges of the native Germans. The fact that white western people will encounter more open doors here and there doesn't mean they are able to understand the choices they have or that they will use these privileges in a way that will change the world into a better place. I have seen more often the so-called underprivileged people making these changes. That a political person is aware of his/her privilege is the least I can expect. But to create a false hierarchy where the privileged people, all of the sudden, pretend to be learning from the oppressed just because they have realized how superior they are, is nonsensical. In my opinion, people should get together when they empathize with others, when they share the reasons to fight together, when they think the struggle is also for them, they should fight not because they feel guilty or have pity. Once I was in a meeting and I wanted to say something about a certain situation involving the refugee struggle and a German supporter said to me, "This is not about you". And I replied that it is about me, it is about all of us. The person laughed. I understood that, at that point, I have lost even my history as a migrant, I was being looked at as a semi-white privileged person, with supposedly had more rights than the refugees in general, but less rights than the refugees when it comes to speaking of my experiences and challenges. I remember asking myself that if it is not because of all of us, what are we doing here? Exercising benevolence? It was a weird feeling and it showed how big our challenge is when it comes to solidarity struggle on a non-hierarchical plane.

MARISSA: Our schools are just as good. Many African families put all of their money into education, but then when you come to Germany nobody acknowledges it. Your degree is not recognized.

Here you can work as a cleaner or in a hotel, making the beds, as a waitress or maybe in the kitchen of a restaurant. And that's because you come from Africa. It is the opposite for a German in Kenya, where he or she becomes a boss because all their qualifications will be accepted. They will be given the best jobs, the best cars and the best houses. There are no Germans in Kenya who want to come back to Germany. They do not have to come back, they live well and they are accepted there.

It is the contrary for us. I come to Germany and whatever education I have is useless. Nobody is interested. It's like, if our schools are not schools, if our education is not education – which is not true. You are treated here as if you have never seen a hospital, as if there are no doctors in your country, whereas from Kenya thousands of nurses are exported to the UK. When South Africa became independent, many nurses and doctors were exported there, as well as to Namibia and Botswana. Sadly the only picture you see here is the wild Africa, the *Massai* in the wild. They do not represent or show anything about our hospitals or other technical professions, our doctors, our lawyers or our education system. But when you come as a doctor here to Germany, nobody will allow you to work here. What I am saying is that we should not be judged by the color of our skin or where we come from. More important is what we are as a person, what we know and what you are doing. I have a friend who was looking for a job and was told by the Germans that, "You are aiming too high".

GABRIELA: At IWS, we try to explain to the supporters that we are not victims, and we do not need someone for help. Help is not what we need. We need people who want to work together with us. Sometimes girls from universities come and say, "How can I help you?" And it ends up that we need to help them finish their studies or PhD theses. We are the living material for their work. Sometimes we say, "Hey, come on, we know who is going to help who here." These are things we have to say over and over again.

There is a lot of racism, and the society has a long way to go if they want to get rid of it. I have lived and squatted in the UK in the late 1980s, and I never faced the racism that I face in Germany, every single day. Consider the language as an example. The postman comes to my house and my name is Ramires-Boll. It is written at the door. Then I choose to sign Boll, because it is shorter. Then the postman asks me: Ramires? If I wrote Boll it is because it is also Boll, but Boll is a German name and he/she gets suspicious because he/she can't make any connections between my looks and my surname.

MARISSA: We went to the Vivantes-Hospital and went to the reception. There was a lady, at the reception, and she was on the phone. She took her time, talking on the phone instead of attending to us. Then she hung up and did something else. Then she picked up the phone again. She was not even looking at us as we were waiting in front of the window of the reception. She just wanted to boil us up and was totally ignoring us. Eventually she spoke to us. We gave her the names and she kept asking, "WHAT, what name?" She only spoke in German and didn't look at us. It was so annoying. She was so rude. You could just see the hate in her attitude, which meant that, "I am obliging myself to talk to you".

MIRANDA: You could clearly see how they treat a white person better than how they were treating our friend who was black. And when the doctor came he talked from a distance. With the white patients he got physically closer. It was so obvious. We went there at 1 pm, and she wasn't admitted until 10 pm. It took 9 hours for her to be attended to by the hospital. And then, when they admitted her, they first put her in isolation or quarantine because she is from Africa, and they were suspecting that she had a very contagious disease or something.

GABRIELA: This also helps us understand our space. The people living in the school are vulnerable, and the school itself is vulnerable. Every day you see it, and every day you start to solve all sorts of problems collectively; this is also why we are there. Since recently, every Friday we have meetings with the district to try and sort out the situation of the school. We know the problems we have there cannot be solved on the district level. We know we are dealing with federal regulations against asylum seekers and migrants.

But there are other problems too, which do not make the situation easier. The black people living here have to face problems daily whenever they step out of the occupied school. So you cannot expect them to come back to the school in the best of moods. There are many young people here who just want to live like other young people around. For instance, when they go to a club, security doesn't let them in. There was a woman who worked in a place, and the boss said, "I am not going to pay you". And to whom could she complain? Since this incident, we participated in a workshop that was organized by another group about the working rights of illegalized people and what they can do when they work and do not get paid (see Filhol Chapter 18, in this volume).

And then, as you said, you have a lot of so-called German white left groups or people who are paternalistic?

GABRIELA: Absolutely paternalistic, they have to stop "el colonialismo metido en la cabeza, asi" [the colonialism in their head]. There are Germans in our group who are conscious about what's going on, but this thing of "I'm guilty, because I'm white" must stop. If they were really willing to struggle against this enormous guilty feeling, they could start by understanding that a woman who came all the way from a far away country with no visa, money, or connections, and ended up finding her way to the occupied school, to the movement is a capable person. This person cannot be treated as a child who is learning about life from scratch.

CARMEN: I think the German left has done little work in the sense of understanding what racism means in reality. There exists the theoretical discourse about racism, but in the base, no real work has been done with people who are victims of racism. Actually the people of color should attend more workshops about racism. There should be such workshops regularly in this country, which underscores it was a colonial power and still is. This maybe could be a way for the German left to fully understand what that means.

This society is based on bourgeois principles and structures, and many on the left come from this bourgeois background and simulate an understanding. But in reality they do not want to give away their privileges and structures. I always remember the Anti-fa, who do politics somehow as a "left sport"; they do not want any real change because they have this middle-class thinking. They have the system somehow in themselves.

I come from South America (Chile), from a family, who were partly members of the Revolutionary Left Movement MIR[9] (Movimiento de Izquierda Revolucionaria), which is a Chilean political organization and former guerilla organization, and for me it is very strange. I am in the Queer Scene and with the Transgeniale-CSD.[10] In 2013, we had a racist incident,[11] and now this community is split because these dominant structures appear again and again, although there exist these anti-racist discourses. When you cannot trust the left, the radical groups here, then could you trust the rest of the society? The left is somehow still saturated in this colonial way of thinking. These are old structures, but many deny it. They do not see that and when you criticize that – either they remain silent or do not want to deal with it right away. But you have to deal with it, you have to work with the people and listen – and yes – it probably hurts.

GABRIELA: It scares me that if the authorities close the school and evict Oranienplatz, refugees will be sent back to the Lager, to the isolation; people will be forgotten and hardly anyone will join the fight anymore. I fear this because this has happened in the past. For a long time, Berlin was quiet about the struggle of the refugees although in other parts of the country there was a lot of activity and work done by refugee- and migrant-groups like *The Voice of the Caravan*. I am afraid that if our fight will be disrupted if we are dispersed, if we don't have an address like the occupied school or the camp in Oranienplatz, the repressive system will devour us.

At least five thousand people still come every year to the solidarity-demonstration for Silvio Meier, who was killed by the Nazis in Berlin in 1992. And this is very important, but how many people come to express their solidarity for Mete Eksi, Ouri Jalloh or Antonio Amadeu or many other non-Germans and/or persons of color who have been killed?

MIRANDA: We were mobilizing to go to the just-opened Heim (refugee-camp) in Berlin-Hellersdorf (suburb of Berlin) in 2013, but many refugees were afraid because of the Nazis. People said that being African, having a dark skin color, would make you a target. Though we wanted to go, there was also the anxiety and fear for our lives because we knew there was a strong Nazi presence. But not just the Nazis, ordinary people came out from their houses, the German people, to demonstrate against the Heim and against, a hundred or so, Syrian refugees who came to this area.

So it's like, if you are black around there, you are just one person, isolated among so many Germans. If you are attacked, you will not even be able to spot who did it as there will be so many of them. I mean, if such a thing happens then who is going to defend you? Who are you going to take to court? So sometimes, there is no point of putting yourself into such problems. For instance, at the demonstration, though many of us wanted to protest against the Nazis, almost none of us, Africans, had the guts to participate.

MARISSA: Many times, I have experienced open racism at the train. When I was working around Lichtenberg (a district of Berlin) and had to take the U1 from Warschauerstraße, wherever I sat no one wanted to sit next to me or close to me or even on the opposite seat. The train could be overcrowded but still no one would come close to me. And this has not happened only once or twice. I became so scared that I said to myself, "Someday this exclusion will kill me!" So I changed my route though it took longer to travel because I got scared for my life.

And what about the experiences with the police?

MIRANDA: Anytime somebody is sick and you call the ambulance, they come together with the police. And when the ambulance comes, they don't take the people. Last time they left somebody who was sick, and the person couldn't even walk, but they said the patient should walk to the hospital and left.

And when the police enter the building, how do they react?

GABRIELA: Last time the police went to every room.

MARISSA: They checked our identification, which scared us.

MIRANDA: They went through the whole house, came in riot gear, checked IDs, held people in their rooms and detained some of us. The police inspected for many hours and blocked every activity until they were done. No one could get out of the building. We had to sit and do nothing. In half an hour, there were a hundred policemen and policewomen. They stormed the Social Center *Irving Zola* (which, meanwhile, was used by the refugees) and photographed all the occupants. This incident happened in December 2013.

Do you have any connections with other anti-racist/women struggles from other countries?

GABRIELA: We hope that with the "March for Freedom" to the European Union from Strasbourg to Brussels in May/June 2014[12] we will be able to connect and interact with other groups, broaden our network and alliance. However, we still have to find out how many of us can technically join the march, because refugee women and men cannot leave the German borders.

How is the relationship with the local people of this neighborhood?

MARISSA: There were some complaints in the beginning by one person, I remember. He wanted to collect signatures against us, because he said we made too much noise and all that stuff, and that the occupied school should be closed. So we had to mobilize against this. Now it subsided. We have neighbors who bring clothing or other stuff or come to ask what we need.

Maybe the biggest enemy is not the neighbor, but rather the yellow press, popularly 'tabloid journalism'?

MARISSA: Yes, they come around and want to interview us. The first question they ask is if we have heard about someone who was raped here in this place. Last time, there was one journalist who even helped to clean up the place, and he did not say he was a journalist. Eventually he asked about six women who supposedly had been raped in this place. I said, "What? Six women, when did that happen?" This is just a newspaper story. I mentioned that we were living here, and we don't know of anyone being raped. Truly, the media flocks around this place with cameras to construct stories.

Do you let the journalists enter the building?

MARISSA: We have been refusing to give interviews, but sometimes they target us when we leave the building. They come and ask if something is true, we might say no, and then there is another question, and you find yourself answering. In the end, you have actually been interviewed without your knowledge.

GABRIELA: Last week, one woman of the house was quoted in the newspaper, but she had not even spoken to a journalist. They want to find stories they can publish and influence the public opinion against us.

MARISSA: They come every day and even want to take photos of the rooms. The entrance is open, so anybody can come in pretending they are just curious as a person and then find someone, who feels seduced to say something. One time the toilets were really fucked, and some people let the press take photos. The next day you could see how the media used these images against us. One journalist came into the Women's Space and opened the doors of the rooms to look around without any respect for our personal space. I asked the person to leave: "This is the Women's Space! What are you doing here? Get out!" This has even happened at night around 10 pm.

MIRANDA: When the incident of rape was posted in Indymedia in May 2013,[13] the mainstream media spotted it, they wrote terrible things, abusive articles, and we had to call a press conference at Oranienplatz to make our position clear. To our astonishment, we met some resistance from some refugee men, who were at the time very dominant at Oranienplatz. The idea that a women's group had organized a press conference to talk about women's issues was alien to them, and the men felt uncomfortable about the fact that the women were being protagonists. The impression was that we would be allowed to speak if we were going to speak for the men, in support of the men. Not being there to do such a job, these men could not comprehend why the women needed to hold the press conference or clarify their stance. On that day, there were many journalists, thirty maybe, and the women started to speak and were interrupted by one man on multiple occasions. The man even said, "Who are you? We've never seen you here working for us." It was really embarrassing for us. We were there to speak against sexism, against prejudice, showing our solidarity to the woman who had published her statement in Indymedia and also to the men in general, but were victimized for our efforts and common struggle.

GABRIELA: I remember that my body was trembling. I could not believe it. We had worked so hard with other men from the movement, to formulate our statements. And there we were waiting for a man to calm down and let us speak. However, we did speak after some other men of the movement were able to calm the agitated man. Surprisingly enough, the next day, the press did not mention the brawl, which was a relief.

What was the clarification to this incident?

MIRANDA: We are constantly there, and we have not seen or heard about rape. It was not just responding to the lady who wrote about being abused. She told about what had happened to her when she was involved with the movement at Oranienplatz, before the school was squatted. People were using this information to connect rape to the school that women were being abused in the school.

What do you think about what was written in Indymedia? How do you see it in the end?

GABRIELA: Our position is that she is telling the truth. A woman usually doesn't like telling a story such as that.

MIRANDA: Anyhow, the newspaper took advantage of this information, and we made the statement to remind the public that sexism and rape or assault or violence on women is everywhere in society. And the school is a part of the society. We wanted to show that as a group we were openly fighting against it. Our concern in this regard, what is the general society doing against sexism or violence against women? How is the general society dealing with everyday patriarchy? Unlike the media broadcast on the rape incident at the school, there are many sexual crimes that are commonplace in the wider white society, then why don't we see any big media coverage on those?

MARISSA: Take, for example, the Oktoberfest at München where every year women are assaulted.

GABRIELA: With the school, we have to be really careful because, in the racist fantasy, black men are sexualized, and black women are potential prostitutes.

What is the situation right now with the school? Do the authorities want you out?

GABRIELA: The talks are still going on. They want to turn the squatted school into a house for projects. They talk about resettling the majority of the refugees some place else. To give space to projects, the living space should be smaller. The majority of the people would have to leave and we know they would opt to send people to the Lager. And this is exactly what we are fighting against.

MIRANDA: The school is full. There are people sleeping even in the corridors under bad conditions. Some sleep next to the toilet if they have just arrived. There are different plans in the rounds of negotiation. The district says only around 20 to 30% of the place could be used for living spaces. And this is impossible because this is one of the prime alternatives to the Lager.

MARISSA: The negotiation meetings are not easy, and it takes forever to agree on anything. Because, where should we all go, when we have nothing? Why should we leave? At the beginning 50 or 60 people were participating in the meetings between refugees, activists and supporters. It takes time to develop something dynamic, something that will work better for all of us.

MIRANDA: The main topic now is the security, and the refugees themselves want to take responsibility for that.

GABRIELA: Now the Senator for Integration in Berlin, Dilek Kolat, has come offering to negotiate with us. The Senate wants to have a list of the people living in the school as well as to know in which parts of Germany people have their cases registered. They say they could transfer their cases to Berlin. There is a group from Oranienplatz and the school who meets with the Senate on this matter. The district won't participate in this meeting. So if the Senate transfers the cases to Berlin, then refugees would be staying in Berlin without fearing the *Residenzpflicht*. Residenzpflicht means mandatory residence. It is a legal requirement affecting specifically applicants of refugee status or those who have been given a temporary stay of deportation. Those affected are required to live within certain boundaries defined by the applicants' local foreigners' office. Residenzpflicht only exists in Germany or within Europe, and several migrants and refugee advocacy organizations have opposed the Residenzpflicht as a violation of fundamental human rights.

MARISSA: But we don't know if we can trust the Senate.

MIRANDA: The Senate interferes because they say the school has not been managed well, that we are devaluating the market value of the school property. The district operates the school, but the district cannot sell the house; to sell the house, an approval from the Senate is required.

MARISSA: There is a lot of pressure coming from the media. All of these bad reports about violence create pressure on the district, and if the police decided there was imminent danger in the school, they would storm in. And the Senate would take control of the situation without having to share decisions with the district.

GABRIELA: Many demands we have are of the responsibility of the Senate of Berlin and of the Federal Government. They have to get involved. The house was not only squatted, it is a part of a movement whose main demands are to stop deportation; close the Lagers; and give permission to the refugees to stay, study and work. This is not some demand only on the level of the district's responsibility, but our demands are addressed to other levels of the Federal German Government.

Maybe some last words for today:

MARISSA, GABRIELA, MIRANDA, CARMEN:

Asylum to women persecuted because of their gender! This is political persecution! We need more recognition on gender-based persecution! End of Patriarchy!

Stop Residenzpflicht! Freedom of movement! Close the Lagers!

We want to live in apartments like everybody else.

We want the right to work, to study and to stay in Germany as long as we want.

And our struggle will continue…

More information

International Women's Space: http://oplatz.net/category/international-womens-space/
Ohlauer School: http://oplatz.net/category/international-womens-space/
Women in Exile, Refugee women get loud! : http://womeninexile.blogsport.de/
Respect, for the labour rights of refugee and illegalized women:
http://www.respectberlin.org/wordpress/
LIA, Ladies International Association München: http://www.lia-munich.de/
Refugee women rights: http://asylumlaw.org/countries/index.cfm

Notes

1 See: http://asylstrikeberlin.wordpress.com/refugee-women/http://asylstrikeberlin.wordpress.com/refugee-women/events/.
2 All the interviews were done by azozomox and Vanessa Diaz.
3 *Heim* or *Lager* refers to the official refugee homes, set up by the German government for asylum-seekers, and often situated very far from any infrastructure; in some cases even located in forests with little access to public transportation.
4 *Franz Schulz* was mayor of the district Kreuzberg from 1996–2000 and mayor of the district Kreuzberg-Friedrichshain from 2006–2013 (July 31).
5 *Women in Exile* is an initiative of refugee women founded in Brandenburg in 2002 by refugee women to fight for their rights. http://women-in-exile.net/ueber-uns/.
6 See: http://de.indymedia.org/2013/05/345257.shtml.
7 The Lager *Eisenhüttenstadt* is a reception camp and also a deportation prison and one of the biggest of its kind in Germany. See: http://lagerwatcheisen.blogsport.eu/category/allgemein/.
8 LARA offers bureaucracy-free help for women and girls who have suffered rape, a sexual offence or assault, or sexual harassment. See: http://www.lara-berlin.de/index.php?id=21&L=6.
9 See the website for more details: http://www.mir-chile.cl/.
10 T-CSD: Transgeniale Cristopher Street - Day (Gay-Pride) is a demonstration and festival, held each year in Kreuzberg, Berlin, to celebrate the lesbian, gay, bisexual, transgender and queer people, but also to protest against racism, gentrification, capitalism and nationalism. The T-CSD so far took place from 1998–2013.
11 One performance of the artist "Miss Pünktchen", which was an official solidarity act of the T-CSD 2013, included racist words – also the n-word. See also, the criticism towards the organizers of the T-CSD 2013 and the declaration and apology of the preparation group: http://transgenialercsd.wordpress.com/.
12 See: http://freedomnotfrontex.noblogs.org/route/.
13 See: http://de.indymedia.org/2013/05/345257.shtml.

Part V

Social centers, radical autonomy and squatting – beyond citizenship and borders

17 Beyond squatting

An autonomous culture center for refugees in Copenhagen

Tina Steiger

Introduction

This chapter depicts the mobilizations for refugees and asylum seekers in Copenhagen, and the emergence of an "autonomous culture house" that provides space for their empowerment. The idea of the Trampoline House emerged as a result of a series of workshops between students and asylum center residents during a time when a center right-wing government, which propagated anti-immigrant and xenophobic rhetoric, came to power in Denmark (Rytter 2013; Johansen 2015). This is placed against the backdrop of Copenhagen's vibrant history of squatter movements, and the incidence of Iraqi refugees and activists squatting a church in the summer of 2009. This chapter sheds light on the autonomous operating structure of the "Trampoline House", drawing its strength from the diverse and broad set of actors involved – academics, activists, artists, students, refugees and civil society joining to actively challenge repressive asylum policies and their consequences for the everyday lives of people. The horizontal and direct-democratic structure, coupled with the involvement of this diverse set of actors, has allowed the house to become a remarkable project, providing refugees with a place of support, community and purpose on multiple levels.

Refugees squat a church in Copenhagen

> Danish officers stormed a church in the capital of Copenhagen to arrest 17 refugees who had been illegally squatting (*Hurriyet Daily News* 14 August 2009)

> Raid in Denmark to Dislodge Iraqi Refugees Leads to Protests and Hunger Strike (Saltmarsh and Contiguglia, *New York Times*, 14 August 2009)

In May 2009, a group of 60 Iraqi asylum seekers sought refuge in a church in Copenhagen's Nørrebro District (see Houston Chapter 14, in this volume). The progressive pastor of the church opened his doors to the group, a phenomenon that has often occurred throughout history, as the local situation of migrants becomes dire. In Danish, this is called *kirkeasyl* and refers to the concept of anyone, regardless of religious affiliation, seeking sanctuary from arrest or political persecution in a church. In Denmark, private and semi-private solidarity organizations have acted to help refugees since the 1960s, and especially young people have been involved in these organizations and networks (Mikkelsen 2011: 228–236).

The group of Iraqis was mostly composed of young men, but also included the elderly, as well as women and children, fearing deportation while Danish officials were stepping into negotiations with the Iraqi government about issues of political asylum.

In the early hours of August 13, 2009, the police raided the church, with the intent to detain those whose asylum applications had been denied (Saltmarsh and Contiguglia 2009). Outside of the church, a group of about 200–300 autonomous activists and refugee sympathizers had gathered, attempting to block the eviction and the transportation of 17 of the refugees to prison, and later the closed section of the Sandholm Asylum Center (de Laine 2009). In a matter of minutes, videos went viral on the Internet showing police beating activists and a brutal use of teargas against demonstrators who attempted to stop the eviction. The incident generated intensive news coverage, sparked demonstrations by thousands and led to a hunger strike by some of the refugees (Saltmarsh and Contiguglia 2009).

During the months leading up to the occupation of Brorson's Church, asylum politics had become one of the hottest and most contested issues in Denmark. Many demonstrations leading up to and after the eviction of the church brought together thousands of people in some of the largest protests against the liberal government and the Danish Folk party's policy since 2001 (Christiansen/Monsun 2010).

The Danish writer and columnist for the country's largest newspaper, *Politiken*, described the anti-immigrant and xenophobic rhetoric propagated by the party during a conference in Germany:

> [T]he Danish People's Party and the radical right in general have made it their trademark to stigmatize ethnic and religious minorities, often by blatant exaggerations and accusations based on prejudices and stereotypes. Misrepresenting facts about immigrants and their cultural heritage while criticizing human rights used to be the game of extremists with little or no significance in public debate, but this has gradually become an increasingly accepted and influential part of Danish politics through the influence of the Danish People's Party (Larsen 2014).

Although the eviction and dislodgement of the Iraqi refugees from Brorson's Church was legal under Danish Law, the government, if they had wanted to, could have allocated residence permits on humanitarian grounds to the arrested refugees (Christiansen/Monsun 2010). The mobilizations surrounding the brutal eviction of the asylum seekers can be understood against the backdrop of Copenhagen's vibrant history of squatter movements, which have claimed spaces for radical autonomous organizing since the early 1970s.

Cycles of squatter movements

By the mid-1970s, Copenhagen witnessed the emergence of the first generation squatter movement called the *Slumstormers*, who occupied empty houses, sites and entire buildings in pursuit of autonomous and radical forms of living. The greatest relic of this time is the *Freetown Christiania*, a self-organized community of approximately 800 people that continues to exist on the city's medieval military moat. A second generation of squatters took over parts of the city by the early 1980s; these were the more radical and militant *BZ Brigades* who claimed various houses in the outer bridge districts. By the end of the decade, the large slum-clearance schemes, which had left many inner city buildings vacant, came to an end, and many of the squatted houses and Social Centers were evicted. Deprived of their central places of convergence, the squatter movement disintegrated and was submerged into various autonomous networks by the early 1990s (Karpantschof and Mikkelsen 2014). It was not until the mobilizations that followed the spectacular clearance of the Youth House in 2007, that a new generation of "street-level activism" was re-ignited, around which rights of refugee and asylum seekers became a central organizing point. Many of the activists who attempted to block the eviction of Brorson's Church were part of the city's autonomous scene and former squatter movement.

Workshops: activists and refugees collaborating

A few months before the occupation of Brorson's Church, a series of workshops organized by the Asylum Dialog Tank (ADT) were held in two of the country's most important asylum centers. The participants included asylum center residents, working together with students of architecture, social work, fine arts and Architects Without Borders, who collaborated to develop ideas and strategies on how to improve living conditions for asylum center residents in Denmark.

The fundamental approach to the workshop was to consider all participants equal, regardless of their legal status. With this they hoped to negate stigmatizing labels, such as *Phase 1, 2 or 3 Asylum Seeker*, given by the Danish Aliens Act, probably the toughest Alien Act of all members of the European Union. They believed that reducing individuals to such labels led to victimization and proliferated the notion of asylum seekers as "unreliable individuals". Therefore, refugees and students received the chance to choose which identity they wanted to adhere, within the realm of the workshop (Goll *et al.* 2009). Over the course of several days, the workshop turned into a social experiment in which the refugees became the experts on asylum law and asylum center living, while the students were invited to help find alternative solutions to the Danish system. By means of video workshops, a reflective dialog and a final presentation, participants analyzed the asylum centers in question and collaboratively developed alternative solutions. What the workshop found was that asylum seekers were generally satisfied with the physical conditions at the camps, and even expressed that "they could endure the disrespect of their human rights, if only there was a set time limit to the situation" (Goll *et al.* 2009).

The participants ultimately concluded that the core problem was structural in nature and set by the Danish Aliens Act, which prohibits asylum center residents from working or pursuing an education. Although the centers were built to accommodate refugees temporarily (no more than 12–24 months), refugees often spent many years in the camps waiting for their application to be processed. As Goll *et al.* (2009) observed, this leaves asylum seekers living in enclosed, parallel worlds, often facing isolation, poverty and mental paralysis – deprived of the basic rights to a self-sustaining life. From the perspective of public opinion, the consequence of the Danish asylum policy caused many Danes to perceive asylum seekers as "costly clients" and "parasites" to the welfare state, rather than considering them potential colleagues and contributors to society.

Upon collaboratively reaching the conclusion that the Danish Aliens Act is the biggest obstacle, workshop participants explored ways of changing voters' perception of asylum seekers, while at the same time empowering them. Workshop participants proposed a self-managed and independent platform where refugees could develop and display identities of their own choice, while empowering themselves, and simultaneously change public opinion and fight for social justice (Goll *et al.* 2009). This is when the idea for the Trampoline House emerged. Rather than offering a mere stepping-stone to refugees, the house should develop into a user-driven space facilitating a jump-start into Danish society.

Toward autonomy by self-organization

In the Danish context, "user-driven" is often used by organizations, projects and spaces to connote that they are both independent from state or private institutions, but also managed by the users themselves. The term implies a horizontal and non-hierarchical operating structure, and is often associated with squats or other activist organizations of the radical left. This is similar to the differentiation made in Spain, between *Centros Sociales Autogestionados* (CSA), Self-managed Social Centers, and *Centros Sociales Ocupados* (CSO), Squatted Social Centers.

Autonomous movements, such as the Danish squatter movement, believe in a "politics of the first person" and are organized along principles of social autonomy, based on mutual aid, collaboration, and direct democracy. Decision-making structures are non-hierarchical, and since there is no belief in strict ideology, internal processes are made by intuition, following what Katsiaficas (2006) calls a "rationality of the heart". This is a rationality based on human reason, dialectically intertwined with passions and emotions. Through this, autonomists believe we can gain back our inner meaning, which has been colonized by the capitalist system (Garland 2007). By adhering to our emotions and simultaneously reason, a society based on equality and free of exploitation can be built. Horizontally organized, social autonomy facilitates discussions and actions by enabling numerous and diverse inputs, whose approval depends on collective consensual agreement. Autonomy is a direct-democratic form of decision-making, creating communities governed by participants and not managerial prerogatives or representative bodies (Katsiaficas 2006).

At the collective level, we may consider autonomy as the "right to self-government" particularly in relation to the state and market, whereas the autonomous *project* can be understood as resulting from the tension between this collective-individual dichotomy – involving a group working together along principles of cooperation, reciprocity, equality, and freedom in order to create alternative ways of living (Katsiaficas 2006; Pickerill and Chatterton 2006). Since squatting has become virtually impossible in Copenhagen since the early 1990s, the city's autonomous movements have found what Martínez (2013) considers anomalous forms of institutionalization. In order to sustain themselves, many squats and formerly squatted Social Centers have found new, often legal ways of self-management, while maintaining as much independence as possible from state or private institutions.

The Trampoline House

> Trampoline House is an independent community center in Copenhagen that provides refugees and asylum seekers in Denmark with a place of support, community and purpose.[1]

In previous years, the Trampoline House described itself as a user-driven culture house, rather than an independent community center. The reason for changing their description to an "independent community center" does not mean that the internal operating structure no longer strives toward direct-democratic forms of decision-making; rather the alteration was made to appeal to donors who may be wary of radical left activist projects.

Although the Trampoline House has a board of directors and five paid staff members who take care of programming, coordination and fundraising, it is based on a horizontal and participative governing structure. Since 2014, about 50% of the funding comes from the Danish Immigration Services, while the remainder comes from the private Oak Foundation, along with other smaller sponsors such as the Roskilde Festival Charity Society. Receiving financial support from these institutions has of course impacted the working of the Trampoline House, since energy and time must be vested into activity reports, communication and the transparent allocation of funds. Also, collaborating with institutional donors has made the project hesitant about publicly supporting or endorsing radical demonstrations and initiatives defending refugee rights. Overall, the content and agency of the project has not been limited by its collaboration with state and private institutions.[2]

All major decisions are discussed during the weekly house meetings, which are attended by approximately 50–60 people – volunteers, activists, staff members, and asylum seekers. This forum not only allows participants to have influence over decisions affecting the direction of the Trampoline House, but has also proven to be an effective forum for letting users speak their mind, contribute and vent their frustrations. The open nature of the house meetings have shown to build up the self-esteem and self-worth of many asylum seekers who find themselves in otherwise despondent mental states, while desperately waiting for their applications to be processed.[3]

Within the space of the Trampoline House, emphasis is placed on neither treating people as victims, nor glorifying them as heroes for the journey they have made. Many refugees who have spent long periods of time within the asylum system are in danger of regarding themselves as helpless victims of an unjust system. This has proven dangerous because it often leads to self-victimization of the individuals in question, blocking them from developing their identity further and making attempts to take control of their lives. These are fundamental cornerstones of the functioning of the house, allowing it to become a collective space where people can help one another. In addition to offering language classes, legal aid, and a range of cultural activities, the house mainly offers space for a community, which aims to increase the self-worth of asylum seekers and refugees.

Although the house is not squatted, in the sense of unlawfully occupying a building, its internal organization has strong similarities to those promoted by radical autonomous movements. By means of a horizontal, participative governing structure, decisions are collectively reached by deliberation. Additionally, the Trampoline House maintains a strong network with other autonomous and squatted projects throughout Copenhagen.

During the Trampoline House's *Asyl Festival*, for example, screen-printed shirts and bags were made in collaboration with the screen-printing workshop of the "Candy Factory", one of the city's most vibrant autonomous spaces. To give insight into the multidimensional approach of the Trampoline House, in merging academia, the arts and civil society around the realities of refugees, the series of the festival may provide insight. The program of the Asyl Festival was comprised of a graffiti workshop, interactive installations about the everyday realities of detention centers, presentations and legal advice from representatives of the Danish Human Rights Council, hands-on workshops for making pirate radio transmitters, a Deleuze and Guattari reading circle, music, movies and a common dinner prepared by the people's kitchen. This integration of activism, art and academia ignited at the Trampoline House may offer insight into local and resilient alternatives to current realities faced by asylum seekers and those who are in solidarity with them. The fusion of actors involved provides an example of bottom-up local resistance to repressive asylum policies prevailing throughout Denmark, and Europe at large.

Notes

1 See: http://trampolinehouse.dk/.
2 The paragraph on the organization and fundraising of the Trampoline House is based on an interview with a coordinator (see Steiger 2011).
3 The paragraph on the decision-making process of the Trampoline House is based on an interview with Reima (see Steiger 2011).

References

Christiansen/Monsun O.W. (2010) *Rydning af Brorsons Kirke bliver mindet*. modkraft.dk progressiv portal. Available: http://spip.modkraft.dk/nyheder/article/rydning-af-brorsons-kirke-bliver.

de Laine, M. (2009) 'Danish police take Iraqi asylum-seekers from church in night raid'. *The Copenhagen Voice*. Available: http://cphvoice.ning.com/profiles/blogs/danish-police-take-iraqi.

Garland, C. (2007) 'Logics of resistance: Autonomous social movements in theory and practice', *Studies in Social and Political Thought*, 14: 81–87.

Goll, M., Hamou, J. and Nielsen, T. O. (2009) 'The Asylum Dialog Tank: A process of empowerment, agency, and social change', *Framework*. Available: http://www.hamou.artcodeinc.com/static/gallery/files/ADT_framework.pdf.

Hurriyet Daily News – Agence France-Presse (2009) 'Danish police storm church to seize refugees'. Available: http://www.hurriyetdailynews.com/default.aspx?pageid=438&n=danish-police-storm-church-to-seize-refugees-2009-08-14.

Johansen, I. V. (2015) 'The Danish People's Party: A journey to the centre', *transform!* Available: http://www.transform-network.net/en/journal/yearbook-2015/news/detail/Journal/the-danish-peoples-party-a-journey-to-the-centre-of-danish-politics.html.

Karpantschof, R. and Mikkelsen, F. (2014) 'Youth, space and autonomy in Copenhagen: The squatters' and autonomous movement, 1963–2012'. In van der Steen, B., Katzeff, A. and van Hoogenhuijze, L. (eds) *The City Is Ours: Squatting and Autonomous Movements in Europe From the 1970s to the Present*. Oakland: PM Press.

Katsiaficas, G. (2006) *The Subversion of Politics: European Autonomous Social Movements and the Decolonization of Everyday Life*. Oakland: AK Press.

Larsen, R. E. (2014) 'How the Danish People's Party challenges the open and inclusive society'. Available: http://www.panhumanism.com/articles/2014-01.php.

Martínez, M. A. (2015) 'How do squatters deal with the state? Legalization and anomalous institutionalization in Madrid', *International Journal of Urban and Regional Research*, 38(2): 646–674.

Mikkelsen, F. (2011) *Transnational identitet under forandring. Indvandrernes sociale, religiøse og politiske mobilisering i Danmark 1965–2010*. København, Museum Tusculanums Forlag, pp. 228–236.

Pickerill, J. and Chatterton, P. (2006) 'Notes towards autonomous geographies: creation, resistance and self-management as survival tactics', *Progress in Human Geography*, 30(6): 730–746.

Rytter, M. (2013) *Family Upheaval: Generation, Mobility and Relatedness among Pakistani Migrants in Denmark*. New York: Berghahn Press.

Saltmarsh M. and Contiguglia C. (2009) 'Raid in Denmark to dislodge Iraqi refugees leads to protests and hunger strike'. *New York Times*, 14 August 2009. Available: http://www.nytimes.com/2009/08/15/world/europe/15denmark.html?_r=0.

Steiger, T. (2011) 'Spaces of autonomy in Copenhagen and Madrid', Unpublished Thesis Université Libre de Bruxelles: UNICA Euromaster in Urban Studies 4 Cities.

18 When migrants meet squatters

The case of the movement of migrants and refugees in Caserta

Romain Filhol

Introduction

The main objective of this chapter is to analyze the case of the "movement of migrants and refugees in Caserta" (MMRC), which presents the peculiarity of being one of the few long-term experiments between Italian squatting activists and migrant workers in Southern Italy. Through my analysis, I provide details on various issues, including:

- How an area characterized by illegality, criminality and lack of rights can accommodate the rise of one of the most important social movements struggling for migrants' rights in Italy;
- How the MMRC managed to obtain significant improvements in the living conditions of migrants in the Campanian Plain and became a legitimate interlocutor with public institutions while still maintaining a highly disruptive capacity;
- How Italian squatting activists and migrant workers intersect in the context of Southern Italy; and
- How the MMRC has contributed to the recognition of migrants as political players.

To do so, I first describe the geographical context in which the MMRC has developed. Then, I explore how the local squatters' and migrants' trajectories have crossed to give birth to this social movement. I also describe the specific organization of the MMRC, linked with a particular repertoire of contention, to understand how they managed to obtain real gains for migrant workers. Finally, I consider how these gains, together with the migrants' participation in the MMRC, have contributed to the recognition of migrants as political players capable of challenging their exclusion, criminalization and precariousness.

This study realized in the Campanian Plain is based on empirical data collected between June 2011 and June 2015. During my fieldwork I conducted several in-depth interviews with the founders of the MMRC as well as with its current activists. I also realized a countless number of observations during the various moments of the MMRC's life, such as activists' meetings, everyday practices, and demonstrations. I also used theoretical documents, various media products and statistical data produced by the MMRC. Finally, my insight about the MMRC could not have been this accurate if I had not actively taken part in its initiatives since 2011.

This analysis intends to be a contribution to the recent literature about the issue of migrants' struggles in Italy, which has been popularized and developed by alternative media, political activists and various researchers (e.g. Brigate di Solidarietà Attiva *et al.* 2012; Caruso 2011, 2015; Outis! 2011; Montagna 2012; Oliveri 2012; Brigate di Solidarietà attiva *et al.* 2012). These writings have underlined the significance of migrants' struggles not only over citizenship issues but more generally over labor issues, enlightening the relationship between migrants' labor exploitation and division of the working-class.

This connection between citizenship and work issues has been particularly straightforward during the protests led by migrant agricultural workers in Southern Italy (especially in Nardò in Apulia and Rosarno in Calabria). However, the analyses of these events have tended not to discuss the fact that these spontaneous revolts have not really been translated into long-term experiments of migrants' struggle. This is why this chapter attempts, through the experiment of the MMRC, to give some insight into the difficulties of building a common movement between local activists and migrant workers in Southern Italy.

The Campanian plain between informality, illegality and job exploitation

Understanding the peculiarity of the MMRC would be impossible without a brief description of its geographical context, characterized by job exploitation, informality and illegality.

A rurban space characterized by low-skilled work, illegality and property speculation

Situated between the Mediterranean Sea and the Sub-Apennines mountains, the Campanian Plain is characterized by one of the highest unemployment rates in Italy (21.5% of the active population in 2014 for the Province of Caserta), while its economy is marked by the weight of low-skilled work in agriculture and construction (officially 5.5% and 8.8% of the total employment, with rates of unreported employment of 25% and 33.7% for the Campania Region; Istat 2015).[1] These two features define a labor market characterized by low salaries (around 30 euros a day in agriculture and construction), precariousness and absence of regular work contracts (Caruso 2013; Filhol 2013).

The Campanian Plain is also characterized by a landscape we can describe as "rurban", composed of a strange mix of agricultural plots, greenhouses and tall buildings (Caruso 2013). This configuration, generated by the extension of the Neapolitan agglomeration in one of the most fertile Italian agricultural plains, has been particularly quick and uncontrolled, especially after the 1980 *Irpinia*

Earthquake, from which the *Camorra* imposed itself as one of the most important "urban planners" of the area.

Finding its origin in the poor neighborhoods of Naples during the nineteenth century, *Camorra* is a mafia-type criminal organization. Historically, we can distinguish an "urban Camorra" specializing in smuggling from a "rural Camorra" acting as a mediator in the food supply chain (Sales 2006). Since the 1970s, its role in the heroin and cocaine trafficking gave it a huge capital, while the reconstruction of the Campania Region after the *Irpinia* Earthquake was the perfect occasion to invest. From this moment, by capturing the market through the creation or purchase of companies linked to the construction sector and by grabbing public tenders, Camorra entered the legal economy in collusion with the local political class to produce the "rurban" landscape described above and made of abusive, unfinished and low-quality constructions (Cesoni 1991; Anselmo 2009).

Squatting to struggle against property speculation, Camorra and political corruption: the case of Caserta Ex-Canapificio

In such a context, squatting is of fundamental importance. As Mudu (2004) argues, one of the biggest achievements by Italian Social Centers has been to present an alternative to property speculation. In the Campanian Plain, it has also been a way to fight Camorra interests in the construction industry and the widespread corruption of the local administrations. In fact, many of the Italian municipalities dissolved for "mafia-infiltration" are located in the Province of Caserta (Rinaldi 2013).

Social Center *Ex-Canapificio* in Caserta provides a good example of the way in which an occupied space constitutes a concrete way to struggle against property speculation, Camorra and political corruption (Filhol 2016). Set up in 1995 in a former slaughterhouse by a group of political activists who succeeded in avoiding eviction, the Social Center began to develop popular cultural and social activities. But in 1998 squatters were asked to leave by the newly elected right-wing municipality, because a local entrepreneur linked with Camorra won a tender on a European project to make a multimedia library from it. Harassed by police, the squatters led a series of public protests against the municipality. Benefiting from popular support, they occupied the streets to pursue their activities and raided the city hall. Having already obtained the funds for the project, the municipality did not let them use the space again. But thanks to their protest, the activists managed to get free use of another abandoned space doomed to property speculation: a former hemp factory (*Ex-Canapificio*) that gave its current name to the Social Center.

From there, the squatters' collective led various struggles against illegality and property speculation in the Caserta area. For example, they pursued a popular campaign to prevent one of the last green places left in the city from being destroyed, and are still fighting to classify it as a public garden unsuitable for construction. They also became an active part of the *Terra dei Fuochi* (Land of fires) movement, which asks for the closure of the abandoned quarries used by Camorra to bury toxic wastes. Eventually, they succeeded in resisting all attempts of eviction since 1998, even while the latter evictions were proven to be linked with Camorra's interests in the construction industry.

A "waiting area" for "undocumented" migrants

The Campanian Plain was one of the first areas of Southern Italy to host international migrants and can be currently considered its largest "migratory hub". Eritrean and Ethiopian migrants were the first to arrive, in the 1960s, followed by migrants from West and North Africa in the 1980s (De Filippo and Morlicchio 1992). Similarly to the rest of Italy, Eastern Europeans now represent the largest migrant community in the Campania region. However, the most particular feature of the Campanian Plain is the presence of thousands of "illegalized" African migrants experimenting with living conditions characterized by social and spatial segregation (Filhol 2013).

The main reason illegalized African migrants have settled there is the possibility of finding a cheap place to stay due to the abundance of low-quality or abandoned houses. Italian landlords have not hesitated to rent these places to migrants, particularly without a legal contract. For those living in the numerous abandoned houses along the Domitian Coastline, the pattern has been different. As one migrant describes, "You can go there, and after a few days someone from *Camorra* will ask you to pay. If you can't, they will come back until you do" (Nigerian male migrant; interviewed by author, 11 November 2013).

Another reason is the possibility of finding a job, even while being illegalized. The growing "liberalization" of the labor market has made possible, every morning at dawn, the growing of informal recruitment practices of manpower on the main roundabout of the Campanian Plain for agricultural, construction, or even industrial jobs. These practices have been particularly hard to stop considering that "when employers know there will be a control, they tell us not to come the next day, and when the police arrive, they only find Italian workers with regular work contracts" (Ivorian male migrant, interviewed by author, 31 March 2015).

This is how the Campanian Plain became a "waiting area" for a lot of illegalized African migrants waiting for the recognition of a refugee status, or hoping for an amnesty as had happened in the past. And what at first looks like a "temporary" solution often becomes a permanent condition. As one migrant explains, "Here, we are like in a hole where they put us. The legal document is a rope that you have to seize, but once you manage to get outside the hole, they won't renew your document or something like that, and you will be back in the hole just like nothing happened" (Ivorian male migrant; interviewed on March 31, 2015).

When migrants meet squatters: why fight together?

Understanding the specifics of the Campanian Plain context enables us to analyze and discuss the way in which Ex-Canapificio activists and African migrant workers have set up an original and successful social movement, defined as a distinct collective process through which actors linked by dense informal networks and engaged in collective action are involved in conflictual relations with clearly identified opponents and share a distinct collective identity (della Porta and Diani 2006: 20).

The meeting between Italian squatters and migrant workers

The way in which Ex-Canapificio squatters have met with the African migrants from the Campanian Plain is represented by a series of events offering an interesting starting point to understand successive dynamics (Filhol 2016).

Located in a popular neighborhood, the Social Center gathered from the beginning a lot of different people, among them Senegalese migrants living in Caserta who had a lot of difficulty getting appointments at the police station to obtain or renew their residence permits. That is why squatting activists started a protest, together with the Senegalese Association of Caserta, which enabled them to conquer the right to get daily appointments at decent hours. From then on, local activists decided with the Senegalese Association to run a "migration help desk" to support migrants with their legal procedure, improving their function of mediation between local administration and Caserta's migrant communities.

This "mediation" role largely increased in 2002 when the "Bossi Fini" law was promulgated (Law 189, 30 July 2002). This law, although heavily restrictive for migrants, provided regularization procedures for "undocumented" migrants who had already lived in Italy for a long time. For the regularization procedures, African migrants from the Campanian Plain had to go to Caserta to present their requests, getting in touch with the Social Center's migration help desk.

Facing an increasing number of requests and cumbersome legal conditions to submit the documentation for being "legalized", squatting activists organized a meeting with the migrants to discuss the situation and decide how to proceed. The main decision was to organize a protest to obtain simplified application procedures. The demonstration led to the occupation of Caserta's main church and was a success in terms of migrant participation, leading to the birth of the "Movement of Migrants and Refugees in Caserta" (MMRC).

Social composition of the MMRC and its insertion in the migrants' networks

As explained above, the MMRC is composed of both local squatting activists and African migrant workers.

Ex-Canapificio's squatters are a collective of about 30 people. Some of them were part of the 1990s Italian antagonist movement (Dines 1999), while others have generally been engaged before with various students' collectives (Italian female activist; interviewed by author, 21 November 2013). Most of them are a gender-balanced mix of middle-class students and precarious workers from the Caserta area.

Approximately 3,000 people participating in the MMRC are migrant workers and generally come from West Africa, with an overrepresentation of Ghanaian people (see Figure 18.1). They are men in 82% of the cases, and their average

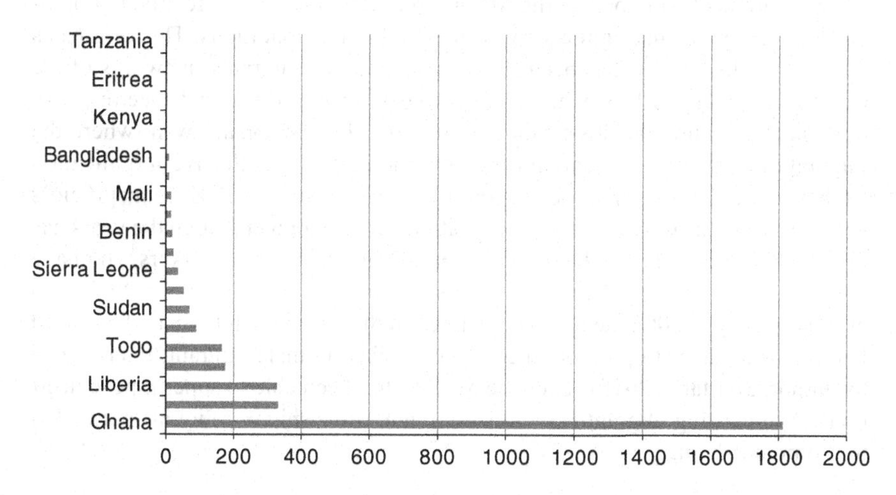

Figure 18.1 Migrant participating to the MMRC by nationality

Source: MMRC, data from October 2014

Table 18.1 Average birth-year of the migrants participating in the MMRC

Period	%
Before 1960	2%
From 1960 to 1970	16%
From 1970 to 1980	39%
From 1980 to 1990	39%
From 1990 to 2000	3%
After 2000	1%

Source: MMRC (survey conducted on 212 migrants taking part in the October 2014 demonstration)

Table 18.2 Average presence in Italy of the migrants participating in the MMRC

Presence in Italy	*Number of individuals*	*Rate*
Less than one year	7	10%
From one to three years	9	13%
From three to five years	25	37%
From five to ten years	16	24%
More than ten years	11	16%

Source: MMRC (survey conducted on 69 migrants attending the weekly meetings in November 2013)

birth-year in 2014 was 1978 (see Table 18.1). They tend to live in the same cities of the Campanian Plain and to work as agricultural workers, masons, or ambulant sellers. They have been in Italy a long time (see Table 18.2). Migrants' participation in the MMRC evolved along the years but has always ranged between 1,000 and 4,000 people.

The migrants taking part in the MMRC represent therefore a relatively homogeneous category, sharing the same spaces and living conditions. This originates by the way the MMRC has been able to penetrate the migrants' networks of the Campanian Plain. In fact, when local activists organized the first meeting with migrants to discuss the Bossi Fini law in 2002, they personally went where the migrants lived, to meet them and promote the idea of a collective mobilization (Italian male activist; interviewed on 13 May 2015). But if Ex-Canapificio's squatters were the trigger of the mobilization, its development along the years has only been possible due to the involvement of various migrant workers who have advocated for it among their compatriots and friends. An example is "P.", who arrived in Italy in 2002; he has always lived in the Campanian Plain and has used his role as a pastor to promote the MMRC (Liberian male migrant, interviewed by author, 30 March 2015). Thus the MMRC has been able to superimpose itself on the pre-existing migrant networks (national, friendship or religious networks) to promote its struggle.

Why struggling together?

In this section I highlight the mutual supporting basis that makes the MMRC strong. For migrant workers, the participation in the MMRC is the expression of a need. To the question, "Why are you taking part into the MMRC?" most of the migrants answer, "Because we need help for our documents". The need for a legal status is therefore what explains the migrants' participation in the MMRC, while satisfying this need is its reason to exist. The recognition of a legal status is the only way to have access to a regular job and benefit from all the welfare state services. This is also the only possibility for migrants to travel back to their countries. A residence permit also offers the opportunity of moving away from a place where the few available jobs pay only 30 euros a day.

For local squatters, the situation is quite different. They don't need any resident permit although they are involved in this struggle. Their commitment can therefore be considered a struggle for the most deprived group. To quote "F.", "When we saw their living conditions, we thought: if there is still humanity in us, we have to fight that" (Italian male activist; interviewed by author, 13 May 2015). This first thought was then articulated with general considerations about the fact that racism and segmentation of the workforce through different legal statuses are two of the main weapons of capitalism to discipline the labor force. Struggling for migrants' legal recognition therefore is not only a struggle for a specific deprived group, but also a struggle against capitalism in the name of all the oppressed workers, including themselves; in brief, re-scaling to connect a particular struggle to the global working-class struggle.

There is an apparent contradiction between migrant workers and local squatting activists about the reasons to take part in the MMRC. Although there is a common struggle, i.e., obtaining the regularization of migrants, their motivations are different. The majority of migrants just need to be supported and want to leave the Campanian Plain after having obtained a legal status, while local activists have engaged themselves in this struggle to radically change the space where they live. And yet, the long existence of the MMRC has proved capable of challenging this tension between the individualism of the migration projects, intended as projects of individual achievement, and the collective dimension of a struggle for all the working-class.

How fighting together can obtain concrete gains

The tension between the migrant individual's need to be supported and the collective struggle promoted by the squatters seems difficult to overcome. If the scale tips toward one side, local squatters risk transforming themselves into mere service providers, and even supporting the trend of withdrawal from the welfare state. But if the scale tips toward the other side, local squatters risk manipulating the migrants' needs to have them conducting a struggle that local workers seem to have abandoned long ago. For this reason, it is important to see how the MMRC has succeeded in finding solutions to this apparent dilemma.

"We don't assist, we struggle together"

Since its beginning in 2002, the MMRC has had to face the fact that it is impossible to fight for citizenship issues without linking them to labor issues.

The fundamental point is that migrants themselves reached this conclusion. After the first meeting, when local activists and migrant workers decided to organize a demonstration, some migrants raised the issue that, "If we go to demonstrate on a particular day, we will risk losing our job. What if other migrants don't go to the demonstration, take our jobs, and obtain regularization thanks to our own sacrifice?" (Italian male migrant; interviewed by author, 13 May 2015). So even if, by demonstrating about citizenship issues, migrants were unable to participate in a proper "strike", they immediately realized the need to find a way to protect their fight from the "strikebreakers" (Ghanaian male migrant; interviewed by author, 14 April 2015).

To respond to this concern, the MMRC took inspiration from a tactic used in the 1970s by the *Movimento dei Disoccupati Organizzati* (Movement of the Organized Unemployed of Naples) (Festa 2003; Italian male activist; interviewed by author, 22 November 2015). They made a list of people attending the demonstrations, with the purpose of ensuring the benefits of the struggle only to those who participated. In other words, the MMRC does not "support" the migrants, providing them assistance and receiving public or private funds to do so, but it "struggles" to help migrants obtain what they need. This approach appears very radical, especially when activists refuse to help migrants who have not taken part in the demonstrations; but "even if it's difficult, this is the fundamental part of the 'contract' that joins together all the migrant workers taking part in the MMRC" (Italian male activist; interviewed by author, 13 May 2015).

This particular approach represents a breaking point with many other social movements involved in struggles with migrants. Critics of this approach generally argue that "we want regularization for every migrant, not only those that struggle with us!" In the context of the division of the Italian antagonistic movement between anarchists and various fractions of post-autonomists (Mudu 2012), the strategy of the MMRC to accept political compromises and promote organization – making it close to the post-autonomist sphere, even if neither migrant workers nor Ex-Canapificio's activists define themselves in these terms – is frequently criticized. But these movements fail to recognize that the MMRC method has set up a successful series of struggles that have been able to last, to obtain significant results, and to respond to the tension between individual migration projects and involvement in a collective struggle.

A specific repertoire of contention: the "lotta-vertenza" method

During their 13 years of activity, members of the MMRC have used the same strategy of struggle that they call *lotta-vertenza*, which translates to "struggle-dispute". Like one of the squatting activists explains, "thanks to our migration help desk, we can see when various migrants have the same problems. So we study the law, then we try to meet with the representative of the public institutions in charge, and if we can't find an agreement to solve the problem, we organize a public protest [struggle] to put pressure on her/him until she/he negotiates and proposes an acceptable solution [dispute]" (Italian male activist; interviewed by author, 13 May 2015).

The success of this method depends on various factors: first, an accurate knowledge of the migratory phenomenon in the Campanian Plain, sustained by the "registration" of the migrants attending the demonstrations; secondly, a knowledge of the norms about immigration, and of the actual power held by the "street-level bureaucrats" who have to interpret it (Lipsky 1980); and finally, an extended solidarity network at the local scale, even related with progressive catholic groups, allowing the MMRC to reach a "critical mass" necessary to make the institutions bend.

The "lotta-vertenza" has made the MMRC able to obtain gains that have concretely bettered the lives of thousands of migrants. Gains range from the small ones, such as obtaining decent hours for appointments at the police station, to the most significant ones, such as obtaining a provincial commission in Caserta instead of the sole national one to examine the numerous asylum requests of the area and give them a positive result, or obtaining the activation of an amnesty (*sanatoria*) (Filhol 2016).

This strategy to be successful needs the promotion of public protests not only in the Campanian Plain, but also in other parts of Italy. In other words, the MMRC has been able to rescale its struggle to get closer to the "centers of power". For instance, it obtained a provincial asylum commission in the Province of Caserta after a demonstration in Rome, where migrants coming from the Campanian Plain participated, ending in a street occupation. This enabled the activists to meet the representatives of the Ministry of Interior who agreed on the requested provincial Commission in Caserta.

The repertoire of contention used by the MMRC is a mix of disruption and negotiations, where the disruption coming from the struggle has facilitated negotiation and has produced real outcomes from the dispute. Tangible outcomes obtained after negotiations have allowed larger mobilizations, because "mobilization grows when there are hopes that some changes can be obtained" (della Porta 2004: 28). Relying on a social legitimacy, both from migrant workers and institutions, the MMRC has channeled the disruptive capacity of migrants to convert it into tangible outcomes for all of them.

MMRC organizational patterns: a flexible structure relying on the Social Center

The MMRC's organizational structure directly reflects a division between daily activities linked to the legal support for migrant workers and particular events usually related to public protests.

On the one hand, Ex-Canapificio activists have managed the legal support that requires an expertise that migrants generally do not have and that squatting activists have progressively acquired since the inception of their first migrant support desk in 1995. This support desk is free of charge to migrants. To fund this activity regularly, on an everyday basis, squatting activists have therefore tried and managed to insert themselves in various public projects, like the SPRAR, a national network of reception centers for asylum seekers and refugees established in 2002 by the Italian Interior. Activists also organize weekly meetings to discuss the various projects and campaigns in which Ex-Canapificio is involved.

On the other hand, the activities linked with the strategy adopted by the MMRC and the organization of public protests are managed by a structure called by the activists "Staff", that gather Ex-Canapificio squatting activists and around 80 migrant workers who are among the most involved in the MMRC once a month. Obviously, migrant workers' participation in the "Staff" and in the decision-making process is fundamental because they are the ones who will ensure the participation of other migrants of the area, and discussion can last for hours before the formalization of any agreement.

The MMRC relies therefore on the Social Center structure and on a network of migrant workers that make the fundamental decisions on behalf of the MMRC. In the last instance, migrant workers decide, even if local squatting activists are generally the ones making the proposals. This can be explained by two factors: first, by the trust that migrants have in the most experienced Italian activists, considered responsible from successful past struggles (Ghanaian male migrant; interviewed by author, 14 April 2015); and second, by the deep knowledge of migration laws and institutional context held by squatters (Ivorian male migrant; interviewed on 31 March 2015). Commenting on this situation, one of the squatting activists explains, "If the MMRC is composed 90% of migrants, it is true that we Italians have a significant "counseling" role, and defining it as "counseling" may not be enough [...] We are conscious that we have an impure form of self-organization" (Italian male activist; interviewed by author, 13 May 2015).

Local activists are therefore aware of their paradoxical weight in the MMRC. But doing differently seems difficult. When I asked a migrant worker activist if the MMRC could exist without Italian activists, he answered, "No, it couldn't, because we as foreigners don't have the power: we don't know the laws" (Ghanaian male migrant; interviewed by author, 14 April 2015). However, "complete self-organization" is one of the objectives of the MMRC that tends to get closer through the years. For example, an increasing number of migrants, generally those who have already been involved in politics in Africa or who have decided to stay long-term in the Campanian Plain, are now part of the legal consultations held by Ex-Canapificio's activists. Yet, "complete self-organization" seems difficult to reach until migrant workers of the Campanian Plain have no possibilities of stable living conditions, education and knowledge of migration laws. I also argue that despite its "impure self-organization", the MMRC is one of the few political organizations where migrants really have a voice and are not only present to "make up the numbers". As a migrant activist states, "I saw a lot of organization for migrants, trade-unions and so on [...] And if the MMRC is different, this is because they truly pay attention to what you have to say, and not only use you to reach agreements you weren't a part of" (Ivorian male migrant; interviewed by author, 31 March 2015).

The construction of migrants' political subjectivities

The statement above perfectly summarizes what the MMRC represents for migrant workers: a way to become a political player, collectively as well as individually.

What does militancy mean for migrant workers?

During my interviews with the members of the MMRC, I realized the vocabulary used by migrant activists was quite different from the one used by Italian activists. It is interesting to highlight the particular meaning of "militancy" for migrant workers, and the conditions in which they become "political players".

First of all, militancy obviously has higher costs for migrants than for Italians. By participating in a public protest, they take extra risks. The first one is the possibility of losing their precarious jobs, considering how easily they can be replaced. The second risk is arrest and, since many of them are illegalized, being immediately sent to a *Centro d'Identificazione e Espulsione* (Center of Identification and Expulsion). The third one is identification and labeling as a "problematic subject" by the police, endangering the renewal of the migrant's legal documents. As one migrant activist relates, "I expose myself a lot within the MMRC, and when I think about it I am scared" (Ivorian male migrant; interviewed by author, 31 March 2015). These particular costs for migrants' militancy surely explain the widespread use of the word "sacrifice" associated with militancy from migrant activists during my interviews, and at the same time the difficulty of mobilizing migrant workers for the long term.

Another point is the fact that the large majority of migrant activists do not have previous political experience or school education, often coming from poor and rural backgrounds. This has three consequences:

- The existence of a strong trust in Italian activists regarding the best strategies for the struggles;
- A concrete bond with the economic and political system they fight. Answering the question, "What is the meaning of your fight?", most of the migrants do not use theoretical conceptualization but a common *feeling* of being oppressed; and
- The importance that the MMRC has had in providing *information* to the migrant workers of the area, about the latest evolution of migration laws, or about the social, economic and political context at the local and national levels. In fact, by holding public meetings every week in Italian and also in English to communicate and discuss the latest news, the MMRC has made it possible for the migrants to develop an active knowledge of the place where they live.

That is why most of the migrant activists think that the MMRC has given them consciousness of their own position, providing the words and concepts to explain it better. To the question, "What have you gained participating in the MMRC?" almost all the migrants answered the same: "I have learned a lot". Through the meetings, through the explanation of the laws and the political context, and through the co-organization of public protests that have obtained concrete gains, Ex-Canapificio squatters have largely contributed to the idea of migrant workers as a class *per se*, self-conscious about class belonging (Marx 2002).

Participation in the MMRC has also allowed migrant activists to be included in local social networks. By managing activities with Italian squatters or by frequenting the Social Center, they have been able to meet local people in another context other than work: protests, social and cultural events like concerts, screenings of documentaries, or social dinners. As one of the migrant activists said, "My life has changed since I became part of the MMRC, I met a lot of people, I even found an Italian girlfriend!" (Ivorian male migrant; interviewed by author, 1 March 2015). Another related, "I remember when we went to the restaurant, all together, Italians and Africans!"(Ghanaian male migrant; interviewed by author, 14 April 2015). Social inclusion has meant, for migrant activists, a real attachment to the place in sharp contrast with the widespread desire of most of the migrants to leave the Campanian Plain as soon as they obtain a legal document. Some migrants have also found jobs thanks to the contacts that Italian activists have provided. In few words, the MMRC has acted as an interface between local people and migrant communities, enabling migrants to imagine themselves as a part of the place where they live and strengthening their will to change it by constructing a different *meaning* of their presence in the Campanian Plain.

The MMRC as a representative collective actor

The militancy within the MMRC has enabled migrant workers to consider themselves part of a wider social category characterized by its own difficulties and its own issues. In this sense we can consider the MMRC as a "representative" of the migrant workers, whose legitimacy has been given both by the numerous migrants involved and by the status of interlocutor with the institutions that the MMRC has gained along the years.

In fact, in a context of socio-spatial segregation and absence of legal and institutional recognition, the MMRC has been able to impose the "migrant workers" category in the political spaces. The public protests have allowed them to enter the "city", leaving aside their usual invisibility (Figure 18.2). The organization of meetings with government representatives has enabled them to be physically present in the public institutions that generally prefer to ignore them and discuss with intermediaries such as associations or trade unions. As a migrant activist summarized, "I have spoken with the mayor, the chief of police, the Prefect, and even the Minister!" (Ivorian male migrant; interviewed by author, 31 March 2015).

The MMRC has therefore allowed migrants to enter the institutional political sphere, but also to join other grassroots mobilizations not directly linked with migrants' issues. Obviously, this does not apply to all the migrants of the MMRC,

Figure 18.2 October 2014: demonstration organized by the MMRC in Castel Volturno

Source: Author

but along the years, many of them have understood how their struggle for a residence permit is connected to wider struggles for social justice. That is why the MMRC is now part of the mobilization for a social salary that has developed in the last months in Southern Italy.

Lastly, the MMRC has represented an empowerment device for migrants, protecting them against the abuses they have suffered at work, from the media, or from the authorities. The MMRC has followed and obtained significant achievements in various legal cases against employers who had not paid their migrant workers who would not have gone to the police because of their illegalized status. In 2013, activists fought for a Ghanaian worker's residence permit because of "labor exploitation", namely a rare judgment which set a precedent (Italian female activist; interviewed on 19 March 2013). In 2010, the MMRC organized the first "black labor strike" during which hundreds of migrants showed up at dawn on the crossroad where they usually wait for a job to express their refusal to work for less than 50 euros a day. This demonstration was massively participated in by migrants all over the Campanian Plain and focused media and political attention on migrants' working conditions. The MMRC also made a big effort to fight against migrants' criminalization perpetuated by the media. For instance, after the Castel Volturno tragedy in which six African migrants were killed by the Camorra, they built an anti-defamation campaign to sustain the migrants' innocence, opposing all the national press linking the murder to drug trafficking. Providing evidence to the prosecutor, the MMRC succeeded in proving the migrants' innocence, while, for the first time in Italy, four *camorristi* were condemned for the aggravated circumstances of "racial hatred". Finally, the MMRC has also protected migrants against various police abuses (Filhol 2016).

Conclusion

My analysis of the MMRC shows how the meeting between local squatters and migrant workers has allowed for the setting up and the growth of a social movement that has challenged migrants' exclusion from the social, spatial, economic and political spheres.

In doing so, they have developed a particular organization and a specific repertoire of contention in order to obtain significant results. Since 2002, the MMRC has obtained more than 8,000 residence permits for migrants whose asylum applications were often rejected elsewhere. At the same time, its peculiar strategies and organization have made the interaction with other antagonistic movements difficult. However, I argue that the particularly hostile context of the Campanian Plain has forced the activists to adopt "extreme" forms of organization to compensate for the specific difficulties of mobilizations they have encountered.

This socio-political experiment underlines the difficulties in setting up self-managed spaces with migrants that represent a social category whose economic, social and cultural capital cannot be compared with those of the local activists struggling with them. However, being conscious of these difficulties is already a step towards trying to solve them, and it is not a coincidence if migrant workers feel more supported by the MMRC than by other organizations dealing with migration issues.

The MMRC's experiment also demonstrates how, through long and patient background work, a "step forward" can be made, shifting from the struggles about citizenship issues, related at least directly just to migrants, to reach wider spaces of struggle regarding labor and economic questions that are issues for the entire working-class. We can therefore consider that the MMRC has helped construct the migrant workers of the Campanian Plain as a class *per se*, with the conscious desire to break into the political spaces that traditionally excluded them.

Note

1 Istat data are available from their website: http://dati.istat.it/?lang=en.

References

Anselmo, M. (2009) 'L'impero del calcestruzzo in terra di lavoro: le trame dell'economia criminale del clan dei casalesi'. In Gribaudi, G. (ed.) *Traffici criminali. Camorra, mafie e reti internazionali dell'illegalità*. Torino: Bollati Boringhieri, pp. 505–537.

Brigate di Solidarietà attiva, Nigro G., Perrotta, M., Sacchetto, D., Sagnet, Y. (2012) *Sulla pelle viva. Nardò, la lotta autorganizzata dei braccianti agricoli*. Roma: Derive Approdi.

Caruso, F. (2011) 'Percorsi di sindacalizzazione del bracciantato migrante meridionale nel distretto della clandestinità: il movimento dei Migranti di Caserta', *Mondi Migranti*, 3: 229–246.

Caruso, F. (2013) 'La porta socchiusa tra l'Africa Nera e la Fortezza Europa: l'hub rururbano di Castel Volturno'. In Colloca, C. and Corrado, A. (eds) *La Globalizzazione delle campagne*. Milano: Franco Angeli, pp. 141–157.

Caruso, F. (2015) *La politica dei subalterni,* Roma: DeriveApprodi.

Cesoni, M. (1991) 'Camorra et politique: démystification du rôle de la drogue', *Cultures et Conflits*, 3: 24–54.

De Filippo, E. and Morlicchio, E. (1992) 'L'immigrazione Straniera in Campania', *Inchiesta*, XXII (95): 40–49.

della Porta, D. (ed.) (2004) *Comitati di cittadini e democrazia urbana*. Soveria Mannelli: Rubettino.

della Porta, D. and Diani, M. (2006) *Social Movements: An Introduction*. Malden: Blackwell.

Dines, N. (1999) 'Centri sociali: occupazioni autogestite a Napoli negli anni novanta'. *Quaderni di Sociologia*, 43 (21): 90–111.

Festa, F. A. (2003) 'L'alchimia ribelle napoletana'. In Cappelli, O. (ed.) *Potere e società a Napoli a cavallo del secolo. Omaggio a Percy Allum*. Napoli: ESI, pp. 381–423.

Filhol, R. (2013) 'Les travailleurs agricoles migrants en Italie du Sud'. *Hommes & Migrations* 1301(1): 139–147.

Filhol, R. (2016) Social Centers in Southern Italy: the Caserta Ex-Canapificio between illegality, migration and rurality. *Antipode* (forthcoming).

Giarrizzo, G. (1992) *Mezzogiorno senza meridionalismo: la Sicilia, lo sviluppo, il potere.* Venezia: Marsilio.

Rinaldi L. (2013) 'Ogni anno si sciolgono venti comuni per mafia' *Linkiesta* (22 October 2013). Available: http://www.linkiesta.it/it/article/2013/10/22/ogni-anno-in-italia-si-sciolgono-venti-comuni-per-mafia/17175/.

Lipsky, M. (2010) [1980] *Street-level Bureaucracy; Dilemmas of the Individual in Public Services.* NYC: Russel Sage Foundation.

Marx, K. (2002) [1847] *La misère de la philosophie.* Lausanne: Payot.

Montagna, N. (2012) 'Labor, Citizenship, and Subjectivity: Migrants' Struggles within the Italian Crisis'. *Social Justice*, 39(1): 37–53.

Mudu, P. (2004) 'Resisting and challenging neoliberalism: The development of Italian Social Centers', *Antipode* 36(5): 917–941.

Mudu, P. (2012) 'At the Intersection of Anarchists and Autonomists: Autogestioni and Centri Sociali', *ACME* 11(3): 413–438

Oliveri, F. (2012) 'Migrants as activist citizens in Italy: understanding the new cycle of struggles', *Citizenship Studies*, 16(5–6): 793–806.

Outis, Revue de philosophie (post)européenne (2011) *Révoltes migrantes, Rivolte migranti.* Milan/Udine: Mimesis.

Sales, I. (2006) *Le strade della violenza. Malviventi e bande di camorra a Napoli.* Napoli: L'Ancora del Mediterraneo.

19 Migrant squatters in the Greek territory

Practices of resistance and the production of the Athenian Urban Space

Vasiliki Makrygianni

Migration in Greece: dialectics of repression and resistance

Ever since the establishment of the Modern Greek state in 1830, the Greek territory has been shaped indelibly by migratory flows. Greece has been shifting constantly from a country of origin to a country of destination and vice versa. Nowadays, due to the country's geopolitical location that indicates it as one of the main gates of Europe, it constitutes a transit space for thousands of migrants heading towards the wealthy North. Before the burst of the contemporary crisis, the Greek terrain was regarded as an intermediate space between war and peace, poverty and wealth, growth and decline. At the moment, it is a crossroad of migratory flows, a destination for many non-Europeans, a transit for those heading north and a point of departure for Greek citizens that try to escape the crisis. Contemporary Greece is a melting pot of conflicts, a space where different expressions of crises come together.

During the last decade the country has been marked by significant changes in migratory flows. There has been a decrease in arrivals from European and Balkan countries and a rapid growth of arrivals mainly from Asia and the Middle East and secondarily from Africa. Currently about 10% of the population is of non-Greek citizenship.[1] Of them, thousands are detained in camps while more and more are crossing the borders every day.[2] The biggest concentration is found in Athens, the capital city right in the heart of the contemporary socioeconomic crisis. There is no precise data on the exact number, but it is estimated that about 350,000 people, that is about 10% of the area population, inhabit the wider Athenian Urban Complex either as permanent residents or as people in transit (Papastergiou and Takou 2012). However, official statistics hardly reflect the actual number of migrants as many of them choose not to record either out of fear of persecution and deportation or due to their temporary status of residency.

But what kind of social relations do these numbers reveal and what sort of spatialities do they generate? Migrant populations face various aspects of crises every day. In fact, migration should be thought of not just as a derivative but also as a detonator or even a permanent state of crisis. Migrants are often murdered, detained or excluded, but they also resist and produce emancipatory spaces, challenging the fear generated by the vicious economic and social apparatus. The spatialities of such practices are indicative. Several spaces of fear, exclusion, racism and death have sprung up. The country is filled with detention camps, a fence is raised along the Greek–Turkish borders (*Evros*), the land and sea border areas have become cemeteries for the "anonymous" strangers, and many islands (such as Mytilene, Kos, Samos, Crete, etc.) are shifting from vacation paradises to overflowing refuges of persecuted migrants. At the same time, several spaces of solidarity, resistance and struggles have emerged. The newcomers have developed notable means of resistance in the Greek territory and have formed new relations with the native population. While the contemporary crisis deepens, the Athenian metropolis becomes increasingly a privileged field of struggle. In this context, squatting as praxis of struggle, resistance and re-appropriation of the deprived means of production and reproduction is gaining ground.

Crises, squatting and migration in Greece: an interlocking triad

A brief overview of the recent evolution of cities in the Greek territory indicates that they are an outcome of race, ethnic, gender and class relations. Moreover it reveals the inextricable link between crises and migration. The production of the urban space is a result of major economic and political transformations and is marked by the continuous movement of migratory populations and their praxis of struggle, resistance and emancipation. Among those praxes, squatting has been drastically shaping the urban terrain.

Hundreds of thousands of people were displaced from their homeland, during the interwar period of the 1920s and the 1930s, that is, after the Balkan Wars (1912–1913), the First World War (1914–1918) and the Greek-Turkish War (1919–1922). Having lost their means of production and reproduction, they were forced, in order to survive, to settle in the periphery of the cities. Although, at that time, a major state project was conducted for the regulation of refugees' settlement, most of them covered their housing needs through arbitrary constructions. During the decades of 1950s and 1960s (after WWII and the Greek Civil War, 1946–1949), approximately 45% of the Greek population, that was more than three million people, were relocated, either to the big Greek cities or to other countries such as the USA, Germany and Australia. As there was no organized social housing plan to absorb the internal immigration, the vast majority of new residents settled into informal and arbitrary makeshift constructions and slums, in the perimeter of the big cities. Plenty of them were arbitrary constructions on squatted terrains. During the decade of the 1990s after the collapse of the Eastern bloc, migrant flows intensified once again, as thousands of migrants mainly from Albania and the Balkans entered the country. Many of them occupied fields and terrains either in the countryside or in the periphery of Athens in order to spend the night. Still, during the 1990s and the early 2000s – the so-called decades of growth – squatting was not a widespread tactic as many of the migrants managed to enter the housing market (Balampanidis 2015). At that time, squatting in Greece was undertaken mainly by political activists who opened several Social Centers. It was the re-appropriation of the political space taken by the socialist government of PASOK during the 1980s. In those spaces, few migrants were involved. From the early 2000s until now, as people started to massively abandon the Middle East, Asia and some African countries, there has been a significant increase of non-European citizens trying to enter Europe. Meanwhile, the current socioeconomic crisis that burst in Greece deteriorated the living standards of the country's residents. The arrival of thousands of migrants along with the rapid marginalization of the population (whether native or newcomers) altered the geography of squatted spaces. New enclaves of occupied spaces sprouted as more and more migrants started to inhabit Athens. Such spaces (mainly abandoned buildings) remain well hidden in the urban fabric. Their ephemeral character and their wide dispersion, along with the informal status of the squatters, make them practically untraceable.

Taking all of this into account, the dialectic relation between capitalism, migration and crises becomes apparent. These processes are either considered a derivative of capital overaccumulation, as several orthodox Marxists point out, or an outcome of civil disobedience, as autonomous Marxism stresses. Yet, it is worth noting how this triad drastically rearranges the urban space.

According to several thinkers, the relation between capitalism and crises lies on the mechanism of the permanence of the so-called primitive accumulation. Several scholars such as Harvey (2003), Chandra and Basu (2007), De Angelis (2007), Caffentzis, (2010), and Holloway (2010) examine the global consequences of neoliberalism and highlight the constant enclosures, thus reinstating the permanent character of primitive accumulation. As capitalism seeks to overcome the crises, it expands the capital relation (i.e., the separation of people from the means of production and reproduction) into new spheres, while at the same time it deepens and intensifies the existing ones. Therefore, in times of crisis new accumulation processes are implemented or existing ones are intensified, for instance, land grabbing or the displacement of populations. Under this scope, resistance and emancipation to such enclosures could mean the reunification with the deprived means of production. Likewise the praxis of squatting can be understood as a response to enclosures, a crack in capitalism's urban continuity, and a negation to capitalist relation, a negation to commercialization and intermediation of the everyday life from the capital.

This dialectic between suppression and resistance lies indelibly in the understanding and the production of urban space. Moreover it allows us to view migrants not only as victims of the capitalist beast, but also as potential agents of resistance and emancipation. It also enables us to consider people's movement as an act of emancipation from the capitalist hegemonic treaty. As stated by Casas-Cortes *et al.* (2015), migration implies a dual status of resistance and emancipation, and this can be defined at different spatial scales – at a hyperlocal and a local level. On the one hand, there is a structural relation between migrants and struggles when crossing borders and actively questioning transnational policies of exclusion, such as the Schengen agreement and the Dublin II Regulation. On the other, emancipatory movements, denials and resistances emerge through the performance of everyday life of the newcomers.

In Greece during the recent crisis apart from detention camps, spaces of fear dominated by fascists group attacks (inspired by the ancient Greek institution of Crypteia) and state pogroms (like the operation Xenios Zeus), several struggles squats and occupations collectives and groups have emerged. This led to the establishment of squats and emancipatory spaces in the urban core that drastically alter Athens' character.

Reappropriating the city space of Athens during a crisis era

As several Marxist thinkers have stressed (see among others: Harvey 2012; Lefebvre 1974; Massey 2005), the urban space is produced through social relations. It is not an empty vacuum filled with people's lives but a vivid formation, open to subversions and alterations. Likewise, squatted spaces frequently derive from solidarity and resistance relations. As the praxis of squatting signifies the reappropriation of the deprived means of reproduction, such spaces question vividly the neoliberal state and open alternatives and cracks to such repressive mechanisms. Since the dominant state policies on migration are based in relations of control, fear and power, squatting comes as a response to spaces of exclusion like detention camps, the so-called "hospitality centers" (κέντρα φιλοξενίας or *Kentra filoxenias*). Migrants themselves, apart from ephemeral shelters they occupy in order to spend a few nights (in parks or squares[3]), appropriate more permanent spaces in order to house their needs. Such spaces are usually of public (state) ownership and include abandoned factories, houses and buildings, or even small parts of land in the fields of suburban areas. Moreover, as squatting refers also to the reappropriation of several aspects of the everyday life, like the social or the political relations, squatted spaces also serve as spaces of encounter and political engagement. In order to understand further their characteristics, their diversity as well as the continuity of the squatting practices before and during the Greek crisis, four indicative examples of squatted spaces in the city center are presented.

The former Court of Appeal (Το παλιό Εφετείο or To palio Efeteio)

In the beginning of 2003, a few months before the 2004 Olympic Games and long before the official debt crisis would burst, several homeless people, some of them drug addicts, squatted an eight-floor building right in the heart of Athens. The district is densely built, very vivid during the day but desolate during the night as the main uses are shops and services and very little housing. The squatted building that used to be a Court of Appeal until 2001 is under state ownership. Starting in 2006 a few migrants moved in, and by 2008 about 500–600 people, mainly migrants, inhabited the place. The squatters were mainly men, from Afghanistan, Pakistan, Maghreb and a few from the Balkan countries Bulgaria or Romania. The space had no facilities (water or electricity), thus the hygienic conditions

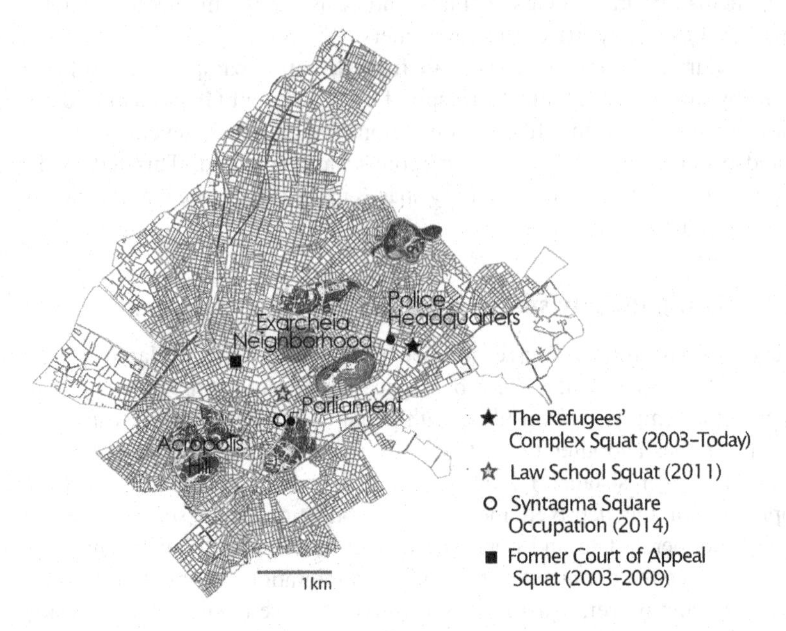

Figure 19.1 Migrants' squats in the municipality of Athens

Source: Author

were extremely bad. There were no strong relations developed with the neighborhood, but several people in solidarity from leftist and antiauthoritarian groups tried to approach the space, especially after the rise of fascist groups that occasionally attacked the squat. The police tried to evict the building several times as it was considered by the state and the media a "hygienic bomb". During these attacks, several people in solidarity approached and defended the squat along with the squatters. Eventually, the police with the support of members of the fascist party "Golden Dawn" in July of 2009 evicted the place. It has remained shattered ever since.

The Refugees (Τα Προσφυγικά or **Ta Prosfigika***)*

In 2003, a complex of buildings known as "The Refugees" was occupied by homeless people, unemployed people and, later on, migrants.[4] The complex was built during the 1930s in order to host the Greek refugees of 1922, and it is one of the few architectural examples of organized building blocks in Athens. In 2015, 150 apartments are squatted by more than 400 people. More than half of the squatters are migrants of more than ten different ethnicities, coming mainly from Turkey, Iraq, Afghanistan or Syria. The exact number and the composition of the inhabitants change constantly as many of the residents, especially the migrants, move very often. Among the squatters are many children and families, while about half of the migrants are women. The "Prosfigika" squat is not just a housing project but an ongoing battlefield, a place of struggle. For the last three years two assemblies have been running, one of the inhabitants and one of inhabitants and people in solidarity, while they have organized collective kitchens, a kindergarten and a barbershop. Some of the squatters also take part in wider Athenian anti-fascist networks. Though the complex is a few meters away from the head police department, no serious attacks or tensions have been recorded. Prosfigika space keeps developing through its contradictions and antagonisms. If anything, the project respects the space's memories and justifies the former purpose of the complex.

The hunger strike of 300 migrant workers

Two exemplifying squats of struggle emerged in 2011 when 300 migrant workers of several nationalities conducted a hunger strike, claiming legalization and rights for migrants and workers living in Greece. The strikers along with those in solidarity squatted spaces in the Athens' Law school and in the Worker's Union building in Thessaloniki. The strike and the squatting of the Law school, in particular, draw a lot of attention, irritated the power mechanisms and became a running sore for the government, as the university space has a specific symbolical meaning for the Greek society.[5] The state attack on the hunger strike was also driven by the willingness to "conquer" and control the university space, thus the migrants' squat in the Law school got in the middle of a very big debate. The squats of Athens and Thessaloniki formed vivid spaces of struggle and counter-information while an international movement of solidarity was raised.[6] Those in solidarity organized medical teams of doctors and students, counter-information groups, meetings and assemblies, 24-hour continuous shifts for defense, escort shifts in hospitals and press conferences. Against the fierce attack of the government and the majority of the mass media, after 44 days of hunger strike, the migrants managed to obtain a six-month residence certificate and to reduce some of the required documents for the legalization for the migrants. Still the state justified once again its vicious role by not keeping its part of the "agreement" later on.

The hunger strike of Syrian refugees in Syntagma square

Right in front of the Greek parliament in Syntagma square,[7] about 150 refugees from Syria started a protest demanding from the Greek state to recognize them as refugees on 19 November, 2014. Five days later, they began a hunger strike that lasted for about a month. The initial group of migrants was reinforced by more Syrians and people in solidarity. It is estimated that a few days later about 500 people were standing opposite the Parliament. The migrants, who dared to make themselves visible right in front of the headquarters of their persecutor, also made two crucial ruptures. They not only broke the spatial contract of the square being only a temporary space of protest, but they also broke the social contract which defines them as poor and helpless, victims of persecution and pogrom, always at the mercy of philanthropy and pity. The presence of the strikers irritated both the government and the municipality (Makrygianni and Tsavdaroglou 2015),[8] as they claimed that the appropriation of the square constituted a shame for the public image of Athens and an annoyance for the consumers of the nearby shops.

The above were only a few indicative examples of migrants' struggles over the appropriation of the city space. In several Athenian neighborhoods (Ilisia, Exarcheia), squats are made in order to host migrants; in a rural area of western Greece, Manolada, several strikes of migrant workers took place, while many migrant collectives occupy public spaces on temporary basis in order to pray, to celebrate or to protest. While such initiatives are harshly attacked by the state, the mass media and fascist groups, a big part of the indigenous population along with squatters and political groups stands in solidarity with them. Several supportive structures have been organized during the last 15 years, such as Greek language lessons, collective kitchens, festivals and demonstrations. Moreover in several cities, some Social Centers have been created in order to stand in solidarity with the migratory population and to enforce the struggle against capitalism and power relations.

Migrants' squats: an endless spatial struggle

Migrants' squats during the last decade redefined the meaning of squatting in Greece. They reminded us that squatting is not necessarily engaged with a fixed political identity. In migrants' cases, the praxis of squatting is not necessarily related only to housing needs and personal space. Thus, it is also associated with the appropriation of several aspects of the everyday, like the public and political sphere, the social and cultural relations or even the space of imagination. Moreover, as in the contemporary squatting struggles various subjects are engaged, it became obvious that migrants do not form a solid and fixed category. Gender, class, race, age, nationality or ethnicity remain strong differentiating characteristics that question their homogenization.

Squatting has strengthened the relations between local political projects (groups, individuals and Social Centers) and the newcomers as they found new common spaces of reference. Migrant squatters have been adopting local ways of struggle and occupation practices. Likewise, they have engaged with local struggles for the university asylum, the protection of the Refugees' complex and its memory, the fight against racism and fascist groups, the workers' rights or the appropriation of Syntagma square and other public spaces.

During these fights, they found a lot of rivals but also many companions in an environment balanced between resistance and fear (see azozomox and Gürsel Chapter 9, Brogstede Chapter 13, Filhol Chapter 18). Their struggles reveal that the social and economic crisis does not only affect the conservation of a big part of the Greek society. They showed that the rapid marginalization of the Greeks made them not only more compassionate but even more open to practices of struggle. Squatting as praxis is gaining ground in a society that is watching the urban and rural territory being dispossessed by capital accumulation processes. Migrants' initiatives confront vividly such neoliberal policies and open a path to the re-appropriation of space.

Moreover, when looking into the migrants' struggles, the connections between the different and simultaneous crises that burst throughout the planet become even more apparent. The continuity of struggles and occupied spaces before and during the Greek crisis of 2008 reveals that the vicious system called neoliberalism produces ongoing multiscalar crises in the urban terrain. In this constant battle between the deprivation of the means of production and the struggle for life and emancipation, these moving populations often appear to be the pioneers.

Notes

1 According to the 2011 national census, the total population numbers 10,815,197 people. Of them, 199,101 are citizens of EU countries, 708,003 have citizenship of other countries, and 4,825 people's nationality or citizenship are not determined.
2 According to the minister for immigration policy, it is estimated that during the first semester of 2015 about 80,000 migrants entered the country. It should be noted that among those who try to enter Europe many die at the borders. According to the IOM, between the years 2000 and 2014, about 22,000 migrants died in the Mediterranean Sea while trying to access Europe. Available: http://publications.iom.int/bookstore/free/FatalJourneys_CountingtheUncounted.pdf
3 An indicative example is the settlement of about 500 refugees mainly coming from Afghanistan and Syria in the summer of 2015 in "Pedion tou Areos", a big park in the center of Athens. In fact, an exemplary wave of solidarity was raised in order to support them.
4 The eight three-story buildings, with 228 apartments in total, have been an issue of debate for a long time as the state tried to expropriate them claiming "compulsory acquisition" in order to gentrify the wider area. Eventually, 177 were expropriated and 51 remain private property.
5 The Law school was one of the occupied university spaces where the students organized their struggle against the junta regime in 1973. Until now, it has constituted a space of struggle and resistance.
6 Finally, after several threats and "negotiations", the strikers moved to another building of private ownership.
7 In Syntagma square (square of the Constitution), several protests have been taking place ever since the formation of the Greek state. In 2011, several people occupied the square following the *Indignados* movement and transformed the square into a constant space of struggle for more than two months.
8 The municipality is governed by a political formation that goes under the name "Right to the City".

References

Balampanidis, D. (2015) 'Geographies of migrants' settlement in the municipality of Athens. Access to housing and interethnic relations in the city's neighbourhoods', *Urban Conflicts, Workshop "Encounters and Conflicts in the city": Thessaloniki*: 342–356.

Caffentzis, G. (2010) The future of 'The Commons': neoliberalism's 'Plan B' or The original disaccumulation of capital? *New Formations* 69(1): 23–41.

Casas-Cortes, M., Cobarrubias, S., De Genova, N., Garelli, G., Grappi, G., Heller, C., Hess, S., Kasparek, B., Mezzadra, S., Neilson, B., Peano, I., Pezzani, L., Pickles, J., Rahola, F., Riedner, L., Scheel, S. and Tazzioli, M. (2015) 'New Keywords: Migration and Borders', *Cultural Studies* 26(1): 55–87.

Chandra, P. and Basu, D. (2007) 'Neoliberalism And Primitive Accumulation In India'. Savyasaachi, Kumar R. (Eds) *Social Movements: Transformative Shifts and Turning Points*. New Delhi: Routledge India, pp. 144–158.

De Angelis, M. (2007) *The Beginning of History: Value Struggles and Global Capital*. London: Pluto.

Harvey, D. (2012) *Rebel Cities: From the Right to the City to the Urban Revolution*. London: Verso.

Holloway, J. (2010) *Crack Capitalism*. New York: Pluto.

Lefebvre H. (1974) *La Production de l'espace*. Paris: Éditions Anthropos.

Massey, D. (2005) *For Space*. London: Sage.

Makrygianni, V. and Tsavdaroglou, H. (2015) 'The right to the city' in Athens during a crisis era. Between inversion, assimilation and going beyond'. In Eckardt, F., Ruiz Sanchez, J. and Sevilla, A.B. (eds) *City of Crisis*. Berlin: Springer-Verlag.

Papastergiou, V. and Takou, E. (2015) Migration in Greece. Eleven myths and more truths. *Rosa Luxemburg Stiftung*. Available: http://rosalux.gr/de/publication/migration-greece.

20 Natural resource scarcity, degrowth scenarios and national borders

The role of migrant squats

Claudio Cattaneo

Introduction

The interconnectedness of global capitalism and border policies is responsible for an intricate relationship between the economic effects of increasing natural resource scarcity; the availability of abandoned real estate properties deemed to be squatted; migrants seeking self-managed enterprises; and the politics from the New Left and the New Right that share some similarities, such as anti-capitalist visions, environmental conservation and support for the opening of squatted social centers. But they diverge with respect to border policies and migration issues. In this chapter, I analyze this interconnectedness, claiming that a degrowth policy from the New Left based on open borders localism will be beneficial to society at large. Learning from the case of Barcelona waste-pickers, it gains insight for urban and border politics.

In the following section, I explore the legacy of the capitalist imperative of infinite growth and the need for resource conservation and scarcity. Secondly, I argue that both the New Left and the New Right share the idea of a non-capitalist future based on degrowth, local economies, and less mega-infrastructures. In particular, reinforcing Latouche's idea, I posit that the future will be degrowth or barbarism. Moreover, similarities in degrowth proposals need a closer look. Thirdly, I explore the differences between degrowth positions to discuss the key explanatory variable; it refers to national borders, migrations, openness and the intensity of cultural identity. For the New Right, a stronger nationalistic/identitarian and even ethno-regionalist approach is associated with an anti-immigration political stance that the left opposes. Fourthly, squatting and Social Centers constitute the direct action arm from both the New Left and the New Right, and their typologies will be explored; while they all agree with an anti-capitalist and material recycling logic, they differ in their approaches towards minorities and migrants. Finally, evidence from Barcelona waste-pickers addresses the interconnectedness of degrowth in urban metabolism, squatting and open border policy, to sustain an economically efficient, ecologically effective and socially just strategy for degrowth. Finally, evicting waste-pickers' squats is a lost opportunity for many benefits, a win-win-win situation that will be explained as a result of the study.

Degrowth and the similarities between the New Left and New Right

There are three critical streams against economic growth: an anthropological critique, a critique of the concept of western development, and an environmental critique.

Caillé (1989) observed that the contentions against economic growth refer to an anthropological critique of neoclassical economics axioms (selfishness, perfect rationality, non-satiation) that do not correspond to human values and behavior. In the 1980s, a group of male French academics started a magazine challenging the imperialism of neoclassical economics, popularized by the Chicago School. Inspired by the anthropological work of Marcel Mauss (2007) on the essence of the gift, they called the magazine "MAUSS".[1] According to Latouche (2001) and Fournier (2008), a re-politicization of the economy is needed to foster democratic citizenship rather than capitalist consumerism. Contrarily, the concept of Economic Anthropology shows that reciprocity, generosity, altruism and communalism are central values in many societies (Sahlins 1977; Malinowski 1978; Mauss 2007), and the market is not a central element in these societies (Polanyi 1944). Furthermore biological evolution is popularized not only by Darwin's "survival of the fittest" (an argument used to justify competitive markets and concentration of market power in monopolies), but also by Kropotkin's (1915) mutual aid.

A second set of growth criticism emerged after World War II with decolonization, which argued the ostensible basis of cultural and economic hegemony of the former colonial nations over the rest of the world as development. Although this argument is not within the scope, this line of critique is particularly relevant to migrations because, among many reasons that force people to migrate to the Global North, there are those related to the detrimental consequences of "development" imposed on the Global South (see the discourses led by Sachs, Illich, Escobar, Esteva, Shiva, and Latouche in the *Development Dictionary*, 1992).

Gilbert Rist (2008) traces a history of development as the core of western imagination, in which it is believed that growth or progress should be able to continue indefinitely and has been constructed within a particular history and culture. Colonialism, which originated in Europe, opposed colonizers and colonized, with a discourse in which the North was seen as "advanced" and "progressive", while the South as "backward", "degenerate" and "primitive". Alternatively, the anti-colonial imperialism that began in the US created instead a new narrative of "developed and underdeveloped nations": it gave an idea of continuity of all nations being on the same path – that of development – rather than a clear-cut separation between civilized and primitive cultures. While the old European colonial discourse was based on cultural primacy, the new narrative of development was based on economic grounds: wealth can be generalized to everyone on earth; injustice and international inequality are therefore still justified as a temporary state of affairs that development will solve. Development was then superimposed, and Gross Domestic Product (GDP) growth became the measure of its success, assimilated to a dogma. Consequently, Rist argues, in line with Sachs, that it is time to deconstruct the imbalance of who has the power to define what the problem is and how to solve it. The root of this power, rather than in the colonizer's military intervention, now lies in the elegant discourses of developed nations which provide a sufficient guarantee of social power to intervene, to transform and to rule, to the detriment of inhabitants of the Global South who are often forced to migrate.[2] Following Rist, proponents of degrowth argue that there are democratic limits to growth, and development and capitalism fail to deliver adequate environmental quality or social justice. Through a repoliticization of citizenship and stronger democratic processes, market capitalism must be replaced (Cattaneo *et al.* 2012).

Finally, there is the environmental legacy of economic growth and increasing natural resource consumption. Defenders of degrowth (Schneider et al. 2010; Kerschner 2010; Kallis 2011; Demaria *et al.* 2013) argue that if ecological sustainability is to be achieved, then economic growth cannot be a priority anymore. Alternatively, defenders of green growth argue that the decoupling of material consumption from GDP growth will be possible with the advent of new technologies and ecological modernization. This is the idea of sustainable development: just like in an ancient Greek comedy, the religion of growth and development believes in the divine arrival of a *deus ex-machina* – a technological god – that will save us all. In any case, resource efficiency must be at the forefront of any economic policy aimed at reducing material consumption and re-using and recycling waste in the most efficient manner. Material intensity will have to decrease drastically in favor of waste regeneration and scrap metal reutilization, aiming for a closed-loop economy.[3] Perfect recycling is not possible because in any process some material is lost, such as the erosion of rubber tyres by the friction with road. Reutilization and recycling can be more efficient and effective if certain labor-intensive processes, like material recovery and separation, are employed – in other words, what waste-pickers do in cities around the world, employing little machinery to recover and segregate all the different useful components from waste through dedicated labor.

The need for degrowth in urban metabolism and its social actors

Most consumption and waste are concentrated in cities. Using an analogy from biology, with social metabolism we understand the performance of economies is not only represented in monetary units, but primarily in terms of flows of physical resources and waste. Material flow analysis shows how physical flows of materials enter, exit or recirculate in a national economy. Likewise any geographical unit of analysis is relevant to describe the amount of minerals and organic resources that are required for the metabolism of a society. To this extent, the analysis of urban metabolism is an important avenue for research because of the amount of waste generated and its potential for recovery in a world of increasingly scarce resources. This has challenging implications for waste management practices and policies in an environmentally effective way.

In this work, both waste and social metabolism are intended in a broader way: for advanced capitalist societies governed by global financial flows, waste has also to do with the material effects associated with financial speculation. Whatever is generated by speculation in the real estate sector should be included; for many of the cities of post-industrial countries, it implies the transformation of their landscapes from industrial spaces back into service and residential spaces, with the consequent abandonment of houses, buildings, and factories. So, abandoned real estate is "waste" that squatters "regenerate". The broad understanding of social metabolism, in line with Toledo and Gonzalez de Molina (2011), considers two components: a hardware, referring to the relationship between the natural and the human system, and a software, referring to the social relationships within society, in particular power distribution, access to information and resources across social classes. For the case at hand, the actors optimizing the urban waste flows differ largely if they are waste-picking migrants or large hi-tech corporations with formal agreements for municipal waste collection. I contend that the latter are not as economically efficient, ecologically effective or socially just as the migrant waste-pickers of the informal economy (detailed in the following sections). It is good to look at physical-chemical processes and cycles from a perspective of environmental limits and scarce resources, and to conceptualize social metabolism as a biological metaphor with the natural world characterized by closed loop cycles. However, it is fundamental to also include the social dynamics of those actors who, like the enzymes of a biological organism, contribute to material recycling in cities, but not without the creation of a social conflict between resourceless pickers and large corporations of the waste industry. It is worthwhile to address this socio-metabolic conflict (Demaria, personal communication with author, January 2015).

A future of local economies?

There are some subtle similarities between the New Left's idea of degrowth and that of the New Right, and here the perspective from the New Right, principally of French origin, is presented. *La Nouvelle Droite* (the New Right), as a concept, originated in the 1970s. The New Right's discourses are associated with a group of intellectuals who formed the Research Group on European Civilization Studies (GRECE), led by Alain de Benoist and inspired by an ancient pagan European reference to identitarian and cultural values in line with those of europeanism and of neopaganism (in connection with nature). These discourses were critical of capitalism, communism and christianism for being too egalitarian, and, in hindsight, censured migrations. In his defense of European values, De Benoist suggested that, like ancient civilizations (Nordic, Celtic, Roman, Greek), the different European cultures should not synthesize but live in a way that maintains their cultural purity; likewise, he denounced the American-style melting pot ideology.

The association between environmentalism and nationalism is increasingly commonplace. Not only among the French, the New Right's defense of the environment is consistently associated with defense of the motherland. In Germany, too, the Neo-Nazi NPD party has leaders who take pride in organic farming and live off of local produce (Thiele and Weiss 2012).

De Benoist's relevant book is "Demain, la décroissance!" (2007), a title copied from the publication of the left-wing degrowthers based on the work of Georgescu-Roegen (1979). It focuses on the ecological crisis and the impossibility of infinite growth, and is also critical of sustainable development. His work is reflected in the magazine *Krisis* published by him, where he defends nature's intrinsic value in line with Naess's visions on deep ecology. In *Krisis*, he also deals with issues of communalism and the voluntary simplicity movement with reference to the work of Etzioni. Spektorovski (2000) claims that the New Right is shifting from an idea of nationalism to that of organic regionalism, ethnicity and organic citizenship, based on the people naturally belonging to a region and opposed to legal citizenship. These notions propagate a non-racist barrier to migrants' integration. "An organic citizenship, by contrast, rejects any possibility of integrating foreigners at the time as it sets the basis for a populist, 'anti-liberal' concept of democracy" (Spektorovski 2000: 359) and is opposed to the liberal citizenship proposed by a capitalist society. Alternatively, Latouche (2006) sets the degrowth economy within a bioregional dimension.

Both Latouche and De Benoist would agree on many issues that not only refer to the material efficiency of social metabolism but are also against the liberal notion of capitalism – a principle that is couched in the French Revolutionary slogan of "Liberté Egalité Fraternité", meaning individual freedom for the bourgeois with no limit to entrepreneurship but economic growth.

The aforementioned discursive articulations are conspicuous on the migration issue: in ecological economics, scholarly approaches from North America defend closed borders as the only way to limit population growth and fast depletion of natural resources. On the contrary, the European approaches are more oriented towards a degrowth vision that embraces the concept of open localism. The difference between the degrowth of the New Left and the New Right is exposed in the following section, which shows openness versus closeness in the localist/regionalist ideas.

Nationalism, migrations and the differences between the New Left and New Right

The lifeboat concept is often used to indicate an extreme situation for ecological salvation in an overpopulated and over-consuming world, where rather than seeking a solution for equal opportunities for all humans, only those who belong to certain territories are entitled as passengers of the boat (no matter how the material lifestyle of the opulent passengers of such boats environmentally impacts other countries).

The North American Carrying Capacity Network (CCN) is one of these organizations with some exponent ecological economists on the advisory board such as Robert Costanza, William Rees, Herman Daly (who resigned after some peer pressure) and David Pimentel who, sided by Virginia Abernethy, are known as the leading figures in the CCN and proponents of his white supremacist ideologies. It defends population control primarily through the reduction of immigration.

The ecological economics argument for immigrant reduction refers to the famous I=PAT equivalence in which Environmental Impact is directly correlated to Population, Affluence and Technological inefficiency. The conflict between the New Left and New Right hinges on the priority given to the PAT elements: left-wing ecological economists and, in particular, degrowthers would claim for a reduction in "affluence" (i.e., personal income) for the population in the most wealthy nations, or for adoption of adequate technologies (either low-tech convivial technologies like bicycles or eco-efficient innovations like e-bicycles), or population control focused on birth reduction and abortion rights. Instead, right-wing degrowthers would primarily focus on immigration limits, based on many ideas that range from ethnicity to white supremacy. For them, the reasons behind migratory fluxes – such as the negative effects of colonization and development – are not relevant, and this constitutes a severely biased position. The roots of their impoverishment are to be found in an ecologically unequal exchange, in the processes of enclosure of the commons, in land grabbing, in privatization to the benefit of, very often, large corporations and NGOs from the Global North. Moreover, while an external financial debt – often illegitimate – is imposed on poor countries, almost no recognition is given to the ecological debt that the North owes to the resource-rich countries of the South. These unequal terms of trade and debt relationships are not addressed in the narrow-sighted vision of anti-immigrant advocates from the New Right.

In addition, anarchist-nationalism with British origins is perhaps the most complex representation of the New Right vision of a stateless, tribal, neo-pagan society embracing a mix of decentralized deep environmentalism with ethnic and racial separatism, whose main figure is Troy Southgate (Graham 2005).

Conversely, a degrowth politics of open localism is maturing within the left-wing degrowth community; this would foster an idea of a culturally open society where cultural differences are highlighted and where, contrary to a vision of fixed tradition and like any biological organism subject to evolution, openness is the opportunity for cultural evolution. This would be based on local traditional culture, such as an inclusive celebration of traditional festivities or even some sort of pre-christian paganism, and could in the long-term contribute to an evolutionary *hybridization* of cultures. Likewise, languages – which are often representative of a culture – are not fixed and evolve in time. The case of the Basque identity is exemplary: *Euskaldun* means Basque speaker, meaning that, irrespective of their origin, one can have a Basque identity if one speaks the language. The ability to learn and speak a language, like a foreigner does to assimilate into a local culture, is a very plausible hypothesis for setting cultural identities in stark contrast to the rigidity of ethnic-based identities and organic citizenship advocated by the New Right.

Squatting and social centers: similarities and differences between the New Left and New Right

Squatting represents the direct action side of both the New Left and, more recently, the New Right, found in Italy and, to a limited extent, in Spain. The next two sub-sections expose the similarities, from the social metabolic and from the anti-capitalist perspectives; the third section shows the practical differences that replicate the respective political ideas.

Material recycling

In a previous section, I explained that, from a purely metabolic perspective, any squat is a place where resource efficiency is accomplished by turning real estate waste into a new use. European cities and, in particular Barcelona, are affected by the speculation phenomenon to the point that empty buildings constitute more than 13% of the housing stock (Sanmartín 2013). The physical resources required to deal with an abandoned real estate vary greatly from putting it on the market, managing it by the public administration, or squatting it, in which case the occupiers make the property functional with minimal restoration, with a preference for repair rather than demolition. Instead, in the former cases, it is likely that the estate will suffer a complete restoration if not complete demolition to make way for a new property obeying legal and social standards. For instance, it is highly unlikely that a local public administration could change an abandoned industrial site into a Social Center without a massive physical reformation. The case of squatting for housing in Rome is relevant here. Mudu (2014) explains that, through the illegal action of squatters, an abandoned industrial site or a public school is transformed into a dwelling for people in need of a safe roof. By doing so, natural resource efficiency is accomplished.

Anti-Capitalism

The legacy between left-wing squatting and resistance to capitalism has been demonstrated as a common denominator of the squatters' movement across Europe (SqEK 2014). Less is known of how right-wing squatting is situated in the anti-capitalist struggle. 'Casa Pound' is a fascist squat that originated in Rome in 2003 and has extensively influenced New Right practices across Italy. Like most squats from the left, it has a clear anti-capitalist and anti-imperialist posture. Not so much interested in environmentalism, Casa Pound seeks the accomplishment of social needs, particularly access to housing, but just for native Italians. It was inspired by Ezra Pound, a strongly anti-American Italian fascist who understood the evils of financial capitalism and its institutions as responsible for the First World War (Cattaneo and Martínez 2014).

As for the similarity between the New Left and the New Right, the left and the right squatted social centers share anti-capitalist ideals that, in the case of squatting, are turned into practice. Nonetheless, the main explanatory factor that sets the difference between them is the openness to socio/cultural diversity and social minorities, particularly migrants.

Minorities and migrants

Left political squatting is characterized by shedding light on how people's diversity and minority populations are excluded, if not repressed, in western societies. Diversity can refer to class issues, ethnic issues, sexual and gender issues, or psycho-physical (dis)abilities. In Berlin, there are cases of left-wing squatting where the issues of minorities and diversity were already tackled 40 years ago (Azozomox 2014). In some cases, squatters have joined around a common working-class identity; in others, around that of being a female migrant. Recently, there was the Irving Zola Haus, a squatted social center thought of and run by people with walking impairments, and the occupation in 2012 of a public square, and later a building, by refugees concerned with the strict German immigration laws.

In the case of Barcelona, although not organized around a migrant identity, squatting stems from the need for waste-pickers (who have mainly migrated from non-European Union (EU) countries, in particular from sub-Saharan countries) for a basic shelter and a storage place for the collected scrap materials. Not that waste-picking is solely related to people who have recently migrated to Barcelona, but the reality shows that – notwithstanding an economic crisis that has left more than 20% of the Spanish workforce unemployed – the economic opportunities related to a waste-picking business have not been considered seriously by its more established inhabitants. The New Right's ethno-centric approach is similarly taken into consideration by Casa Pound or Hogar Social Ramiro Ledesma (HSRL), the fascist squat in Madrid.[4] Both Casa Pound and HSRL squatted in other cities after their first squatted social center and strengthened the discourse on housing and social rights, but only for the citizens of their respective nations. In particular, neo-Nazi activists from HSRL give away food to impoverished Spanish citizens. However, it is not clear how radically these new fascist groups have been able to break with their activism from the recent past, when, prior to the housing crisis that hit Spain, fascist activists beat up homeless people, as they still do with migrants (Garrido 2014).

It is doubtless that fascist activists have always had little concern for diversity and minorities, so how can it be that they are now opening social centers? Does the social activism of fascist squatters constitute a genuine ideal, a sort of conversion towards social justice, or is it only political opportunism in the wake of a systemic crisis of a very uncertain future? Because of the very recent emergence of these experiences, a clear analysis is difficult – Casa Pound, the more established one, is only 10 years old. Also, the young activists from the New Right are different from the previous generation of fascist activists. Movements like that of nationalist autonomists give the idea that a certain fusion is taking place between the typical far-right nationalists of identitarian ideology and autonomous and Do-It-Yourself practices that are more in line with left squatter movement. The aesthetic of New Right activists is also different from the traditional one represented by the skinhead and gets close to that of the anarchist black block. The Italian national anarchists and national-autonomists[5] are clearly anti-capitalist, anti-development and even anti-speciesists; paradoxically, equality between animals and humans is above equality between national and foreign inhabitants.

The race separatist and anti-immigrant stance and the lack of any serious consideration of the reasons behind migrations make right-wing squatters' practices just as problematic as the ideologies associated with the New Right.

The lack of an in-depth analysis that considers the global situation in a historical perspective of why people have to migrate from poor to rich countries is per se a problematic issue for which New Right race separatism and anti-immigration positions need to be rejected. However, there are also practical reasons for rejection that will be analyzed in the case study for which open borders and the consequent input of workers willing to undertake certain ecologically viable functions for urban social metabolism are an opportunity not to be lost.

Case study: waste-pickers from Barcelona

The a priori rejection of open border policies defended by the New Right needs to be assessed from the empirical evidence of waste-pickers in Barcelona who positively contribute to the economic efficiency, ecological effectiveness and social justice perspectives of Barcelona's social metabolism.

The case at hand reveals that over the course of several years, as the value of natural resources, in particular, metal, has increased due to their greater scarcity, an opportunity to recycle scrap metal from urban waste has risen. Notwithstanding the economic crisis that hit Europe in 2008, the opportunity to make a living out of recycling scrap metal has primarily been undertaken by citizens who have migrated recently or relatively recently from impoverished areas outside the EU, 60% from Africa. Only 30% of Barcelona's full-time waste-pickers are of Catalan origins (Cantero 2015). Traditionally an activity undertaken by *gitanos* (gypsies), who represent another minority present all over the Spanish territory, metal recycling has grown to the point where more people participate using simpler means. Moreover, due to the crisis, several Spanish citizens who recently received residency permits but belong to non-European ethnic minorities did not hesitate in beginning a waste-picking business since they lost their job or their former business went bankrupt.

The case reveals that there are cultural barriers towards undertaking activities of scrap metal and recycling for which it is more likely that a long-term unemployed person born in Spain (or elsewhere in the EU) will be less interested in this business than a former entrepreneur born outside the EU area who recently migrated to Spain. Cultural barriers take time to overcome since a cultural shift is needed. And in this case, it is significant that scrap metal recycling can be considered a dignified activity. Due to the urgent natural resource crisis and the need to dramatically improve the material efficiency of our economies,[6] any worker willing to undertake scrap metal recycling by creating a value from material resources, which would otherwise be wasted, should be welcomed. The previous two statements imply that the resource crisis urges workers like those who have recently migrated from non-EU areas to undertake scrap metal recycling because they are able to adapt to new business opportunities.

Commonly waste-pickers operate in Barcelona by pushing supermarket trolleys along the streets and loading them with valuable materials they find in waste containers or next to them. When trolleys are fully loaded, they either sell them to metal dealers or store them in places that they have occupied, mainly in the former industrial district of Poble Nou, where abandoned industrial hangars are available. They normally squat large places where many of them come together, each one managing their own space within the larger property. It is common for them to also create a dwelling in these places.

The activity of scrap metal recycling is not new: prior to waste-pickers pushing trolleys, *chatarreros* (scrap metal dealers) have been operating in a similar way for many years, but with two main differences. Rather than push a trolley, chatarreros drive small lorries where they can load more weight, and rather than storing metal in squatted estates, they do it in their houses, which likely are not squatted. They are normally citizens born in Spain or former migrants from rural Spain; gitanos represent most of their cultural make-up. In recent years, some non-EU-born waste-pickers have upgraded their businesses by buying their own lorries to transport scrap metal.

In general, both waste-pickers and chatarreros sell their resources to larger buyers, normally within the city itself, who have direct contact with industries with plants located in the city's periphery that re-process the metal into raw material that will serve as input in industrial processes. Finally, there is an institutional channel where urban waste is managed: when the size or the weight is too large for disposal in a container, citizens drive to "green points", specific locations run by the city administration where waste is separated according to its typology (toxic, electronic, wood, metal, organic, general). Moreover, paper, glass, plastic and, to some extent, organic waste is also managed institutionally through the containers located in the streets.

Economic efficiency

From an economic perspective, waste-pickers represent a solution at zero institutional cost who operate in efficient market conditions, quite similar to the conditions for perfect competition – perfect information, no barriers to entry, great number of agents, price takers that cannot influence the market. A market for waste management helps reduce the work load normally undertaken by the public administration at minimal to no overhead cost. In Barcelona, waste-picking is mainly limited to metals. However, there is an emerging market for paper recycling. Specifically, in cities, like Delhi and in wider Global South, plastic and organic waste is recycled. It is more than likely that as natural resources become scarcer and more expensive, business opportunities in non-metal waste recycling will increase. There are different typologies of operators in the waste management sector. So the plurality of agents will eventually make waste-picking an economically efficient sector. However, there are vested interests that, with the excuse of regulating this market, want to create local oligopolies or monopolies; large corporations with interest in the waste management business are lobbying the city administration in order to obtain concessions to operate in the collection and resale of urban waste. If this were to occur, the informal sector of waste-pickers might suffer an important loss because they would not be able to access and manage waste so easily (Demaria, personal communication with author, January 2015). This is occurring both in Barcelona and in Delhi where a corporate alternative to waste-picking is waste incineration. The monopolist-environmentally detrimental manner at which waste incineration is carried out is unsustainable.

Ecological effectiveness

Waste-pickers – in particular, those squatting abandoned industrial warehouses, pushing trolleys and separating materials by hand – are extremely effective converters of material waste into useful resources. It can be said that their entire activity is effective. By pushing trolleys, they operate with little energy consumption. An average human metabolism has an energetic requirement of 2500kCal per day, which is equivalent to a quarter of a liter of gasoline fuel. The human body is an extremely efficient energy converter.

Also those who use a lorry, load and unload it manually, so they are more energy effective than the large trucks for waste management used by the local administration.

By squatting abandoned warehouses, they also contribute to the previously introduced effectiveness of putting back into use properties that would otherwise be left to decay (with future material costs for managing the abandoned wreck) or be subject to real estate speculation, transformed into other uses through a demolition and further reconstruction, which is an extremely resource-intensive activity. So by squatting these spaces, they contribute with minimal resource input to maintain a space that is useful both as a dwelling and as a storage place.

Finally, because they separate materials manually, they manage waste in a more effective way than the institutional method adopted in the green points. There, if a piece of waste is composed of more than one material, rather than being dismantled by separating the different materials as waste-pickers would do, it is thrown to the general waste container. By doing so, an opportunity to return a scarce natural resource to the economic system is lost, and more waste must be physically managed (requiring transportation and a place for disposal such as a landfill or an incinerator).

Social justice

Although waste-picking is seen as a source of income for many citizens, it is not clear whether or not they consider this as a dignified source of employment, and whether or not they would continue to operate in this sector if other job or business opportunities were available. Also, it is hard physical work, and almost all waste-pickers are young and strong people. Depending on the value of metal, income can be extremely low, so it is highly important that a storage facility is used against the volatility of metal prices. Also, the living conditions within these abandoned hangars are highly inadequate, often lacking access to water or electricity. As squatters, their situation is further complicated as they cannot sign contracts to access these facilities. However, these people work in this sector because there is no other employment alternative. By living in these places, they can at least self-manage their lives, which would otherwise depend on institutional aid, and they would probably be worse off in terms of social dignity. The present institutional situation is highly problematic due to the barriers that the administration and certain lobby groups are posing. As mentioned earlier, waste-picking undertakes the political economy perspective of social metabolism where the role, type and function of the actors are relevant. The fact that the waste-pickers' warehouses are evicted complicates their business. In July 2013, an industrial hangar in Poble Nou was evicted. Here, 300 people, half of whom lacked permits of residence, were living and used it as their storage place. In total, 700 people were estimated to live in 62 squats, mainly in the Poble Nou area. In spring 2015, another squat, close to the Glories square (now a hub of the futuristic Barcelona of architectural fashion and design) was evicted. The hypocrisy of a city administration that sells the idea of a smart sustainable city simultaneously evicts places where environmentally sustainable methods are operated and managed – is ironic. Correspondingly, the monopolist alternative to waste management with a corporate operator under an institutional permit gives the idea that, although waste recycling is important, it also has to be done in a socially just manner, one that provides dignified employment to many. From Barcelona to Delhi, perspectives and narratives illustrates that it is highly significant to recognize waste-pickers' social function. If the problem is the lack of formal institutionalization, then the creation of a waste-pickers syndicate would be a more sustainable alternative than the concession of waste management operations to a corporate actor.

Conclusions

From a purely economic standpoint, the extraction of natural resources has negative externalities that should be paid. If these resources were properly taxed to pay for their social and environmental costs, then the value of scrap metal and other urban waste would be higher and the income for waste-pickers better. At the juncture of climate crisis and intense exploitation by capitalist initiatives, it is a missed opportunity to not consider these entrepreneurs who do an important and urgent task as dignified. Thereafter, in a world of scarcer resources, this activity represents a typical degrowth opportunity that could enhance economic efficiency and ecological effectiveness. The waste-picking sector should be valued together with the main operational actors who are mainly people not native to the EU.

There is the social need to improve the material living standards of these people. There is also the need for creating a union of waste-pickers that would receive some sort of institutional protection. This would give them the opportunity to, at least, get a permit to secure warehouses among the many abandoned buildings which, in turn, would serve as a protection against the impediments that are derived from their repeated evictions. Hence, Barcelona's administration must amend the humanitarian crisis perspective that waste-pickers as migrants or refugees or asylum-seekers need aid for their sustenance, to one that views waste-pickers as dignified and ecological laborers whose income-generating activities are highly critical for the ecological and economic sustainability of any city.

This symbolizes a triple win: 1) formalization of an economic sector that operates close to the efficient situation of a perfectly competitive market and away from that of a corporate oligopoly or, at worst, a monopoly; 2) from the ecological and material perspective, the recycling capacity of urban metabolism in the present situation of a degrowth future; 3) the opportunity to dignify a self-managed sector that would drastically reduce the public administration's cost of waste management (including abandoned buildings) as well as the public cost of social aid. In sum, since the majority of operational waste-pickers in Barcelona are migrants who are significantly contributing towards a progressive eco-sustainability of the city, the claims of the right-wing ideologists of race separatism and restrictive immigration laws would enormously undermine the benefits Barcelona enjoys from the waste-picking activity.

Notes

1 For details on the Movement Against Utilitarianism in Social Sciences see: http://www.revuedumauss.com/.
2 Although this text defends a position in favor of open borders and migrations, it has to be acknowledged that migrations are not always good, particularly from the perspective of desperate people that, in fate of development policies imposed in their regions, fall into misery or desperation and have to move to the Global North in search for a better life.
3 Material intensity refers to the amount of resources required to generate one GDP unit. Trends show that material intensity in post-industrial nations has been decreasing in the past decades, primarily due to the growth of the service sector – while industrial production and its heavy material and polluting burden have been transferred to developing nations.
4 This squat was occupied in summer 2014 (and evicted in May 2015) by members of the Social Republican Movement, a far-right political organization named after a Spanish activist and ideologist of the early twentieth century inspired by Italian fascism and German Nazism.
5 On the issue of national anarchists and national-autonomists consider two websites: http://nazionalanarchismo.jimdo.com/ and http://autonominazionalisti.blogspot.com.es/; http://machorka.espivblogs.net/2015/01/03/conoscerli-per-isolarli-nazional-anarchismo-resistenza-nazionale-e-autonomi-nazionalisti/.
6 The material efficiency of our economies is the unit of useful service provided per amount of natural resource extraction, or also unit of useful service provided per amount of non-valuable waste generated.

References

Azozomox (2014) 'Squatting and diversity, gender and patriarchy in Berlin, Madrid, Barcelona'. In SqEK, Cattaneo C., and Martínez, M.A. (eds) *The Squatters' Movement in Europe. Commons and Autonomy as alternatives to Capitalism.* London and New York: Pluto Press.

Caillé, A. (1989) *Critique de la raison utilitaire – Manifeste du Mauss.* Paris: La Découverte.

Cantero, M. (2015) Espigolar la ferralla. La Directa (21 May 2015). Available: https://directa.cat/espigolar-ferralla.

Cattaneo, C., D'Alisa, G., Kallis, G. and Zografos, C. (2012) 'Degrowth futures and democracy', *Futures* 44(6): 515–523.

Cattaneo, C. and Martínez, M.A. (2014) 'Squatting as an alternative to capitalism, an introduction'. In SqEK, Cattaneo, C. and Martínez, M.A. (eds), *The Squatters' Movement in Europe. Commons and Autonomy as alternatives to Capitalism.* London and New York: Pluto Press.

De Benoist, A. (2007) *Demain, la décroissance! - Penser l'écologie jusqu'au bout.* Paris: Edite.

Demaria, F. Schenider, F., Sekulova, F. and Martinez-Alier, J. (2013) 'What is Degrowth? From an Activist Slogan to a Social Movement', *Environmental Values* 22(2): 191–215.

Fournier, V. (2008) 'Escaping from the economy: the politics of degrowth', *International Journal of Sociology and Social Policy* 28(11–12): 528–545.

Garrido, J.M. (2014) Los nazis siembran el pánico en el barrio obrero y multicultural de Tetuán: comienzan las palizas a los inmigrantes. *Elplural.com* (5 September 2014). Available: http://www.elplural.com/2014/09/05/los-nazis-siembran-el-panico-en-el-barrio-obrero-y-multicultural-de-tetuan-comienzan-las-palizas-a-los-inmigrantes/.

Georgescu-Roegen, N. (1979) *Demain la décroissance.* Lausanne: Pierre-Marcel Favre.

Graham, M.D. (2005) 'Co-opting the counter culture: Troy Southgate and the National Revolutionary Faction', *Patterns of Prejudice* 39(3): 301–326.

Kallis, G. (2011) 'In defence of degrowth', *Ecological Economics* 70(5): 873–880.

Kerschner, C. (2010) 'Economic de-growth vs. steady-state economy', *Journal of Cleaner Production* 18(6): 544–551.

Kropotkin, P. (2004) [1915] *Mutual aid: a factor of evolution.* London: Kessinger Publishing.

Latouche, S. (2001) *L'invenzione dell'economia.* Casalecchio: Arianna Editrice.

Latouche, S. (2006) Abajo el desarollo sostenible! Viva el decrecimiento convivencial! In Colectivo Revista Silence. (2006) *Objetivo Decrecimiento.* Barcelona: Leqtor.

Malinowski, B. (1978) *Argonauti del pacifico Occidentale.* Roma: Newton Compton.

Mauss, M. (2007) [1924] *Essai sur le don. Forme et raison de l'échange dans les sociétés archaïques.* Paris: PUF.

Mudu, P. (2014) *Ogni sfratto sara' una barricata. Squatting for housing and social conflict in Rome.* In SqEK, Cattaneo, C. and Martínez, M.A (eds) *The Squatters' Movement in Europe. Commons and Autonomy as alternatives to Capitalism.* London and New York: Pluto Press.

Polanyi, K. (1944) *The Great Transformation: the Political and Social Origins of our Time.* Boston: Beacon Press.

Rist, G. (2008) *The history of development: From Western Origins to Global Faith.* London: Zed Books.

Sachs, W. (ed) (1992) *The Development Dictionary: A Guide to Knowledge as Power*. London: Zed Books.

Sahlins, M. (1977) *Stone Age Economics*. London: Tavistock.

Sanmartín, O.R. (2013) 3,44 millones de casas cerradas a cal y canto, muchas de ellas de la burbuja. *El Mundo* (18 April 2013). Available: http://www.elmundo.es/elmundo/2013/04/18/suvivienda/1366275980.html.

Schneider, F., Kallis, G. and Martinez-Alier, J. (2010) 'Crisis or opportunity? Economic degrowth for social equity and ecological sustainability. Introduction to this special issue', *Journal of Cleaner Production* 18(6): 511–518.

Spektorovski, A, (2000) 'Regionalism and the Right, the case of France', *The Political Quarterly* 71(3): 352–361.

SqEK, Cattaneo, C. and Martínez, M.A. (2014) *The Squatters' Movement in Europe. Commons and Autonomy as alternatives to Capitalism*. London and New York: Pluto Press.

Thiele C. and Weiss, M. (2012) Idylle in Grün-Braun. *Süddeutsche Zeitung* (13 April 2012). Available: http://www.sueddeutsche.de/politik/unterwanderung-des-biolandbaus-durch-rechtsextreme-idylle-in-gruen-braun-1.1332321.

Toledo, V. and Gonzalez de Molina, M. (2011) *The Social Metabolism A socio-ecological theory of historical change*. New York: Springer.

21 Euro trash in Loïsada, New York

Hans Pruijt

In most cases, squatter migrants move from a poor place to a rich place. The opposite is possible too. This is a micro case study of Europeans squatting as migrants on the Lower East Side of New York in the late 1980s. At the time, this was still a relatively poor neighborhood. The main source of information is an in-depth interview with Anna (the name has been changed for anonymity), a Dutch woman who, after arriving as an art student in New York in 1988, tried to join local squatters and then formed a group to open up their own building. In the Netherlands, she had lived in a large squat, but her first encounters with the Lower East Side squatter scene were alienating:

> When I came to New York in 1988 there were maybe five or six big squats. I did not have a home and visited all these squats to try to get a room. This did not work at all. One building said: "we already have enough women". The other squat said "we already have enough Europeans". Yet another group said "we already have enough white people". Thus as a white, female European, they did not want me.

Therefore she searched for an empty building that she could squat herself, found one and recruited some fellow students to join in the squatting action. This worked out well. However, Anna found out that the Americans in her own group found it difficult to justify the presence of white American and European squatters in the neighborhood. This became apparent when she started discussing the ideal size of the group. Her opinion was that:

> it would not be a good idea to get the building full of people, because as soon as the group becomes larger than fifteen of so, it becomes very difficult to organize. Then, someone has to take control. Who will check on what everyone is doing, and whether everybody is paying the monthly dues? I did not want that someone would be the leader, I wanted to keep it democratic, that the group, during the house assembly, decides what will be done.

The Americans in the group did not agree:

> They had what I call "white American guilt", they felt guilty because black and Porto Rican people were homeless, and that we as a group – we had one black guy and a Spanish guy – were mainly white. They were upset about this, and wanted more people of color. I said, I don't care if someone is Black, or Hispanic, or White, I think that it is interesting that we are all artists, to create an artist's squat. Well, they found this very elitist. They did not understand my viewpoint or argument at all.

Reactions in the neighborhood were hostile:

> During the first week the Porto Ricans in the neighborhood – it is a Porto Rican neighborhood – discovered that we were in there. They did not like it at all. Then they threatened us with violence. [...] In the beginning, those Porto Ricans yelled at us, and threw things at us.

However, the hostility did not seem to be caused by discrimination. When the squatters explained to the neighbors that it was a shame that the house had been standing empty while so many people needed a home, and that they were not junkies, and were fixing up the building, and moreover backed this up by constantly bringing in building materials, "their anger turned into interest".

Nevertheless, there were political attempts to discriminate the squatters on the Lower East Side as "Euro trash" (Van Kleunen 1994: 306). Porto Rican politician Antonio Pagán campaigned against the squatters in a conflict over a proposal to raze squats for new housing (Pruijt 2003). He put up posters that suggested that the presence of European squatters caused American homeless people to camp out in empty lots:

> SQUAT YUPS GET FREE RENT, SPACE, ELECTRIC, GAS, WATER.
> WHAT GET YOU?
> EURO TRASH IN THE SQUATS.
> AMERICANS IN THE LOTS. (Van Kleunen 1994: 306)

Frank Morales, longtime squatter on the Lower East Side, commented that this was:

> in fact a way of attempting to denigrate our efforts, as though we weren't indigenous to the neighborhood, but simply invading "eurotrash" – which we actually in some cases took on as a badge of pride, cognizant of the squatter movement throughout Europe!" (e-mail communication with author, July 2015)

Because this is only one case, and largely based on the experience of one person, the conclusion can only be tentative. Anna's story highlights a very natural strategy: identifying a shared identity and networking on the basis of this. It also shows an extreme case in which local squatters were troubled by self-accusations and accusations of being privileged, and that therefore it was inappropriate for them to squat (refer also to azozmox and IWS chapter 16, in this volume on privilege and guilt). Their solution was to apply bureaucratic quota, which hampered shared identity-based networking.

Finally, Anna's story showcases another relevant strategy: taking the initiative to start an autonomous squatting action, using skills and experience gathered in her home country. This can be seen as the royal way to join the local movement.

References

Pruijt, H. (2003) 'Is the institutionalization of urban movements inevitable? A comparison of the opportunities for sustained squatting in New York City and Amsterdam', *International Journal of Urban and Regional Research* 27(1): 133–157.

Van Kleunen, A. (1994) 'The squatters: a chorus of voices ... but is anyone listening?' In Abu-Lughod, J. L. (ed.) *The Battle for Tompkins Square Park*. Oxford: Blackwell, pp. 285–312.

22 Squatting and the undocumented migrants' struggle in the Netherlands

Deanna Dadusc

> Here in the Netherlands our existence is structurally denied. But this does not mean that we do not exist. We are here. We are living on the streets or in temporary shelters. We are living in a political and legal vacuum – a vacuum that can only be filled by the recognition of our situation and our needs. Our lives have been put on hold because we don't have papers, but we refuse to have our existence denied any longer. We refuse to remain invisible. We refuse to remain victims. We demand a structural solution for anyone who is in our situation and for all others who might find themselves trapped in the same political and legal vacuum. We demand recognition of our existence. We demand our existence to be acknowledged in official policies and laws. We are here and we will remain here. (We Are Here Manifesto)[1]

All over Western Europe groups of undocumented migrants are organizing themselves to resist and protest against the current migration regime and border system. In this chapter it will be argued that while the migration regime aims to push undocumented migrants into invisibility, to silence their voices, to tame their bodies and to let them live in a constant state of fear, this does not deprive migrants of their capacity to rebel and to struggle. For years, sewing the lips, hunger strikes, setting fire to one's body have expressed acts of protest occurring daily both in foreign detention centers and on the streets of Europe. Most of these practices relate to one's body: when every aspect of one's life is criminalized, the only weapon undocumented migrants have left is their own body, and often, the power over their own death.

However recently a new mode of protest started emerging in many Western European countries, including the Netherlands, Germany and Italy namely collectively squatting unused buildings. This marks an important shift in the undocumented migrants' modes of struggle that goes from isolated acts of protest to a collective mode of resistance that affects the everyday lives of undocumented migrants. Indeed, in this chapter it will be argued that living in squatted buildings has been used by undocumented migrants as a tool of protest and to gain visibility, but also to open collective spaces where it becomes possible to organize their struggles in a systematic manner, to intervene in the way they are supposed to experience their everyday lives, and to take their basic rights in their own hands, thereby overcoming the structural denial of juridical rights.[2]

"We Are Here" to stay: squatting and undocumented migrants' struggles

In the Netherlands, the "We Are Here" movement has been one of the first attempts of collective mobilization by a group of undocumented migrants. The people involved in this recently formed movement come from a variety of countries and live in a juridical and political limbo. Their asylum requests have been rejected, but they are unwilling, unable, or not allowed to go back to their countries of origin. This is due to a number of reasons, ranging from bureaucratic procedures that do not allow them to access the refugee status, or international laws that prohibit the Netherlands from deporting people to specific countries where their lives would be in danger, or because the very countries of origin refuse to accept their return. In a political system where possessing legitimate identification papers is a basic condition for the exercise of freedom, every aspect of life of an undocumented migrant is considered illegal (Palidda 2008; Aas 2011).

In this context, since the beginning of their mobilization, the strategy of the "We Are Here" movement has been to occupy pieces of land or squat empty buildings in order to live together, construct networks of solidarity among migrants and organize a common struggle that would let them exit the stigma entailed by the juridical label of "illegal migrant". The "We Are Here" movement started their campaign in September 2012, when a small group of undocumented migrants settled in the garden of a protestant church in Amsterdam. This was the beginning of a large mobilization that attracted hundreds of undocumented migrants and asylum seekers from all over the Netherlands. In November 2012 the "We Are Here" group squatted a courtyard in Osdorp, a neighbourhood at the outskirts of Amsterdam West. This settlement attracted much support from civil society, and they became visible both in local and national media. After the eviction of this camp, they squatted several buildings in different parts of the city: an empty Church *(Vluchtkerk)*, empty office spaces *(Vluchtflat, Vluchtkantoor, Vluchtgebow, Vlucht-toren)*, a government-owned building (*Vluchtgarage*), a former hospital and a school.

In these spaces undocumented migrants have been creating the possibility of exercising their basic rights instead of simply waiting for their rights to be granted by the government. These squatted buildings have been used by undocumented migrants for housing, but also as social, cultural and political centers, where different groups of people can encounter and mobilize for further political action. Indeed, in the buildings squatted by the "We Are Here" group, several activities emerged: debates, music, dance, educational projects, legal information and assistance by activists with experience in migration law, medical aid, and the creation of networks of support. As undocumented migrants are criminalized in every space of European society, creating these social and political projects has been a step further in the struggle, as what is needed is not only a roof for surviving, not only basic human rights, but also the possibility of living a decent life, which includes the possibility of socialization, communication, and recreation. By taking action into their own hands, the "We Are Here" group have been able to go beyond the social and political attitudes that criminalize them and that aim to turn undocumented migrants into fearful and passive subjects.

Indeed, in the Netherlands during the past decades, the situation of undocumented migrants has increasingly been framed as a "problem" that needs to be fought through stricter surveillance and security techniques. In 1993 it became compulsory to carry identification documents. In 1998 access to social services became dependent on a residence status, and it became compulsory for the migrant to collaborate with her own identification and deportation (i.e., the Law on Identification and Benefit Entitlement Act). In 2010 it became clear that the Dutch government intended to address the so-called "migration problem" by means of criminal law. Since then, undocumented migrants found on the Dutch territory are handed a notification to leave the Netherlands within 48 hours. Violating this ban would entail imprisonment in foreign detention centers for a term of up to six months.

Foreign detention centers are not supposed to have punitive goals, rather they were built with the scope of identification purposes. However, in the Netherlands, as in most of the other European countries, the way they are organized, the way they function and the rules that govern them make them resemble a prison. Hence, undocumented migrants are literally treated as criminals. However, contrary to the regular prison system, the inmates of foreign detention centers have no access to the juridical system: there is no court case nor procedure to establish the duration of the "imprisonment". People are held until further notice, which can be days, weeks, or months. Information about their legal procedure and status is often in languages that they do not understand. Contact with the outside is forbidden, and visitors are allowed under very restrictive circumstances. The treatment is austere, the rooms are very often over-populated, and the tendency for suicide, self-lesionism, panic attacks and other both psychological and psychical problems is very high. Despite European regulations, not only adults, but also children are detained for long terms. Currently, while no-border movements are challenging the existence of these institutions, new centers are being build for detaining and deporting families with children, such as the foreign detention center in *Zeist* (Anarchistische Anti-deportatie Groep Utrecht 2015).

After detentions, undocumented migrants can either be deported or thrown back on the streets with the imperative to leave the country on their own initiative. Many will not leave, and will go back into the circle of "illegality", invisibility and exploitability, under the threat of being found by the authorities and being detained once more. The legal production of illegal people is a political technique to obtain governable and exploitable subjects, outside of the government's responsibility. Indeed, while the concept "illegal" in Western juridical systems tends to refer to specific acts or practices, the label of illegal migrant invests one's whole life, body and subjectivity. This condition entails the complete exclusion from the system of rights, from the "right to have rights", and from access to basic needs such as medical treatment, housing and education (Ruhs 2010). At the same time these subjects are governed through strict laws, deportation threats and confinement in foreign detention centers (Andrijasevic 2010).

Thus, when a person, and not simply an action, is defined as illegal, each aspect of her life is subject to criminalization. This entails the production of fearful subjects, who are forced to hide and to be silent, pushed into invisibility and isolation, thereby often subject to abuse and exploitation (Anderson 2010). For instance, many undocumented migrants who manage to be employed without a contract do not have access to any labor rights and, if they protest against their working conditions, the employer often threatens to reveal his or her status to the immigration police. This is often the case of undocumented domestic workers. While the invisibility of domestic work makes it a safe way to work without legal status, this can also be turned against the workers, who find themselves in a double condition of invisibility and isolation (Triandafyllidou 2013). This mode of "illegalization" of migrants' life makes ungovernable flows governable and productive, through the strict interrelation of coercion over their bodies, and "works at distance" on their subjectivities and on their living conditions. Thus, in Europe and in the Netherlands, both symbolic and physical borders are being reinforced not simply to exclude and reject people, but to create the conditions for government exploitation of migrants' lives.

In the current critical debate around migration, borders are often defined as "a tool of exclusion" that aim to "demarcate a coherent inside from a chaotic outside" (Rajaram and Grundy-Warr 2007: x), and Europe is defined as a *fortress* (Geddes 2001). However, borders are not walls. Concepts such as "exclusion" and "fortress" allude to a clear cut between inside and outside that runs the danger of missing the complexity of border dynamics, and of obscuring the way borders work as tools not simply to exclude, but to control and govern the migration flows in a way that makes them productive for capital (Mezzadra and Neilson 2013). Indeed, borders are not completely closed; migration is not blocked all together. Rather they have a certain porosity, where flows of migrants can transit, although by illegal means (Bigo, 2005). This is not simply due to a lack of the security apparatus, but it configures as a violent technique to channel and regulate the migration flows and to govern the life of the undocumented once the border is crossed, by keeping them in a condition of illegality, precariety and exploitability (Anderson 2010). Therefore, in the current neoliberal Europe, the struggles around borders and migration regimes work as zones of visibility of the relations of power that circulate around citizenship and rights, and where the relation between coercive modes of power, disciplinary power and biopower find their higher intensity (Mezzadra and Neilson 2013).

In this context, the political goal of the "We Are Here" movement is to protest this structural denial of basic rights, to exit the loop of invisibility, and to challenge their criminalization by state and European politics: they aim to place their conditions with which they are confronted every day in the political agenda, to reformulate migration policies to grant full access to basic rights to all undocumented migrants. The name of the movement "We Are Here" explicitly addresses a struggle aimed at emerging from the circuits of invisibility and isolation, and at making their condition visible in the eyes of the government and of civil society. By creating these platforms for collective action, undocumented migrants also resist the way their everyday lives are governed by the migration regime. Indeed, these modes of collective organization had a strong effect on the subjectivities of those involved: from silent subjects, who need to hide and who used to live under conditions of constant fear, to active and powerful collectives, able to organize their own struggle, to share the joy of marching on the streets, of telling their stories without being afraid, and of opening free spaces of both contestation and socialisation.

Responses of the government

Squatting in the Netherlands used to be tolerated, but in 2010 it became illegal. Moreover, as argued above, the law is by definition against illegalized people, especially when they need to undertake criminalized actions such as squatting in order to survive. However, the balance between the law and the interests of the urban authorities has been shifting, according to the political agenda of the moment. While in certain cases the Mayor has brutally evicted the buildings, in other cases the authorities have decided not to intervene with the criminal law as they would do with regular squats, and they tried to negotiate and to find a temporary agreement between the owner and the occupants. Indeed the movement managed to attract political and social attention: under this pressure, negotiations worked as a tool to keep the movement under control, to keep an open channel of communication, and to de-escalate the conflict. The Mayor did not miss the chance of using negotiations to monitor and register the refugees involved. He required a list of the names and nationalities of the 159 migrants involved; such a list primarily served the interests of the state in identifying, registering, individualizing, and observing the refugees involved. Overall, while the "We Are Here" aim has been both to be together and to be visible, the government has responded by trying to break down the group, to separate them, to offer so-called "solutions" that would let the group divide and eventually dissolve.

As a response to this first occupation, the Major of Amsterdam, Van der Laan, offered to move them to an asylum seeker camp for the winter. The group refused this offer because they did not want to be separated from each other, and because they wanted to continue protesting until they got a stable solution, not a temporary accommodation. As a consequence of failed negotiations, the Major issued the order evicting the Osdorp camp "because of safety issues". The eviction took place on an early morning in December. A group of activists and supporters that were blocking the access to the gate of the yard were violently removed by the police, while all the migrants were arrested and the protest camp dismantled. The same night most of the migrants were released by the police, in the middle of nowhere, in the cold, without a place to go.

Two days after the eviction of the camp, a large group of squatters supported the migrants in occupying an abandoned church in Amsterdam West (the so-called *Vluchtkerk*), where they managed to stay for more than six months due to an agreement with the owner. Since then, the collaboration between the "We Are Here" group and the squatters' movement has become more intense, and multiple buildings have been squatted and evicted. On May 31, 2013, the group, now composed of 180 migrants, squatted an empty office building owned by the housing corporation De Key: the Vluchtflat, on Jan Tooropstraat 29. During the legal defense of the building, the lawyer representing the movement, Rahul Uppal, argued that although squatting in the Netherlands has been a crime since 2010, the decision of evictions must be evaluated as proportionate to the interests of the squatters and the ones of the owners. The lawyer argued that because the group is composed of "uitgeprocedeerde asielzoekers", who are not allowed to be in the Netherlands because their asylum requests were rejected but cannot be deported to their country, they are not allowed to access basic rights such as housing. Their only alternative to squatting is living on the streets or in detention centers, so it would be disproportionate to evict them. However, the Public Prosecutor argued that the fact that the squatters are "uitgeprocedeerde asielzoekers" is not relevant, because it is in the interest of the state to prevent unlawful squatting, and that the owner has the right to access their properties. The building was evicted on September 30.

Also on this occasion, the Minister of Foreign Affairs Teeven declared that the migrants could find shelter in the asylum seeker center in Ter Apel, but the "We Are Here" group replied that this was not considered a solution. Indeed, it immediately became clear that as a response to their mobilization, politicians started framing and treating undocumented migrants not only as criminals but also as victims. The attitude brought forward discourses such as "we should do something about this, and who cares about what they want". Approaching migrants as vulnerable victims dismisses their capacity to articulate their own opinions and thoughts, and leads to a paternalistic approach towards them that makes them dependent on charity or governmental projects. This way of approaching the issue differed from the framework that addresses undocumented migrants as a "security problem", yet, it framed it as an "humanitarian emergency" and led to proposals for projects aimed at solving the perceived problem at an individual level: namely supporting migrants with the basic facilities for surviving and keeping them under the governmental control, rather than addressing the broader picture and challenging how the overall system of rights and privileges circulates through borders, documentation and citizenship.

After refusing the offer by the Mayor, the "We Are Here" group squatted another office building in the heart of the city, just in front of the Rijks Museum: the so-called *Vluchtkantoor*, owned by the German pharmaceutical company Bayer. Christmas was approaching, the city was filling up with tourists, and the spectacle of the undocumented migrants' struggle had to be placed in the shadows again. To avoid damaging the image of Amsterdam as the city of freedom and tolerance, the Mayor tried to vacate the building as quickly as possible. In order to avoid escalation, visible protest and bad publicity, he promised an alternative for the whole group: shelter for six months in an abandoned prison on the

Havenstraat, on the condition that each undocumented migrant would cooperate with their own deportation to the country of origin. Most of the members of the movement were about to refuse the offer, as accommodation in a former prison with restricted access and strict monitoring of their activities was not considered an acceptable achievement in their struggle.

However, confronted with the threat of being thrown on the streets in the middle of the winter, and of being immediately evicted if they squatted yet another building, the group eventually accepted the offer and voluntarily left the *Vluchtkantoor*. Only once the agreement was signed by all the members of the group, the Mayor made clear that the offer was valid only for those who were listed in the spring 2013. However, since then, the composition of the movement had been subject to many variations. In this way, the Mayor finally managed to divide the group and their range of supporters: those on the list were sheltered in the former prison, and the other half were left on the streets. Once more it was December, and once more the authorities simply turned a blind eye to this part of the group and kept on congratulating themselves for providing a solution for the other half.

The part of the group that stayed on the streets started regular protests in front of the municipality, in the streets, and in front of human rights associations. When the group occupied the entrance of the city hall, they were told that it was illegal and that they would be arrested. When they claimed they would camp in front of the municipality, they were told that this was also illegal and that they would be arrested. When they said they would sleep on the streets, they received the same answer. The government simply ignored the problem and waited for the supporters to find an alternative solution. Through the groups of supporters, the migrants managed to find temporary accomodation in churches and in already-existing squats. It was only after two weeks on the move that the group managed to squat a government-owned building in the outskirts of the city (*Vluchtgarage*), where they managed to stay for more than one year, although under terrible living conditions. The building was outside of the city, therefore quite hidden from media and public visibility: this made the occupation easier for urban authorities to tolerate. Moreover, the building was owned by the municipality, which left the local governments more space for negotiation.

Therefore, the government has been using the squatters' movement that supports undocumented migrants as an informal service, which has taken a heavy social and political burden off its shoulders. Urban authorities do not want groups of undocumented migrants on the streets, and do not want to provide them with shelter, as this would imply carrying a political burden that they need to avoid. Often, after the eviction of buildings squatted by undocumented migrants, the police encouraged the group to occupy a new place. Moreover, during a court case related to the lack of basic human rights for undocumented migrants, the judge argued that these groups have practical access to the right to a roof, and the squatters' movement is providing them with a roof. Thus, while with one arm the urban authorities have been trying to demotivate the movement to squat by dividing the group, evicting their buildings, and using both physical and symbolic violence, with the other arm they have been encouraging this practice, as it is configured as a practical solution to a problem they should be taking care of.

Undocumented migrants (and) political activism

While using the groups of supporters as an informal service, the government also criminalizes those who struggle together with undocumented migrants. In Europe, laws against smuggling, trafficking and organized crimes are being used to prosecute and convict those providing basic support to undocumented migrants. In Austria, Germany, Italy, France and the Netherlands, supporting or helping "illegalized" migrants is considered a crime and is often prosecuted under the organied crime law. The latest example in the Netherlands is a new law effective since the 1 March, 2014 (Rijksoverheid 2014). This law puts an end to the right of domestic privacy: when the police, the immigration police (IND), the military or other state agencies suspect that so-called "illegal aliens" are present, then they are allowed to enter any house, search it, and arrest people without any warrant. The persons present at the address may also be body-searched, including the contents of bags and clothing, personal correspondence and any other personal data such as that carried in mobile phones. Furthermore, the police are allowed to ask for identification of people on the street that raise the "suspicion of illegality" (by appearance, or by speaking a foreign language, for example), and to arrest them (Advocaat van de Duivel 2014). This is not simply a tool to criminalize migrants and supporters: in the Netherlands, as much as in the rest of Europe, the war on migration is being used as an excuse to suspend rights, to extend police powers, to surveil and to control any part of the population that is questioning and threatening to the "normal running of things".

The supporters of undocumented migrants are not simply criminalized by the law, but are also stigmatized by the discourses of politicians, who try to divide undocumented migrants and their supporters. In particular, Amsterdam's Mayor Van der Laan has tried to create a differentiation between migrants and what he has defined as "Dutch political activists" (van Acker 2014). In his statement, he reinforced racist discourses, implying that migrants are passive objects, victims or criminals, and as such not capable of political action. Van der Laan indeed stated that while he intends to "help the migrants within the boundaries of the law", he is standing against so-called "Dutch political activists" whose aim is to change migration laws. According to the Mayor, so-called "Dutch political activists" have been using the undocumented migrants in order to pursue their own battle against the government, while the politicians are willing to help undocumented migrants within the boundaries of the law. This is clearly a paradox, as the only existing laws are aimed at criminalizing undocumented migrants.

Creating such a clear cut between migrants and 'political activists' is not only a way to separate the groups, but also to depoliticize and victimize the group of migrants. All over Europe movements of migrants are organizing themselves to protest the structural denial of basic rights, to gain visibility, and to challenge their criminalization by state and European authorities. Thus, migrants are themselves political activists. Hence, there is not such a distinction between migrants and political activists. The only existing distinction is the one between documented and undocumented people, a division created by European laws, borders and xenophobic discourses. The differentiation between documented and undocumented is the heritage of colonial relations and the prerequisite for relations of exploitation. As such, it needs to be challenged not only by addressing the politics and the culture perpetuated through borders, but also by struggling in the fields of our everyday relations. From this perspective, documented activists and squatters have not simply been supporting the migrants' struggles. Rather, they aim to resist the relations of power that produce the social-economic circumstances through which documented people's rights and privileges circulate.

However, these struggles are fought from different standpoints. On the one hand, for undocumented activists the political aim is to obtain "papers for all", and to be included in the system of rights. This often leads to a mode of struggle that leads to negotiations with the government, legal battles, and political compromises. On the other hand, for activists that have the privilege of being documented and of having access to these rights, the aim is often to create a world with "no papers at all", where one's life is not defined and confined by the possession or lack of documents. These different strategies are determined by different perspectives, but they do not exclude one another. Rather, they need to go hand in hand to achieve both short-term solutions and long-term transformation to the migration regime and the border system. Marching on the streets, engaging in legal battles and squatting houses are not changing the situation, but these are little steps and tactics to open up common spaces of contestation, autonomy and solidarity, where life and social relations are not defined by the possession of papers, and where both undocumented and documented activists can cooperate and overcome both criminalizing and victimizing attitudes.

Notes

1 See: http://wijzijnhier.org/.
2 This chapter is based on the author's experience as researcher and activist within both the "We Are Here" movement and the squatters' movement in Amsterdam.

References

Aas, K.F. (2011) 'Crimmigrant'bodies and bona fide travelers: Surveillance, citizenship and global governance', *Theoretical criminology* 15(3): 331–346.

Advocaat van de Duivel (2014) 1 maart 2014. Het begin van het einde? *wordpress.com* (23 february 2014). Available: http://duivelsadvocaatje.wordpress.com/2014/02/23/1-maart-2014-het-begin-van-het-einde/.

Anarchistische Anti-deportatie Groep Utrecht (2015) Bouwbedrijf De Vries en Verburg gaat nieuwe gezinsgevangenis Kamp Zeist bouwen. AAGU (July 2015) http://aagu.nl/2015/De_Vries_Verburg_bouwt_gezinsgevangenis.html.

Andrijasevic, R. (2010) 'From exception to excess: detention and deportations across the Mediterranean space'. In De Genova, N. and Peutz, N. (ed) *The Deportation Regime:*

Sovereignty, Space, and the Freedom of Movement. Durham: Duke University Press, 147–165.
Anderson, B. (2010) 'Migration, immigration controls and the fashioning of precarious workers', *Work, employment & society* 24(2): 300–317.
Bigo, D and Guild, E (eds) (2005) *Controlling frontiers: free movement into and within Europe.* Surrey: Ashgate Publishing.
Geddes, A. (2001) 'Immigration and European integration: towards fortress Europe?', *Refugee Survey Quarterly* 20(2): 29–72.
Mezzadra, S and Neilson, S. (2013) *Border as Method, or, the Multiplication of Labor.* Durham: Duke University Press.
Palidda, S. (2008) *Mobilità umane: introduzione alla sociologia delle migrazioni.* Milano: Cortina Raffaello.
Rajaram, P.K. and Grundy-Warr, C. (eds) (2007) *Borderscapes: hidden geographies and politics at territory's edge.* Minnesota: University of Minnesota Press.
Rijksoverheid (2014) Teeven breidt mogelijkheden vreemdelingentoezicht uit. www.rijksoverheid.nl (11 February 2014). Available: https://www.rijksoverheid.nl/actueel/nieuws/2014/02/11/teeven-breidt-mogelijkheden-vreemdelingentoezicht-uit.
Ruhs, M. (ed) (2010) *Who needs migrant workers?: labour shortages, immigration, and public policy.* Oxford: Oxford University Press.
Triandafyllidou, A. (ed) (2013) *Irregular migrant domestic workers in Europe: who cares?.* Surrey: Ashgate Publishing.
van Acker, J. (2014) Van der Laan hekelt politiek actvisme. *IndyMedia.nl* (8 March 2014). Available: https://www.indymedia.nl/node/21717.

23 Migration, squatting and radical autonomy

Conclusions

Pierpaolo Mudu and Sutapa Chattopadhyay

In this chapter, we have set ourselves the task of summarizing the main questions that are covered substantively in this book by thirty authors, some of whom are members of various collectives. This book is both a political intervention and a summary/analysis of political actions and research findings. This book is neither solely on migration nor only on squatting, but on the relation between the two, which ultimately intersects in resistance and destabilization actions against racist regulatory policies. The book is not a detailed analysis of how badly capitalism or the nation states or the local political parties treat migrants or criminalize squatters. Already quite a lot of scholarly and non-academic articles and books have been devoted to the atrocities of the state and the global misery (Bourdieu 1993; Chomsky 1999; Comité invisible 2014), or the specific national and international situations on repression of migrations (Palidda 2013), or the extreme racialization of criminal justice and the rise of the prison-industrial complex directly linked to the expansion of global economy, proto-capitalism, eurocentrism, white supremacy and the bombings of resource-rich countries in the name of democracy and development (Dent and Davis 2001; Davis 2003; Engel Di-Mauro 2012). Although there is a lot of recent literature on the possibilities of building new societies and utopias from different perspectives (Negri and Hardt 2009; Holloway 2010; Wright 2010; Graeber 2013), few books consider the possibility of building right now a different life (not in the unpredictable future of promised revolutions, or changes after the death of capitalism), not governed but emancipated and self-managed. Self-management is at the core of any radical political project, in particular, against fake promises by state institutions, which have co-opted it into a basis for legitimating ridiculous "participatory" forms of urban governance (Brenner *et* al. 2012; Mudu 2012).

Self-management carried out by migrants and squatters means building political subjects that create a dissensus, "a dispute about what is given, about the frame within which we see something as given" (Rancière 2004: 304). This means to attack the symbolic power relations that constitute "[...] the given by utterances, of making people see and believe, of confirming or transforming the vision of the world and, thereby, action upon the world, and thus the world itself [...]" (Bourdieu 1991: 170). Following Rancière (2004) we can affirm that migrants and squatters put two worlds together: the world where rights are valid and the world where they are not. They put the two worlds in one through practices that represent an incubator of different global relations opposed to any form of exploitation.

Currently the struggles presented in this book depict some of the most advanced frontier of class contestation. Class contestation is happening through the rebuilding of solidarity networks that do not represent welfare from below but potential alternative social patterns. This is very obvious to all governments. In fact, the supplemental role of welfare provision that exists, indeed provided by many squatting experiences, is not enough to ensure a niche of survival for squatting experiences that could be supported by the state or even legalized. Autonomous solidarity patterns are hardly conceivable and virtually impossible in neoliberal regimes. A century ago, the construction of the welfare state cannibalized the mutualistic experience of the working class to maintain workers' dignity and efficiency outside working time. Welfare state privatization, its passage to big corporations, is now a multiple powerful force of reshaping spaces of living. The state just represses autonomous solidarity forms from below, even if they are sometimes the only way to provide shelter, food and decent space for migrants. The issue with squatting, and migrant squatting even more so, is that these experiences are carried out autonomously based on self-management, with the result of rejecting securitized humanitarianism and destabilizing notions of inclusion and exclusion that need to be constantly updated to hide the vested class privileges behind the convergence of security, police, social welfare and migration policies. Great suspicion must be granted to those who "recite standard references to Europe's respect for human rights, democracy and history of welcoming refugees" because this is just "a prelude to introducing ever harsher immigration and asylum laws" (Anderson's foreword in this volume).

Increased border crossing can be radically re-conceived as a peculiar kind of social movement that contests nation-states, national boundaries, identities, and inherited privileges. Migrants and squatters pioneer the frontiers and borders of the political actions in a very broad sense. The struggles that are carried out by migrants-squatters by occupying spaces imply, at least, four global issues: 1) squatters and migrants are treated as dangerous social exceptions to the state of exception that runs neoliberalism, restricts democracy and regulates citizenship; but 2) migrant-squatters, housing movements and Social Center activists are able to form and perform some sort of solidarity against discrimination and segregation mechanisms; 3) migrant-squatters unsettle, disturb, and represent a new kind of cosmopolitan disobedience; 4) key sites for struggles are not only cities but many different places where explicit politics of scale are built.

The global scale of struggle is performed when squatting is a way to proclaim one's existence, materially, visibly and directly, not just to reclaim a generic "right to the city" but a "Right to Inhabit" the space on the planet without being segregated in suburban degraded areas, and illegalized in camps. There are many potential and real tensions between migrants and "indigenous" social movement activists involved in squatting and solidarity initiatives. Certain assumptions/claims that are sometimes made by activists and migrant solidarity groups are re-examined and critically investigated, offering the reader the possibility of reconsidering how much people in these particular social movements themselves are reflecting upon and working through these contradictions. It would be politically dangerous to see such contradictions as representing insurmountable obstacles to develop the foundations for societies that dismantle oppression mechanisms. In fact, we build radical autonomy projects on the possibility of self-management of our lives when we consciously address the mechanisms of the reproduction of power privileges in order to disarticulate them.

References

Bourdieu, P. (1991) *Language and Symbolic Power*. Cambridge: Polity.

Bourdieu, P. (1993) *La misère du monde*. Paris: Seuil.

Brenner, N., Marcuse, P., and Mayer, M. (eds) (2012) *Cities for people, not for profit: critical urban theory and the right to the city*. London: Routledge.

Chomsky, N. (1999) *Profit over people: Neoliberalism and global order*. New York: Seven Stories Press.

Comité invisible (2014) *À nos amis*. Paris: La Fabrique.

Davis, A. (2003) *Are Prisons Obsolete?* New York: Seven Stories Press.

Dent, G. and Davis, A. (2001) Conversation: Prison as a Border: A conversation on Gender, Globalization and Punishment. *Signs* 26(4): 1235–1241.

Engel Di Mauro, S. (2012) Prison Abolition as an Ecosocialist Struggle. *Capitalism Nature Socialism* 23(1): 1–5.

Graeber (2013) *The Democracy Project: A History, a Crisis, a Movement*. New York: Allen Lane.

Holloway, J. (2010) *Crack Capitalism*. London: Pluto Press.

Mudu, P. (2012) 'At the Intersection of Anarchists and Autonomists: Autogestioni and Centri Sociali', ACME, 11(3): 413–438.

Negri, A. and Hardt M. (2009) *Commonwealth*. Cambridge, MA: Harvard University Press.

Palidda, S. (2013) *Racial Criminalization of Migrants in the 21st Century*. Farnham: Ashgate.

Rancière, J. (2004) 'Who Is the Subject of the Rights of Man?', *The South Atlantic Quarterly* 103(2/3): 298–310.

Wright, E. O. (2010) *Envisioning Real Utopias*. London: Verso.

Index

Printed in the United States
By Bookmasters